THE TISSUES OF THE BODY

THE TISSUES OF
THE BODY

BY

W. E. Le GROS CLARK, F.R.S.

FELLOW OF HERTFORD COLLEGE
PROFESSOR EMERITUS OF ANATOMY
IN THE UNIVERSITY OF OXFORD

'The problem of organization is the central
problem of biology . . . the riddle of form
is the fundamental riddle.'

J. NEEDHAM, *Order and Life*

FIFTH EDITION

OXFORD
AT THE CLARENDON PRESS

Oxford University Press, Ely House, London W.1

GLASGOW NEW YORK TORONTO MELBOURNE WELLINGTON
CAPE TOWN SALISBURY IBADAN NAIROBI LUSAKA ADDIS ABABA
BOMBAY CALCUTTA MADRAS KARACHI LAHORE DACCA
KUALA LUMPUR HONG KONG TOKYO

FIRST EDITION 1939
SECOND EDITION 1945
THIRD EDITION 1952
FOURTH EDITION 1958
FIFTH EDITION 1965
REPRINTED LITHOGRAPHICALLY IN GREAT BRITAIN
AT THE UNIVERSITY PRESS, OXFORD
BY VIVIAN RIDLER
PRINTER TO THE UNIVERSITY
1967

PREFACE TO THE FIFTH EDITION

IT will be generally accepted that an introductory textbook should be not only informative; it should also endeavour to be intellectually stimulating. Experience in teaching has shown that the latter intention can most nearly be realized by bringing the student from the earliest days of his career up to vantage points on the advancing front of scientific adventure, and giving him preliminary glimpses of current problems which are exciting the attention of scientists. Hence, when this book was first written, an attempt was made to combine a record of elementary knowledge with short discussions of recent advances in the study of the tissues of the living body. Since the fourth edition appeared seven years ago, further significant advances have been made in certain aspects of tissue structure and function, and a new edition has given the opportunity of incorporating references to some of these. Parts of the book have therefore been rewritten, but the temptation to expand and elaborate has been avoided as far as possible. Some new illustrations have been added, and a few of the figures in the fourth edition have been replaced.

The author would like to thank many friends who have offered suggestions for improvements; it is due to their kindly advice that certain crudities of phrase and statement which appeared in the previous editions have been amended.

W. E. Le G. C.

Oxford 1965

CONTENTS

FIGS. 43 *and* 44 *are reproduced in colour facing pages* 96 *and* 97 *and* FIGS. 80 *and* 81
are reproduced in colour facing page 230

ACKNOWLEDGEMENTS

GRATEFUL acknowledgement is made to the authors, editors, and publishers of the following works and journals for permission to reproduce or utilize illustrations:

J. S. P. Beck and B. N. Berg in *The American Journal of Pathology*, vol. 7, for Fig. 96; J. Dixon Boyd, *Contributions to Embryology*, No. 152 (Carnegie Institution of Washington Publication No. 479, 1937), for Fig. 79; J. C. Brash in *The Edinburgh Medical Journal*, vol. 41 (Oliver & Boyd Ltd.), for Fig. 45; *The British Journal of Surgery* (John Wright & Sons Ltd.) for Figs. 56 and 57 (D. Stewart in vol. 24), and 68 (A. G. Timbrell Fisher in vol. 10); E. J. Carey in *The Anatomical Record*, vol. 21 (The Wistar Institute of Anatomy and Biology), for Fig. 64; E. R. and E. L. Clark in *The American Journal of Anatomy*, vols. 38 and 53 (The Wistar Institute of Anatomy and Biology), for Figs. 85 and 86; G. W. Corner, *Contributions to Embryology*, No. 112 (Carnegie Institution of Washington Publication No. 394, 1929), for Fig. 12; Cunningham's *Text-book of Anatomy*, 9th edition, edited by J. C. Brash (Oxford University Press), for Figs. 47, 87, 88, 122, and 123 and Figs. 43 and 44 (in colour); H. Gray's *Anatomy*, 27th edition, edited by T. B. Johnston (Longmans, Green & Co. Ltd.), for Figs. 7 and 77; C. H. Heuser, *Contributions to Embryology*, No. 138 (Carnegie Institution of Washington Publication No. 433, 1932), for Fig. 10; C. J. Hill in *The Philosophical Transactions of the Royal Society*, B, vol. 215, for Fig. 78; *The Journal of Anatomy* (Cambridge University Press), for Figs. 32 and 40 (R. W. Haines in vol. 68), 35 and 48 (P. D. F. Murray and J. S. Huxley in vol. 59), 41 (Sir Arthur Keith in vol. 54), 61 (E. Gutmann and J. Z. Young in vol. 78); *Human Embryology*, vol. 2, edited by F. Keibel and F. P. Mall (J. B. Lippincott Co.), for Fig. 92; L. B. Arey, *Developmental Anatomy* (W. B. Saunders Co.), for Figs. 58, 59, and 75; A. Maximow and W. Bloom, *Textbook of Histology* (W. B. Saunders Co.), for Fig. 83; P. Weiss, *Arch. Entwickl. Mech.*, vol. 116 (Springer, Berlin), for Fig. 22; H. H. Woollard, *Contributions to Embryology*, No. 70 (Carnegie Institution of Washington Publication No. 277, 1922), for Fig. 76; H. H. Woollard, *Recent Advances in Anatomy* (J. & A. Churchill Ltd.), for Fig. 15; J. Gross and F. O. Schmitt in the *Journal of Experimental Medicine* (Rockefeller Institute for Medical Research), vol. 88, for Fig. 19; M. H. Draper and A. J. Hodge in the *Australian Journal of Experimental Biology and Medical Science* (University of Adelaide), vol 27, for Fig. 53; H. E. Huxley in *Endeavour*, October 1956, for Fig. 54; and R. E. Billingham in the *Journal of Anatomy* (Cambridge University Press), vol. 82, for Fig. 101; V. T. Marchesi and J. L. Gowans in *The Proceedings of the Royal Society*, vol. 159, 1964, for Fig. 82.

THE STUDY OF TISSUES AND THEIR CELLULAR BASIS

1. THE CELLULAR BASIS OF TISSUES

TISSUES are the matrix of the body, the stuff out of which are constructed all those organs and systems which fit together to make up the machinery of the complete living individual. For this reason, they form the basic study of the science of anatomy.

Clearly, one of the most important factors determining the disposition and form ultimately assumed by all anatomical structures, whether they are muscles, nerves, or glands, &c., must be the nature of their constituent materials. The study of these materials is also important for the consideration of anatomical problems such as the adaptation of structure to functional requirements, the capacity of various organs for growth, regeneration, and repair, and their morphogenesis during embryonic and evolutionary development. Hence tissues must first be studied as tissues before it is possible to appreciate the morphology of the composite structures to which they contribute.

All tissues are composed of cellular elements and their derivatives. The cells may be closely packed, held together along their junctional margins by the adhesion of their surface membranes or in some cases by direct protoplasmic (syncytial) connexions, or they may be rather widely scattered through an intercellular ground substance containing tissue fluid, fibrous elements, deposits of inorganic material, and so forth. Even these non-cellular elements, however, are ultimately the products, directly or indirectly, of cellular activity.

It is customary in textbooks to define a tissue as an assemblage of cells, all of the same type, arranged in a regular formation. Such a definition is apt to be misleading, for it implies an abstraction which does not exist in the living body. It is true that many kinds of tissue are composed *predominantly* of one type of cell, specialized for the particular functions served by them, but as part of a working unit they also contain other kinds of cellular element which are ancillary and even essential to the maintenance of their nutrition and functional efficiency. Indeed, these other elements are in many cases ultimately responsible for the organized arrangement of the specialized cells which is characteristic of a tissue, from the structural and also from the morphogenetic point of view.

Mere aggregations of muscle cells or nerve cells cannot by themselves form muscle or nerve tissue. Muscular tissue is composed of more than muscle fibres, for it contains intrinsic elements such as connective-tissue cells and fibres, a special arrangement of blood vessels, and so forth. Similarly, nervous tissue is composed not only of nerve cells and their processes, but also of other cellular elements without which the nerve cells themselves are unable to preserve their vitality or to function as an organized tissue. While, of course, it is appropriate to study the chemical, physical, and anatomical properties of isolated muscle and nerve cells as the basic functional elements of muscular and nervous mechanisms, the study of these cells in the bulk, i.e. as organized tissues, must necessarily take into account the other elements which enter into the composition of these tissues and which condition their normal activity in the living body.

A tissue, then, may be defined as an assemblage of cellular and fibrous elements, in which one particular type of cell or fibre usually predominates, organized to form the material basis of one of the functional systems of the body. Vascular tissue, therefore, includes not only the essential endothelial lining of all the blood vessels, but also the organization of muscular and fibrous elements which are fundamental to the normal activity of arteries and veins. Again, a study of the skin as a tissue involves not only a consideration of the elementary structure of the epidermis, but also of its substratum, the dermis, its peculiarities of blood and nerve supply, and its immediate derivatives such as hair, nails, sweat glands, &c.

Tissues are, in summary, *organized* living material, and it is from such a point of view that they will be considered in the following pages.

2. THE ANATOMICAL STUDY OF TISSUES

The science of anatomy is concerned with the study of living organisms primarily from the point of view of their structure. Its ultimate aims include the interpretation of the structural organization of the body in terms of the morphogenetic factors which determine the composition, form, and disposition of the various constituent organs and systems, the variations which they show in different phases of growth and activity, and their adaptation to functional requirements.

Of all the technical expedients employed in the study of anatomy, that of simple dissection with scalpel and forceps is the most time-honoured. By this method the gross organization of the body can be defined in considerable detail. It is an essential approach to the study of the topographical anatomy of the human body in relation to the subjects of medicine and surgery. It provides the basis for all physiological studies, for the study of structural variations, whether individual or racial, and for the study of comparative anatomy. Further, the study of racial variations in man leads on to the wider field of physical anthropology, while comparative anatomy is the starting-point for the investigation of problems of organic evolution.

Dissection is always open to the criticism that, since it is usually practised on dead tissues, it does not necessarily give accurate information regarding their appearance during life. This objection is a real one, and must always be taken into account in anatomical studies. For while, no doubt, the appearance of many gross structures in the dead body, such as muscles, nerves, blood vessels, &c., reflects fairly closely their appearance in the living body, some tissues and organs (particularly the viscera) may become considerably altered in shape and texture after death.

With the use of appropriate technical methods, simple dissection can certainly supply much information regarding the constituent tissues of the various organized units of the body. Moreover, the naked eye can be supplemented by a dissecting microscope with comparatively low magnifications, or even by the highest powers of the compound microscope. Microdissection, as an anatomical technique, involves the use of exceedingly fine glass needles, micro-spatulas, micro-pipettes, and so forth, which are fixed to a rigid holder provided with a control mechanism whereby the points of the instruments can be manipulated in the field under the microscope. In recent years new types of micromanipulator have been developed in which the use of a single 'joystick' control greatly simplifies the actual process of microdissection. Microdissection has been further facilitated by the introduction of new optical methods for increasing the working distance between the objective of the microscope and the tissue to be dissected. Indeed, it is now possible for the organs and tissues of a living animal to be examined *in situ* at a working distance which permits operations under quite high magnifications. By means of modern micromanipulators single cells can be isolated and dissected without the need of great technical skill, and it is also possible to study the effects of injecting chemical substances into the cytoplasm, or the results of experimental mutilation of the cell.[1]

While gross dissection can do little more than give information regarding the gross organization of the body, it does provide the material for the study of morphogenetic problems, and of problems involving a consideration of the relation of form to function. The ultimate solution of these problems, however, requires the use of the microscope. Hence, microscopic anatomy, or histology as it is also called, occupies the central position in the realm of anatomical studies.

Microscopic anatomy can be divided into two main fields of inquiry, the study of tissues fixed and stained after death, and the study *in vivo* or *in vitro* of living tissues. In the first case, the anatomical composition of a tissue in terms of its cellular components may be explored by making film preparations, or by preparing extremely thin sections which are cut with a microtome after the tissue has been embedded in paraffin wax or celloidin, or after it has been frozen. Clearly, the preliminary treatment required for preparing microscopic sections is certain to distort to some extent the appearance of a tissue as it exists in life. It is interesting to note, however, that studies of living cells by the specialized techniques now available have shown that the inferences drawn by the older anatomists from dead material, regarding the structure of tissues and of the individual cells of which they are composed, have often been found to be remarkably correct.

The use of stains in the microscopic study of tissue anatomy is of fundamental importance. Many stains are selectively taken up by particular types of cell, or by particular structures within each cell. They therefore serve to differentiate the cellular elements which enter into the composition of a tissue, and they also bring into view intracellular structures which may be invisible in the unstained cell by reason of the fact that their refractility

[1] See the chapter on micrurgical technique by R. W. Chambers and M. J. Kopac in *McClung's Handbook of Microscopical Technique*, edited by R. M. Jones, 1961.

coincides with that of the immediately surrounding medium. The stains used are solutions of dyes prepared from animal or vegetable extracts, or chemical compounds of the aniline group. Their affinity for the particular elements which they colour depends on processes of absorption, actual chemical combination, or simple solubility.

Stains may be used as microchemical (or histochemical) reagents for the identification of certain chemical constituents of cells and tissues, and for determining their distribution and their relation to different phases of cellular activity. This field of *histochemistry* (as it is called) is evidently very relevant to anatomical studies. Though it is by no means a new subject, it has undergone considerable development in recent years with increasing efforts to find selective stains which may serve as indicators in specific chemical reactions. A simple example of a histochemical stain is seen in the affinity for basic dyes (*basophilia*) shown by the nucleus, and also by certain constituents of the cytoplasm of some types of cell. The basophilia is due to the presence of acid substances which have been identified as nucleoproteins and which consist of a protein combined with a nucleic acid. Further, on the basis of their chemical constitution, a distinction can be made between ribose nucleic acid and deoxyribose nucleic acid, of which the latter is found mainly in the nucleus and the former in the cytoplasm. There is good evidence that deoxyribose nucleic acid may be selectively stained by a technique known as the Feulgen method, the principle of which depends on the fact that, by hydrolysis in weak acid, an aldehyde group is freed and can then be demonstrated by a colour reaction with Schiff's reagent (fuchsin–sulphurous acid). Ribonucleic acid may be demonstrated by means of an enzyme, ribonuclease, the application of which abolishes the basophilia. Spectrographic absorption methods have also been used for studying these and other chemical substances in the cell, and have provided a means for estimating them quantitatively and for studying the changes in their distribution which occur in relation to cycles of cell growth and activity.

Besides staining in the strict sense of the term, certain tissue elements can be studied microscopically by the method of impregnation with metallic salts. In this case the elements are brought to view, not by direct coloration with a soluble dye, but by the deposit on them of particulate material. It is not possible, however, to make a clear distinction between staining and impregnation, for some reagents depend on both processes for their results. The whole theory of staining in its application to microscopic anatomy is a particularly complicated subject, and for its discussion, as well as for the practical details of histological technique, reference must be made to textbooks which deal specially with them.[1]

Histological technique forms the basis of embryological studies. The progressive development of the tissues and organs of the body can be followed by preparing serial sections cut systematically through embryos of different ages. Each series of sections then provides material from which the detailed anatomy of the embryo at one particular phase of development

---

[1] For a brief but excellent account of the rationale of cytological technique see J. R. Baker, *Cytological Technique*, London, 4 edn., 1960.

can be studied, and, by linking them up in order, it is possible to see a picture of progressive change as the structure of the embryo gradually becomes more elaborated. At first this involves the differentiation of tissues of various kinds from embryonic cells of a generalized type (*histogenesis*), and later on the progressive building up of the various organs to their definitive form (*organogenesis*).

To facilitate the study of the anatomy of minute embryos, and to permit their examination in a three-dimensional plane, enlarged models are commonly reconstructed from the serial sections. This is accomplished by making a large-scale model of each individual section, in which the particular organs or tissues which are being studied are plotted out, and then, by fitting together in series the individual models, an enlarged replica of the entire embryo is obtained. The method of studying intrinsic anatomical structure by the use of serial sections and reconstruction models is by no means confined to embryology; it is also widely used for the investigation of the finer details of the anatomy of mature organs and tissues.

It will be realized that anatomical methods involving the study by dissection or microscopic sections of dead and preserved tissues can extend beyond the examination of what may be termed 'normal' material. The study of pathological material, or of material which has been subjected to experimental procedures during life, is extremely important for the elucidation of problems relating to the growth, differentiation, and adaptational potentialities of different tissues and organs. The field of experimental anatomy, in particular, offers possibilities for this type of investigation.

Experimental anatomy, as a method of scientific inquiry, includes all those experimental procedures which are directed towards the interpretation of organic structure. The analysis of growth mechanisms, of the factors which initiate and control the anatomical differentiation of tissues and organs, of the adaptation of structure to function, and so forth, must all be ultimately pursued by the method of experiment, and, in discussing the tissues of the body in the following pages, we shall note how prominent a part is taken by this in modern anatomical research. By way of example, we may note here such studies as the investigation of the intrinsic anatomy of the nervous system by observing the structural and functional changes which ensue after lesions involving groups of nerve cells or tracts of fibres; the investigation of the morphological specificity of nerve-endings in relation to different types of sensation by seeking to correlate the results of sensory tests with the microscopical study of different types of nerve-ending in the skin and other tissues; the endocrine control of the growth of bone, of the cyclical changes in the sexual system, or of the development of secondary sexual characters, &c.; or the mechanical and chemical factors which induce or modify tissue differentiation.

Methods of studying the anatomy of living cells and tissues *in vivo* and *in vitro* are clearly of great importance, since they provide a means of eliminating post-mortem artefacts, and also allow a continuous study of structural changes the details of which can often only be inferred indirectly by the study of dead, preserved material. Cellular organization can be studied directly in the living animal under high powers of the microscope in the

B

case of tissues which are sufficiently transparent to allow examination by transmitted light. For example, by spreading out the thin mesentery of an anaesthetized animal over a glass plate, the minute structure of capillary blood vessels and lymphatic vessels, as well as their cellular content, can be made visible. Living cells which float freely in a fluid medium, such as blood corpuscles or germ cells, can also be directly studied by the examination of small droplets of the fluid freshly removed from the body.

During the last forty years considerable progress has been made in the technical methods of examining tissues *in vivo*. The differentiation of blood vessels, muscle fibres, and nerves, for example, has been intensively studied in the translucent tails of larval amphibians, and it has been possible by such a method to follow the reactions of these tissues to mechanical, chemical, or other stimuli. Special mention should be made of the technique which enables mammalian tissues to be studied by constructing transparent chambers in the ears of rabbits and other animals. Briefly, this consists of punching a hole in a rabbit's ear (under aseptic conditions) and enclosing it between thin plates of mica or celluloid, or some transparent plastic material. The plates are kept firmly in position by screws so that the area is effectively sealed off while the blood supply to the adjacent parts of the ear is not disturbed. The transparent chamber formed in this way is filled with exudate and quickly becomes invaded by connective tissue, blood vessels, lymphatics, and nerves. A permanent preparation is thus provided in which the cellular elements of these tissues can be watched and studied under high magnifications from day to day (and even for many months or years). In some cases, instead of punching a hole, the skin is carefully dissected off on each side over a circular area and a thin transparent layer of the original tissue left behind which can be studied directly. Various modifications and improvements of the transparent-chamber technique originally devised by Sandison have been made from time to time, and for many years now the method has been employed for the most intimate study of tissue growth, differentiation, and activity in the living body.[1]

It will be realized that the microscopic appearance of fresh cells by ordinary transmitted light will not show all the cellular structures which are revealed in fixed material by appropriate selective staining. Many of these details, however, can be brought to view by special optical methods. One of these is the method of *dark-ground illumination*. The principle of this microscope technique depends on illumination by light rays which are directed obliquely towards the object of examination, allowing intracellular structures to be shown up by the refraction and diffraction of the light rays at their surfaces of contact with the surrounding medium (see Fig. 1). The method of dark-ground illumination even permits the visual demonstration of ultra-microscopic particles which can be perceived indirectly by the diffraction haloes about them. Another method of the greatest importance which has been developed in recent decades is termed *phase-contrast microscopy*.

---

[1] J. C. Sandison, 'The transparent chamber of the rabbit's ear', *Amer. Journ. Anat.* **41**, 1928; E. R. Clark *et al.*, 'Recent modifications in the method of studying living cells in transparent chambers', *Anat. Rec.* **47**, 1930; R. H. Ebert, H. W. Florey, and B. D. Pullinger, 'A modification of the Sandison–Clark chamber', *Journ. Path. Bact.* **48**, 1939.

This makes use of an optical principle whereby comparatively small variations in the refractivity of intracellular elements can be converted into variations in the intensity of transmitted light so that the latter can be detected by the eye (or by a photographic plate). Thus, in the living cell intracellular particles which are invisible by ordinary transmitted light (because in their refractive index they differ so little from the surrounding medium) can be brought to view as clearly as in material which has been specially stained to demonstrate them. The use of phase-contrast microscopy in the examination of undamaged living cells has made it possible to distinguish artefact from reality with a degree of certainty not previously possible in cytological studies.[1] Other methods are also available for revealing structural details which are beyond the powers of optical resolution by the highest powers of the microscope as ordinarily used. These include the study of living tissues by means of ultra-violet light, polarized light, X-ray diffraction methods, and electron beams. With techniques such as these, it is actually possible to obtain information regarding the pattern and arrangement of the aggregated molecular constituents (*micellae*) of living protoplasm in different parts of a cell, or in cells of different types.

In the electron microscope (which allows a magnification of as much as 200,000 diameters or more), the electron beams are focused by means of an electro-magnetic field, and the image of the section is recorded photographically or observed directly on a fluorescent screen. While the theoretical resolving power of the microscope is rather less than an Ångström unit, in practice the maximum resolution of biological material is about 20 to 50 Å.[2] The resolving power depends among other things on the thickness of the sections, and these are cut with a glass knife at a thickness of something less than $0.05\,\mu$ (i.e. less than 500 Å). The tissue to be examined is commonly fixed in osmic tetroxide (which also functions as a stain) and embedded in a thermoplastic. Since the section has to be placed in a vacuum for examination it will be understood that the opportunities for distortion and artefact in the course of preparation of material from living tissues are very great indeed. Thus much of the study of electron microphotographs necessarily involves the question of how the appearances which they show should be interpreted. However, the difficulty is beginning to be overcome by the differential elimination of known chemical substances following the application of appropriate enzymes.

An important method used for demonstrating the anatomical details of living cells is that of *vital staining*. Certain non-toxic dyes are available which will stain selectively living tissue without affecting its normal functions or vitality. They can be used either by injecting solutions of the dye into an animal during life (*intra-vital staining*), or by the immersion of fresh tissue immediately after its removal from the body in solutions of the dye *in vitro* (*supra-vital staining*). These methods have been specially applied to the study of certain types of connective-tissue cells (see p. 49), of nerve

[1] R. Barer, 'Progress in practical microscopy', *Journ. Roy. Micr. Soc.*, Dec. 1951. See also the chapter by the same author on 'Phase, interference and polarizing microscopy', in *Analytical Cytology*, ed. by R. C. Mellors, 2 edn., McGraw–Hill, 1959.

[2] The symbol Å refers to the Ångström unit of measurement which equals $\dfrac{1}{10000}$ micron.

fibres and nerve-endings, and of cytoplasmic contents of the cell such as mitochondria, secretory granules, and so forth. They have also been used to good effect in experimental embryological studies, for it is possible to 'mark' certain cells in the developing embryo by vital staining, and then to follow them to their ultimate destiny in the subsequent differentiation of the tissues (see p. 31).

The study of living and growing tissues *in vitro* has been made possible by the development of the technique of tissue culture.[1] This valuable technique was initiated by an American anatomist, R. G. Harrison, in 1907, in his studies of the formation and growth of nerve fibres. Expressed simply, the method consists in placing a minute fragment of fresh tissue in a suitable nutrient medium and preserving it (under aseptic conditions at a suitable temperature) in a sealed chamber on a microscope slide, or in a watch-glass, flask, or some other glass receptacle which enables it to be examined microscopically.

In Harrison's original technique, embryonic tissues were placed in a hanging drop of lymph over the depression of a hollow-ground slide. Blood plasma was later substituted for lymph, as it proved to be a more effective culture medium. It forms a solid coagulum over the surface of which cell proliferation can readily proceed. Nutrient fluids are required for the continued survival of a culture, as well as frequent washing to remove metabolites. To stimulate growth and differentiation of cells *in vitro* (apart from simply maintaining their vitality), embryonic extracts are now also used as a common constituent of tissue-culture media, for it has been found that these extracts (such as the juice obtained from crushed chick embryos) are particularly rich in growth-promoting substances. Similar substances are also present in certain organs of adult animals; for example, in the cultivation of mammalian tissues extracts of spleen may be employed.

With the use of the tissue-culture technique, it has been possible to study the growth and differentiation *in vitro* of a great variety of tissues, such as muscle, nerve, and epithelium, and even to cultivate the embryonic rudiments of whole organs such as the eye and internal ear, or a limb bud. Perhaps the most remarkable result of this work has been to demonstrate the amazing powers of self-differentiation of embryonic tissues when they are allowed to continue their development completely isolated from their normal environment.[2]

In all studies of the tissues of the body and of the functional units in which they are organized, it is of the greatest importance to realize that the fabric of which they are composed is in a continual state of flux during life. Yet they maintain their morphological identity in spite of the unceasing removal and substitution of their material elements which occur in the process of metabolic interchange. Strictly speaking, there is nothing static about any anatomical unit in the living body. As we shall see, even the

---

[1] For a general account of the technique of tissue culture see R. C. Parker, *Methods of Tissue Culture*, 3rd edn., New York, 1961.

[2] For a short account of the results which have been achieved by the technique of tissue culture see the chapter on 'Histogenesis in Tissue Culture', by Honor B. Fell in *Cytology and Cell Physiology*, 2nd edn., Oxf. Univ. Press, 1951, and the article by the same author on 'Recent advances in organ culture' in *Science Progress*, No. 162, 1953.

inorganic constituents of bone are being continually removed and replaced. The material of a bone in life, therefore, is not the same from one month to another. Notwithstanding this fact, however, in its contour and architecture the bone outwardly preserves its identity. Again, the skin of the hand, which appears to be fixed in position with a permanent pattern of lines and creases, is in a continual process of removal and replacement. Thus a cross-section of skin as seen under the microscope is also a cross-section of time—it represents an arbitrary point in an uninterrupted process of change. But not only does this process of change affect anatomical units as a whole, it involves the very molecules out of which living matter is constituted. In the phenomenon which has been called 'transamination', nitrogen is being continually shifted on in living tissue from one organic molecule to another in an everlasting chain of 'hand to hand' transport. Thus, not even the ultimate elements of living substance are stable. Yet the pattern to which they conform remains the same. It seems almost to demand the postulation of a pre-existent system of physical forces with a stable spatial organization to whose pattern the molecules of organic substances must perforce adapt themselves as they become assimilated into living matter.

It is as though, when we look at the living body, we look at its reflection in an ever-running stream of water. The material substratum of the reflection, the water, is continually changing, but the reflection remains apparently static. If this analogy contains an element of truth, if, that is to say, we are justified in regarding the living body as a sort of reflection in a stream of material substance which continually passes through it, we are faced with the profound question—what is it that actually determines the 'reflection'? Here we approach one of the most fundamental riddles of biology—the 'riddle of form' as it has been called, the solution of which is still entirely obscure.

3. THE ANATOMY OF THE CELL

It has been noted that all living tissues have a cellular basis. The study of tissues as organized living material must therefore be preceded by a consideration of the anatomy of the cell.

The essential anatomical features of the living cell may be described by reference to a generalized type of cell found in embryonic and adult connective tissue—commonly termed a *fibroblast* (Fig. 1). Such a cell is composed of a minute fragment of a jelly-like substance called protoplasm, in the middle of which is a rounded globule, the nucleus. The protoplasmic basis of the nucleus is termed the *karyoplasm*, and that of the cell body in which it is embedded, the *cytoplasm*. Since protoplasm is the material substratum of all living processes, its structural composition has provided one of the central problems of cytological study.

For many years, the interpretation of protoplasmic structure had been vitiated by the fact that it was mainly studied in dead tissues preserved with fixatives, often dehydrated by treatment with alcohol, and prepared for sectioning by embedding in paraffin wax or some other substance. Evidently, treatment of this kind is likely to disturb considerably the finer microscopic details of cellular structure and to distort the actual anatomical appearance of the cell as it really exists during life. Even the direct examination of pieces of fresh tissue immediately after they have been removed from the body does not avoid the possibility of post-mortem artefacts. For, unless the cells are kept alive by the specialized technique of tissue culture, they rapidly undergo degenerative changes, with coagulation of the protoplasm. In this case, microscopical examination gives a picture of a dying or

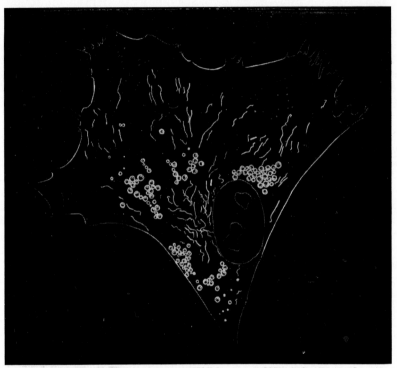

FIG. 1. Drawing of a living cell (fibroblast) as seen under high magnification with darkground illumination in a culture of embryonic tissue. The nucleus is outlined by a fine nuclear membrane and contains two nucleoli. Scattered in the cytoplasm are many filamentous mitochondria, and near the nucleus are clumps of highly refractile fat globules. × 1000 (approx.). (Drawing by Dr. H. B. Fell.)

a dead cell but not the true appearance of a healthy living cell. Finally, it is to be remembered that, even if their vitality and growth are maintained artificially, all tissues removed from their normal environment in the living body for examination *in vitro* must for this very reason be abnormal, and the cytologist is required continually to take this fact into consideration in his attempt to construe the normal anatomy of tissue elements.[1]

Cytoplasm. On the basis of the examination of fixed and stained tissues, it was at one time supposed that cytoplasm has a definite reticular structure, being composed of an exceedingly fine network of relatively solid or viscous

[1] For recent statements on the structure of the cell and its constituent elements, reference may be made to various articles in *Cytology and Cell Physiology*, op. cit.

fibrils enclosing in its meshes a more fluid substance. It is now recognized that this appearance is the result of the formation of a coagulum or precipitate in damaged cells. The suggestion that cytoplasm has essentially a granular structure has also been disproved, for the fine granules often seen in cells with the light microscope can be displaced by centrifuging, leaving a clear substance whose functions as living protoplasm are not disturbed.

In a healthy, living cell of a generalized type, microscopical examination under the highest powers of the light microscope gives cytoplasm an appearance of being optically structureless. It seems, in fact, to be a homogeneous watery fluid of viscous consistency, containing in colloidal solution a number of chemical substances such as various proteins, carbohydrates, and lipoids, as well as crystalloids in true molecular solution. This is not to say, however, that cytoplasm is homogeneous in its physical and chemical constitution; indeed, there is evidence that this is not the case. But such differentiation as exists is not necessarily static; its distributional pattern may be supposed to fluctuate continuously with varying physiological conditions. In certain types of cell, intracellular differentiation of cytoplasmic structure may reach visible definition and appear relatively fixed. Examples of such a differentiation are the striations seen in voluntary muscle fibres, the fibrillae of nerve cells, and also the characteristic spindle formation that occurs during cell division. Although it was at one time doubted whether such structural appearances (as seen in fixed preparations) in all cases accurately reflect morphological reality, it is now clear from studies with the electron microscope that they do have an objective existence.

That the cytoplasm of most cells is a fluid of very low viscosity is shown by the free movement of particles within it, the diffusion of soluble substances which have been introduced by micro-injection, and the effects of centrifuging. Towards the surface, however, it is usually more of a gel in consistency, and this zone is sometimes termed the ectoplasm. But there is no sharp boundary line between the ectoplasm and endoplasm, and both are capable of local and rapidly reversible sol–gel transformations in the course of normal cell activity. When a cell is injured or killed, the cytoplasm as a whole becomes converted into a relatively stiff gel. The surface membrane of a cell is called the *plasma membrane* which is estimated from electron microphotographs to be somewhat less than 100 Å thick, and is believed to consist fundamentally of a lipid film of molecular dimensions to which is adsorbed on either side a layer of protein molecules. It is of the utmost importance in the consideration of the metabolic activity of a cell in relation to its immediate environment in living tissues. This becomes obvious when it is realized that all interchange between the cell and the surrounding medium must take place through the cell membrane, whether this involves the transfer of nutrient substances into the cell or the discharge of metabolites and other products of cell activity. This membrane, indeed, is the anatomical and physiological boundary between the cell and its environment, and for this reason its precise constitution has been one of the most important objects of cytological study in recent years.[1] For

[1] See the chapter by J. F. Danielli on 'The cell surface and cell physiology' in *Cytology and Cell Physiology*, op. cit.

example, electron microscopy has revealed that, apart from its capacity for absorbing fluids by direct diffusion, the cell membrane also shows a remarkable feature in transferring water or other fluid material into the cytoplasm by a process called *pinocytosis*. In this process the cell membrane becomes invaginated as a deep recess into the cytoplasm, engulfing the fluid; then, the deep part of the recess becomes 'nipped off' from the rest of the cell membrane to form an enclosed vesicle (Fig. 2). Such vesicles transport their contained fluid to various regions of the cytoplasm or the nucleus, and in some cases (for example, in the transference of fluid plasma

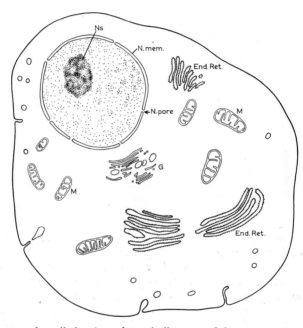

FIG. 2. Diagram of a cell showing schematically some of the structures made visible by the electron microscope. The nucleus, with its nucleolus (*Ns*), is seen enclosed by the nuclear membrane. The latter is a double membrane perforated by nuclear pores. Structures in the cytoplasm include the Golgi apparatus (*G*), endoplasmic reticula (*End. Ret.*), and mitochondria (*M*). Also to be noted are the scattered small vesicles formed by invaginations of the cell membrane that become 'pinched off' and thus separated from the latter.

from capillary vessels to the surrounding tissues) right across the cell to be liberated again at its opposite surface. The ingestion of solid particles by certain kinds of cell (*phagocytosis*) is effected in the same way, i.e. by the preliminary enclosure of the particles in recesses of the cell membrane and the subsequent conversion of the recesses into closed vesicles.

It seems likely that cell membranes in different types of cell have specific properties, and that such distinctive properties play an essential role in the cohesion and adhesion of cells all of the same type in the formation of organized tissues. Thus it has been observed that if dissociated and isolated cells of many different types are mixed at random in a tissue culture suspension—epithelial cells, mesodermal cells, and so forth—the epithelial cells appear to 'seek out' one another to the exclusion of other cell types and then aggregate and cohere to become organized in an epithelial sheet. It has been suggested that this phenomenon of 'sorting out' of cells of like type depends on a selectivity in the mechanism of adhesion.[1]

In addition to its more general functions, the nature of the cell membrane and its properties of differential permeability in relation to ionic transfer have particular reference to the electrical potential changes associated with the propagation of the effects of stimuli along such specialized cells as those of nerve and muscle tissue.

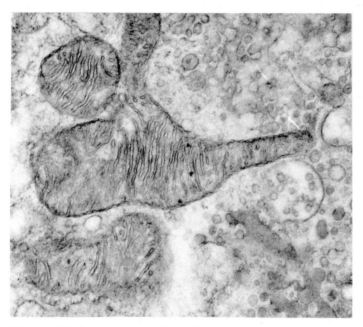

FIG. 3. Mitochondria from a plasma cell, showing the partial septa projecting into their interior. × 40,000. (Microphotograph supplied by Dr. G. A. Meek.)

Within the cytoplasm of cells there are generally found certain essential elements which are believed to be of considerable importance, even though their function remains obscure. These include the *mitochondria*, the *Golgi apparatus*, and networks of flattened membranous sacs called endoplasmic reticula. All these structures are evidently important components of the living cell (Figs. 2, 3, and 5).

Mitochondria are minute filamentous or granular bodies usually scattered throughout the cytoplasm, but in some cases, particularly in secretory cells, congregated in close relation to the nucleus (Fig. 1). In the living cell they give the appearance of being actively motile, and in cinematographic records of cells grown in tissue culture they are observed to be in continual

[1] M. S. Steinberg, 'Reconstruction of tissues by dissociated cells', *Science*, **141**, 1963.

writing movement and to be capable of multiplying by division. If the cell is damaged the movement slows down and may cease. In number they range up to several hundreds in one cell. Beyond the fact that they are composed of proteins and lipoids in varying proportions, little is known of their chemical composition, and not a great deal is yet known of their functional significance. It was at one time suggested that they become actually converted into secretory granules in glandular cells, and form the basis for the differentiation of such structures as the hair-like processes of ciliated epithelium and the myofibrillae of muscle cells. However, this conception of such a direct transformation has for long been completely abandoned, and it

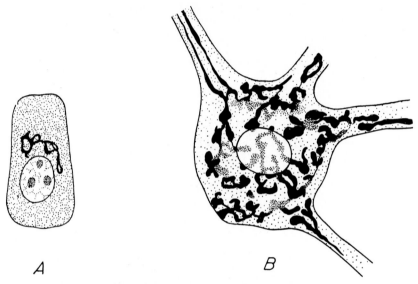

FIG. 4. Diagram of an epithelial cell of the thyroid gland (*A*) and a motor nerve cell from the spinal cord (*B*). These illustrations show the typical appearance of the Golgi apparatus as it is seen in fixed and stained material.

is now clear from electron-microscopic studies that mitochondria are in some way closely associated with the elaboration of products of cell activity, probably through the activity of intracellular enzymes which are known from centrifugation experiments to be concentrated at their surface, or on the partial membranous septa (*cristae*) which appear (from electron microphotographs) to occupy their interior. Since these enzymes include the so-called respiratory enzymes, it has been suggested that the mitochondria constitute the main centres of oxidative processes within the cell. A close topographical relationship between mitochondria and the cell nucleus has also been observed at certain phases of mitotic division, and this has suggested another possible role for these puzzling structures: that they are perhaps concerned with the formation or dissolution of the nuclear membrane. What can be quite certainly stated in general terms is that the mitochondria in the cytoplasm constitute the essential elements of an intracellular chemical 'factory' which is responsible for a great variety of intracellular syntheses and metabolic activities.

The Golgi apparatus has been described as a protoplasmic reticulum common to all types of cell (Fig. 2). Usually situated around or in close proximity to the nucleus, it undergoes a characteristic displacement in glandular cells in relation to their secretory activity. In other cases it may become dispersed throughout the cytoplasm in the form of fine granules, and it undergoes rapid disintegration in degenerative conditions. The Golgi apparatus, which can be demonstrated histologically with such reagents as silver nitrate and osmic acid, has been regarded by some observers as an

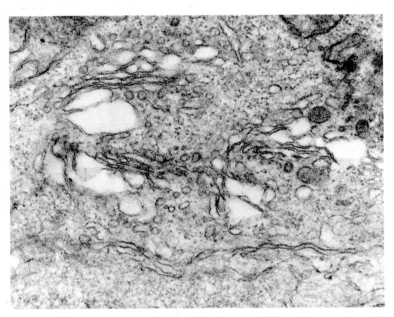

Fig. 5. The Golgi apparatus of a plasma cell. × 50,000. (Microphotograph supplied by Dr. G. A. Meek.)

artefact. On the other hand, its existence as a definite structural entity seems to be confirmed by the observation that it can be displaced without losing its characteristic discrete appearance in cells which are submitted to great centrifugal force in an ultracentrifuge,[1] and the fact that it is displaced centripetally shows that it is composed of a substance less dense than the surrounding cytoplasm. There has been much argument to and fro on whether the reticular appearance of the Golgi apparatus seen in fixed and stained preparations is an accurate representation of this cytoplasmic inclusion as it occurs in the living cell, or whether it is in some degree an artefact due to the precipitative action of fixatives on the lipoid material which it contains. According to the latter view, the apparatus is essentially composed of a local concentration of small vacuoles (demonstrable by

[1] H. W. Beams and R. L. King, 'The effects of ultracentrifuging upon the Golgi apparatus', *Anat. Rec.* **59**, 1934.

staining with neutral red) embedded in a matrix of lipoid substance which is not always clearly demarcated from the surrounding cytoplasm. It has been argued, further, that the reticulum seen in fixed and stained sections is the result of the precipitation of silver or osmium on the surface of lipoid droplets which in the undamaged cell are discrete and separate. However, the more conservative view seems to be gaining ground from recent studies of the living cell by phase-contrast microscopy and also from studies with the electron microscope. This view affirms the reality of the Golgi complex as a definite entity that consists essentially of groupings of vesicles, double-layered membranes packed in parallel formation and clusters of granules (Fig. 5).[1]

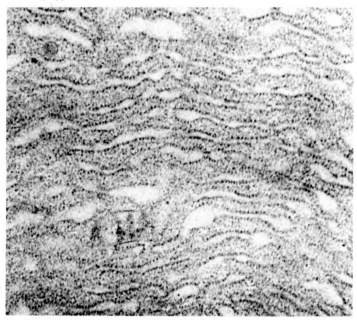

FIG. 6. Endoplasmic reticulum of a pancreatic cell. × 45,000. (Microphotograph supplied by Dr. G. A. Meek.)

In spite of much intensive study, it must be admitted that the precise role played by the Golgi apparatus in the physiology of the cell remains an enigma. The relation of variations in the appearance of the apparatus to cycles of cellular activity, as well as the appearance of the vacuoles often associated with it and the changes which they undergo, suggests that it is an important factor in cell metabolism. In glandular cells it is almost certainly concerned with secretory functions, perhaps acting as condensation

[1] For contending views on the Golgi apparatus see J. R. Baker, 'The structure and chemical composition of the Golgi element', Quart. Journ. Micr. Sc. 85, 1944; A. J. Dalton and M. D. Felix, 'Cytologic and cytochemical characteristics of the Golgi substance', Amer. Journ. Anat. 94, 1954; J. R. Baker, 'The Golgi controversy'; and D. Lacy and C. E. Challice, 'The structure of the Golgi Apparatus in vertebrate cells examined by light and electron microscopy', in Mitochondria and Other Cell Inclusions, Camb. Univ. Press, 1957.

membranes for the segregation of material used in the elaboration of the secretory products, and by its vesicles serving to transport the secretory products of glandular cells towards the plasma membrane at the surface of the cell (Fig. 2). Possibly, also, it plays a more direct chemical role in intracellular syntheses.[1]

Also situated in the region of the Golgi apparatus is to be seen in appropriately stained cells a minute dot, the *centriole*, which is sometimes double. This marks a point of protoplasmic condensation, and its significance lies in the part it plays in the process of cell division. Hence the centriole is usually absent in those specialized types of cell, for example, mature nerve cells, which are not capable of dividing.

Endoplasmic reticula consist of strands of double membranes disposed sometimes in a meshwork and sometimes heaped up in parallel formations (Fig. 6). The double membranes appear in most cases to be flattened sacs, portions of which may be separated off to form minute vesicles. In some reticula tiny granules of ribonucleic acid (ribosomes) are closely aligned along the outer surface of the membranes; similar granules are also to be found dispersed, either sparsely or densely, throughout the cytoplasm. It is under the influence of ribonucleic acid that protein synthesis within the cell proceeds.

The nucleus. The nucleus of a generalized type of cell occupies an approximately central position in the cytoplasm, and it is enclosed in a definite membrane, the *nuclear membrane*. This is a double membrane and is not complete, for it may be interrupted at frequent intervals by small gaps, *nuclear pores*, that permit the transport of nuclear material into the surrounding cytoplasm, and of cytoplasmic material in the reverse direction. A nucleus usually contains one or two small, round, and highly refractile particles, the *nucleoli* (Figs. 1 and 113). In living cells, the nucleoli may show a rather irregular contour. Like the cytoplasm of the cell body, the karyoplasm of the nucleus appears in ordinary transmitted light to be a structurally homogeneous fluid in the normal living cell. In fixed and stained preparations, however, it shows a fine network interspersed in which are minute granules. The substance of which the network is composed is commonly called *chromatin*, because it shows a marked affinity for basic dyes, and in its chemical composition it is characterized by its rich content of nucleoproteins. Phase-contrast microscopy has now demonstrated that the filaments and granules of the chromatin network visible in the stained nucleus actually exist (at least in part) as morphological entities in the living cell. They are invisible under the ordinary light microscope only because their refractive index is almost the same as that of the medium in which they are embedded. However, the fact that, by micromanipulative methods, the nucleolus may be moved about quite easily within the nucleus has been taken to indicate that the filaments can hardly form a rigid meshwork, though, of course, these methods are perhaps open to the criticism that they may effect rapid liquefactive changes in the normal constitution of the chromatin of the undamaged nucleus by mechanical injury. Of considerable

[1] See the chapter by G. H. Bourne on 'Mitochondria and the Golgi complex' in *Cytology and Cell Physiology*, 2nd edn., Oxf. Univ. Press, 1951.

interest is the recent discovery that a sex difference can be recognized in the chromatin content of the nuclei of many types of somatic cell, for in the female of some mammals (including man) a special chromatin mass, the sex chromatin, is usually to be seen as a small compact body in close contact with the nuclear membrane or (in the case of nerve cells) as a little 'satellite' related to the nucleolus. A similar structure is only rarely to be detected in the nucleus of males.[1]

In typical cells there is but one nucleus, but in some there are two or more. This may be the result of the incomplete separation of cells after division, or the fusion of originally separate cells.

The nucleus is the controlling centre of all cellular metabolism. Not only do the various functional activities of the cell depend on its integrity, but if it is damaged and destroyed the whole cell undergoes rapid dissolution.

Cell inclusions. Within the cytoplasm are frequently to be found particles of inert, non-living material which are either taken up secondarily by the cell from the surrounding medium or else are products of its own activity. They are to be distinguished from the organized living contents of cytoplasm which have already been described (e.g. mitochondria, &c.) and which are commonly termed *organelles*. Examples of inclusions are fat globules (which show a tendency to collect in proximity to the nucleus), the secretory granules of glandular cells, pigment granules, and carbohydrate in the form of glycogen (as in the liver). In the case of cells which are actively motile or amoeboid, the cytoplasm may contain a variety of ingested material as, for instance, the remains of cell debris removed in the process of repair after tissue destruction, or foreign particles such as bacteria.

4. CELL DIVISION

All cells which have not undergone extreme morphological specialization are normally capable of proliferation by dividing. The usual mechanism of cell division is the rather complicated process of *mitosis*, but the cells of some tissues (e.g. cartilage cells under certain conditions) have been stated to divide occasionally by the simpler method of *amitosis*, which involves a direct constriction and splitting of the nucleus without any apparent preliminary anatomical changes in the latter. On theoretical grounds it may be doubted whether true amitosis ever does occur in normal mammalian tissues; such cases as have been reported are perhaps those in which an intranuclear rearrangement of chromatin is not readily detected by usual histological methods.

The process of mitotic division is essentially a mechanism whereby the hereditary units or *genes* in the 'mother cell' are passed on in their full complement to the 'daughter cells'. The genes are carried by discrete filaments or rod-shaped bodies within the nucleus, called *chromosomes*. The latter are represented in the resting nucleus by the filaments and granules of chromatin, and there is indirect evidence that the pattern of spatial distribution of genes in each chromosome is preserved in this network

---

[1] M. L. Barr, 'Sexual dimorphism in interphase nuclei', *Amer. Journ. of Human Genetics*. **12**, 1960.

between one cell division and the next. The number of chromosomes is constant for any one species; in human cells there are (according to the most recent studies) forty-six. In some animals, also, the individual chromosomes in each cell show characteristic differences in size and form. In the first stage of mitotic division (*prophase*), the chromosomes become visible under the light microscope as the result of the deposition of refractile material (nucleic acid and protein) on the threads of chromatin. They appear as fine, convoluted threads which are already split longitudinally into two parts, the *chromatids* (Fig. 7 ii).[1] Microdissection of the living cell

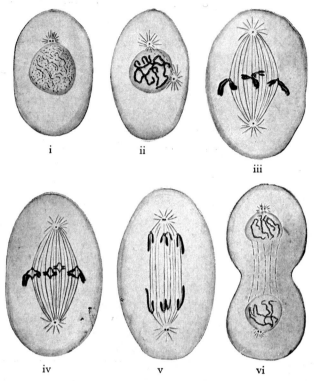

FIG. 7. Diagram showing the different phases in the mitotic division of a cell. i. and ii. prophase; iii. metaphase; iv and v. anaphase; vi. telophase. It will be understood that these phases are arbitrarily defined for the purpose of descriptive convenience. They are not sharply separable, since they represent stages in a continuous process. (From H. Gray.)

has demonstrated that the chromosomes in the prophase of mitosis have a solidity which allows them to be pulled out in loops and stretched with a fine needle point.

As the chromosomes become apparent, the nuclear membrane begins to disintegrate and soon disappears altogether. At the same time the centriole (if it is not already double) divides into two and the latter move to

[1] At one time it was supposed that the chromosomes at this stage are united end-to-end, to form a continuous coiled thread called the 'spireme'.

opposite sides of the nucleus. Between the two centrioles is formed a zone of modified cytoplasm termed the *spindle*. In fixed preparations it is seen to be traversed by a system of fine fibrillae, the *spindle fibres*, which extend, like a series of lines of force, from one centriole to the other. These fibrillae are discernible in living cells examined under polarized light, and are evidently the visible expression of an orientation of protein molecules between the centrioles. As the centrioles move apart, the nuclear elements come to lie in the central region of the spindle, the chromosomes arranging themselves in a radiating formation around its equatorial plane (Fig. 7 iii). At this stage of mitosis, which is termed the *metaphase*, the chromosomes become shorter and thicker, and actual dissection of the living cell shows they have a viscous consistency and are apparently capable of active movement. In the next stage (*anaphase*) the two chromatids which form each chromosome are dragged apart and move in opposite directions towards the extremities of the spindle (Fig. 7 v). The central portion of the spindle now elongates, and as a consequence the two sets of chromatids (which now become the daughter chromosomes) are separated still further. In the last stage of mitosis (*telophase*) each group of daughter chromosomes is enclosed by the formation of a new nuclear membrane, and the individual chromosomes appear to undergo dissolution (Fig. 7 vi). What actually happens, however, is that they swell (presumably by the imbibition of water), and their charge of nucleic acid and protein becomes more diffuse and dilute so that finally they are no longer demonstrable as individual units by the usual histochemical tests. At the same time, the cell body itself divides by a constriction which narrows down to a fine thread and ruptures as the daughter cells draw themselves apart.[1]

The whole process of mitosis, as seen in cultures of mammalian tissue, takes from thirty minutes to two or three hours to complete. The duration of mitosis in the living body has been calculated by indirect methods. It has been shown, for example, by correlating the incidence of mitotic figures with the growth rate of regenerating liver, that in this tissue each mitosis lasts on the average forty-nine minutes.[2] There is evidence to show that, under normal conditions of light, there is a twenty-four-hourly rhythm in the mitotic division of animal cells, a rhythm which is well known to occur in plants. It is interesting to note, also, that this periodicity is abolished by continuous exposure to light, but not by continuous darkness.[3]

Mitosis has been recorded cinematographically, and the films give a very vivid impression of the dynamic energy which is expended by a cell when it divides. The active motility of the chromosomes, their lining up and drawing apart in the equatorial plane, and their mass movement to opposite poles are expressive of the changes going on within the cell. More striking still is the movement which involves the cell as a whole. The last

---

[1] For a study of the living cell in process of mitosis, see A. F. W. Hughes, *The Mitotic Cycle*, London, 1952. See, also, D. Mazia, 'Mitosis and the physiology of cell division', *The Cell*, **3**, 1961.

[2] A. M. Brues and B. B. Marble, 'An analysis of mitosis in liver restoration', *Journ. Exp. Med.* **65**, 1937.

[3] A. Carleton, 'A rhythmical periodicity in the mitotic division of animal cells', *Journ. Anat.* **68**, 1933–4.

phases of division are accompanied by the continuous protrusion and re-traction of pseudopodial processes which, as seen in a speeded-up film, suggest a sort of effervescence or 'bubbling', increasing in intensity as the daughter cells struggle to separate themselves. Finally, when the cells move away from each other, drawing out a fine thread-like connexion of protoplasm, the 'effervescence' calms down and the cells resume their normal appearance.

During the maturation of the sex germ cells (ovum and spermatozoon), a type of cell division occurs which in general resembles mitosis but is much more complicated. It is called *meiosis*. The process of maturation involves two of these highly modified cell divisions, which allow for the interchange of the hereditary material between homologous chromosomes and also reduce the chromosomes to half the number found in somatic cells. This reduction (which occurs in the first meiotic division) is a necessary preparation for fertilization, so that when this occurs the mingling of the male and female chromosomes reproduces the normal number. For the details of meiosis, reference should be made to standard works on genetics.[1]

[1] For an account of the process of meiosis see E. B. Ford, *Genetics for Medical Students*, 5th edn., London, 1961; M. M. Rhoades, 'Meiosis', *The Cell*, **3**, 1961.

II

THE DEVELOPMENT OF TISSUES IN
THE EMBRYO

1. THE EARLY DEVELOPMENT OF THE HUMAN
EMBRYO IN BRIEF OUTLINE

THE rudiments of the main tissues of the body are laid down in the human embryo during the first few weeks of development. In this section a brief sketch will be given of the structural transformations whereby the fertilized ovum is converted into an organized embryo in which these tissues assume a definite morphological plan.

Fertilization of the ovum by the sperm occurs during its passage down the oviduct (Fallopian tube) from the ovary to the uterus. The initial stages in development therefore take place in the oviduct, for it is probably not till three or four days later that the segmenting ovum reaches the uterine cavity where it becomes implanted in the uterine mucosa. The entry of the sperm into the ovum provides a direct stimulus which initiates the process of development. The nature of this stimulus is probably of a simple type, for in lower vertebrates (e.g. amphibia) cell cleavage in a parthenogenetic egg can be induced by simple mechanical stimulation. The sperm also contributes the paternal quota of hereditary factors, and the path which it takes inside the ovum is one of the factors which helps to determine the plane of cleavage when the ovum first divides.

The rapid cleavage of the fertilized ovum soon transforms it into a ball of loosely packed cells which appear anatomically similar except for slight differences in size. This is the *morula* (Fig. 8 A). The cells of the morula (*blastomeres*) next become arranged so as to form an outer layer or capsule, and a central core. The capsule, which is called the *trophoblast*, is concerned with the nutrition of the early embryo but takes no part in its actual formation; it need not therefore be considered further here, except to note that it becomes elaborated to form villous processes (chorionic villi) which penetrate deeply into the uterine wall. Within the trophoblastic capsule a cavity appears, filled with a gelatinous matrix. As a result, the central core of cells becomes pushed to one side, and forms a localized clump, the *inner cell mass*, which is attached at one point to the inner surface of the trophoblast. The whole structure is now termed the *blastocyst* or *blastula* (Fig. 8 B).[1]

Next, the inner cell mass becomes differentiated to form two hollow vesicles whose walls are each composed of a single layer of cells. The upper or dorsal vesicle is the *amniotic cavity*; the lower is the cavity of the primitive gut or *archenteron* (Fig. 9 B). The cells lining the amniotic cavity are

[1] For a statement on the earliest stages in the development of the human ovum, see W. J. Hamilton and J. D. Boyd, 'Phases of human development', *Modern Trends in Obstetrics and Gynaecology*, London, 1950.

partly concerned in forming the skin (epidermis) of the embryo and are termed *ectoderm*. Those lining the archenteron will later provide the lining epithelium of the alimentary tract and its derivatives, and comprise the *entoderm*.

Lastly, in the gelatinous matrix filling the rest of the blastula, there appear loosely arranged spindle-shaped cells. These contribute to the formation of embryonic connective tissue and are collectively termed the *primary mesoderm*. At this early stage, therefore, the three primary germ layers of the embryo are established.[1]

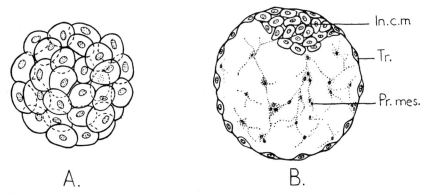

A.   B.

Fig. 8. Diagram illustrating the transformation of a morula into a blastocyst in a human embryo. *A*. Morula. *B*. Blastocyst, showing the inner cell mass (*In.c.m.*); the trophoblast (*Tr.*); and the primary mesoderm (*Pr. mes.*) which fills the cavity of the blastocyst.

The organization of the embryo itself occurs at the area of contact between the amniotic and archenteric vesicles. Here, where the ectodermal and entodermal layers of cells are pressed together in close contact, a flattened, oval, bilaminar plate appears—the *embryonic disk*. The dorsal surface forms the floor of the amniotic sac, while the ventral surface is equivalent to the roof of the primitive gut. As development proceeds, the flattened disk becomes converted into an elongated cylindrical embryo. This occurs as the result of the expansion of the amniotic sac and the growth of the embryonic disk itself. The latter bulges upwards into the amniotic cavity and is gradually 'pinched off' from the upper surface of the archenteron (Fig. 9 c). During this process a part of the archenteric cavity becomes included within the body of the embryo to form the *intra-embryonic gut*; the greater part remains outside as the *extra-embryonic gut* or yolk sac. The restricted communication between the yolk sac and the intra-embryonic gut is the vitelline duct, and this becomes narrowed down until it is finally obliterated.

The caudal end of the embryonic disk is attached to the wall of the

---

[1] It should be noted that the germ layers are not such distinct morphological entities as the older embryologists conceived them to be. This is particularly evident in the formation of the secondary mesoderm, which undergoes differentiation in such intimate relation to ectoderm and entoderm that it is hardly possible to follow with certainty the destiny of the three elements in this region.

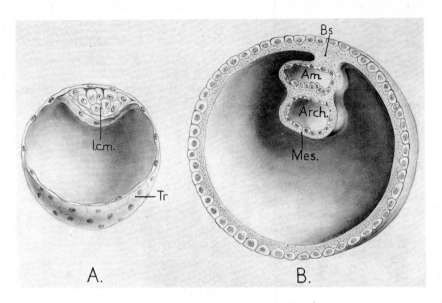

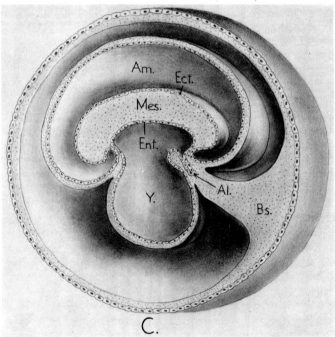

FIG. 9. Diagram illustrating the formation of the human embryo in early development.
A. Blastocyst, showing the inner cell mass (*I.c.m.*) enclosed within the trophoblast (*Tr.*).
B. The formation of the amniotic sac (*Am.*) lined by ectoderm, and the archenteron (*Arch.*)
lined by entoderm, is indicated. The epithelial walls of these sacs are covered by a layer of
mesoderm (*Mes.*) which is continuous by the body stalk (*Bs.*) with the mesoderm lining
the trophoblast. *C*. The amniotic sac has now expanded and the embryo begins to take
shape. It is covered dorsally by ectoderm. The archenteron has become divided into
the intra-embryonic gut and the extra-embryonic gut or yolk sac (*Y*.). The entoderm
of the intra-embryonic gut is separated from the ectoderm by secondary mesoderm except
in front and behind, where they are in contact, forming respectively the bucco-pharyngeal
and cloacal membranes. A tubular entodermal diverticulum from the hind end of the gut,
the allantois (*Al.*), extends back into the body stalk.

blastula by a solid mass of mesoderm, the *body stalk*. This provides the basis for the development of the umbilical cord, and through it blood vessels come to extend from the embryo to vascularize the developing placenta.

At about the fifteenth day of development, the embryonic disk begins to undergo important changes. Viewed from its dorsal aspect, an opaque

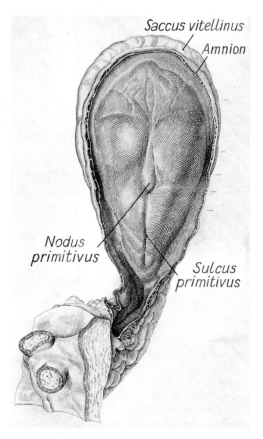

FIG. 10. Dorsal view of a presomite human embryo after removal of the amniotic sac. The line of the primitive streak is marked on the surface by a longitudinal groove (sulcus primitivus) at the anterior end of which is the primitive knot (nodus primitivus). Caudally is seen the body stalk by which the embryo is attached to the chorion. × 50. (From C. H. Heuser.)

median streak comes into view, extending antero-posteriorly in the caudal part of the disk (Fig. 10). This is the *primitive streak*, a structure of great significance since it marks the site of the first active differentiation of tissues. The primitive streak, in essence, is a linear zone of cell proliferation in the upper (ectodermal) surface of the embryonic disk. Along it, the proliferating cells sink in and then extend forwards and laterally, becoming sandwiched in between the ectodermal and entodermal layers of the disk. This newly formed tissue is the *intra-embryonic mesoderm*.

From the anterior end of the primitive streak (which forms a knob-like thickening—the *primitive knot*) the mesoderm extends forwards underneath the ectoderm in close contact with the roof of the gut. This is termed the *head process*, and it becomes moulded into a solid rod of cells, the *notochord*, which extends along the median axis of the embryo. The notochord provides the basis for the future development of the vertebral column.

The lateral extension of the cells of the primitive streak forms a longitudinal mass of tissue on either side of the midline, the *axial mesoderm*, and also spreads out further to form the mesoderm of the body wall.

As we shall see, the proliferating cells of the primitive streak not only become themselves differentiated and arranged to form embryonic structures, they also elaborate an organizing substance of a chemical nature which is able to influence surrounding regions of the embryonic disk and to induce changes in them which lead to the differentiation of a number of other structures. One of these is the *neural tube*—the forerunner in the embryo of the brain and spinal cord.

The developing nervous system first becomes evident as a thickened plate of ectodermal cells, the *neural plate*, extending forwards along the mid-dorsal line from the front end of the primitive streak. The neural plate sinks in along its median axis to form the neural groove, while its lateral margins rise up as the neural folds. The lips of the neural folds then become approximated and fuse together. The fusion of the neural folds first occurs in the middle of their length, and subsequently extends forwards and backwards until the whole of the open neural groove is converted into a closed neural tube. Finally, the neural tube sinks below the surface and becomes separated off from the ectoderm (from which it was initially derived) by the interposition of mesodermal tissue (Figs. 11 and 122).

Even before the neural groove becomes completely closed, its front end shows a conspicuous enlargement which marks the commencing differentiation of the brain. This expansion becomes still more obtrusive when the formation of the neural tube is completed. The greater part of the tube, however, retains a fairly even calibre, and forms the spinal cord (Fig. 12).

Attention may now be turned to the mesoderm. The axial mesoderm (derived, as we have seen, from the cells of the primitive streak) undergoes a segmentation which is the primary basis of the segmental character of the vertebrate body. It thus forms a series of rectangular blocks, or *somites*, ranged on either side along the developing neural tube (Figs. 11 and 12). The somites are for the most part composed of compactly arranged cells, and form the *myotomes* which give rise to some elements of the somatic or voluntary musculature of the body. Other portions of the somites contribute to the formation of the axial skeleton and connective-tissue elements of the skin, and here the constituent cells are loosely arranged in a fairly abundant intercellular matrix. This type of embryonic mesodermal tissue (which is also found in the body wall and elsewhere) is commonly termed *mesenchyme*.

Further laterally, the mesoderm becomes split into two layers by the appearance of a cavity—the *coelomic cavity*.[1] The latter represents the first

[1] This cavity actually first appears in the human embryo as a cleft in the primary mesoderm of the blastocyst.

appearance of the body cavities which persist as the peritoneal, pleural, and pericardial sacs. The outer layer of mesoderm provides a substratum to the skin of the body wall and is termed the *somatopleure*. The inner layer, or *splanchnopleure*, covers the entodermal lining of the primitive gut. A portion of the splanchnopleure is incorporated in the wall of the yolk sac and is not concerned, therefore, with forming any part of the embryo itself; the part which contributes to the wall of the intra-embryonic

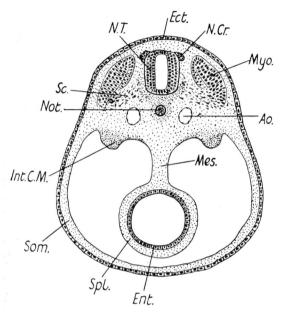

FIG. 11. A schematic cross-section through the body of an early embryo showing its main features. *Ect.* Ectoderm. *N.T.* Neural tube. *N.Cr.* Neural crest. *Myo.* Myotome. *Sc.* Sclerotome. *Ao.* Aorta. *Not.* Notochord. *Int.C.M.* Intermediate cell mass. *Mes.* Mesentery. *Som.* Somatopleure. *Spl.* Splanchnopleure. *Ent.* Entoderm.

gut provides the basis for the development of the visceral musculature and connective-tissue elements of the alimentary tract.

We have now reached the stage in our description in which the embryo is beginning to assume its distinctive shape. It is covered on its outer surface by embryonic skin or ectoderm; it contains a cavity lined by entodermal cells, which marks the beginning of the alimentary tract; the nervous system is represented by a closed tube of ectodermal cells beneath which lies the notochord; and the mesodermal somites are in position on either side of the median axis. Further details regarding the histogenesis of various tissues will be described in other chapters, but brief mention may be made here of some of the ensuing stages of development of the embryo as a whole, in order to provide a basis of reference for the problems which are discussed later.

The anterior end of the neural tube expands to form a series of three vesicles from which the fore-, mid-, and hind-brain are respectively

developed, and the precocious development of the brain leads to an early enlargement of the head of the embryo. The eyes first appear as bilateral outgrowths from the fore-brain vesicle, which approach the surface ectoderm.
The notochord becomes surrounded by a condensation of mesenchyme

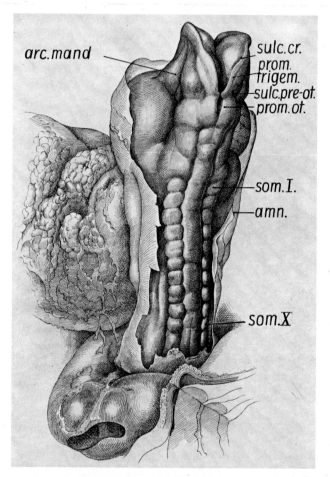

FIG. 12. Dorsal view of a 10-somite human embryo. The amniotic sac has been opened
to expose the dorsal surface of the embryo. The yolk sac is seen to the left. *arc. mand.*
Mandibular arch. *amn.* Amnion. *prom. ot.* Otic prominence. *som. I, som. X.* First and
tenth mesodermal somites. *sulc. cr.* Sulcus cristae (marking the line of origin of neural
crest cells). *sulc. pre-ot.* Pre-otic sulcus. *prom. trigem.* Trigeminal prominence. × 40. (From
G. W. Corner.)

(sometimes termed the sclerotome) which is derived from the mesodermal
somites, and which later undergoes chondrification and ossification to form
the bodies of the vertebrae and the basal elements of the skull. The noto-
chordal tissue itself eventually becomes obliterated except for vestigial
remnants which are still to be found in the adult.

The intra-embryonic gut is at first a closed cavity except for its communication with the yolk sac. Its front part (i.e. the anterior end of the fore-gut) forms the pharynx, in whose walls are developed a series of arches which in lower vertebrates provide the basis for the development of the gill arches. These arches are demarcated by intervening grooves on the outer and inner surface of the pharyngeal cavity, the *external* and *internal pharyngeal clefts*. The anterior extremity of the pharyngeal cavity comes into contact with an ectodermal depression which sinks in from the surface, the primitive mouth or *stomodeum* (Fig. 9 c).

At first, the stomodeal pit is separated from the cavity of the gut by a thin membrane—the *bucco-pharyngeal membrane*—formed by the direct contact of ectoderm and entoderm, but this soon breaks down so that the alimentary cavity establishes a communication with the exterior through the opening of the mouth. Similarly, at the caudal end of the embryo the cavity of the hind-gut is separated from a surface depression, the *procto-deum*, by the *cloacal membrane*: The latter also breaks down, and in this manner the anal and genito-urinary apertures become established.

It should further be noted that from the region of the hind-gut a small diverticular process grows out into the mesoderm of the body stalk (Fig. 9 c). This is the *allantois*, a structure of considerable importance in the early development of lower vertebrates, but whose functional significance in the human embryo is doubtful.

When the intra-embryonic gut is first formed, it is a simple elongated cavity stretching from the stomodeum to the proctodeum. It rapidly becomes lengthened and coiled, however, and here and there it sends out diverticula which form the basis of glands whose ducts open into the alimentary tract of the adult.

The heart first appears as a bilateral tubular structure in the splanchnopleure in close relation to the fore-gut, and its rapid enlargement gives rise to a conspicuous prominence in the early embryo between the pharynx and the stalk of the yolk sac. From the heart, blood vessels (which are differentiated in mesenchymal tissue) extend out into the body wall of the embryo, to the wall of yolk sac, and also through the body stalk to the developing placenta.

The basis of the genito-urinary apparatus is laid down in mesodermal tissue situated at the junction of the somatopleure and splanchnopleure, ventrolateral in position to the somites (Fig. 11). This tract of mesoderm is termed the *intermediate cell mass*. Here the gonads (ovary or testis) become differentiated, and a series of ducts and tubules also appear in it, which form the basis for the development of the duct of the testis, the oviduct, the secretory tubules of the kidney, and other associated structures.

During the fifth week of human development, the limbs appear as buds growing out from the body wall. The limb buds consist at first of a central core of undifferentiated mesenchyme enclosed in a covering of ectoderm. Condensations appear in the mesenchyme, marking out the skeletal elements of the limbs, and these later undergo chondrification and ossification. Other elements such as muscles, blood vessels, and joints make their appearance, and nerves grow down from the developing neural tube to innervate these structures.

By the end of the second month of intra-uterine life, most of the organization of the adult individual is laid down. In the third month, the human embryo becomes recognizable even by the untrained eye as a human being, and maturation of the various tissues and organs proceeds fairly evenly during the later weeks until birth takes place.

It should be emphasized that the initial stages in the development of the human embryo take place very rapidly. Developmental phases, which in lower vertebrates are fairly simple to follow, thus become compressed and overlapped to such a degree that they are not easy to decipher. Moreover, well-preserved early embryos are only rarely procured for anatomical study; hence many details of early human development have to be inferred by analogy from those which have been followed in the development of other mammals, particularly mammals which are closely related to man (the Primates).

2. MORPHOGENETIC FACTORS IN EARLY EMBRYONIC DEVELOPMENT

In the preceding section a brief and somewhat bald description of the structural changes in the early development of the human ovum has been given, showing the derivation and initial differentiation of tissues such as the nervous system, muscle segments, notochord, and the entodermal lining of the alimentary tract. In higher vertebrates the fundamental processes involved in the transformation of the segmented ovum into an organized embryo are to some degree obscured by adaptations related to the nutrition of the developing organism. Hence it is necessary at this point to refer back to lower forms of life in which these processes are much simplified. Moreover, it is on lower vertebrates that most experimental work has been carried out in the attempt to elucidate morphogenetic factors in early embryological development.

The initial stages of development, as seen in many invertebrates and in a simple form such as *Amphioxus*, may be expressed simply (and somewhat diagrammatically) as follows. The segmented ovum or morula first becomes converted into a blastula, a hollow sphere whose wall is composed of a single layer of cells.

Next, one side of the blastula becomes pushed in (or invaginated) towards the centre of its cavity so that the latter finally becomes obliterated. The whole structure now forms a hollow sac, the archenteron, whose wall is composed of two layers of cells. The outer layer is the embryonic skin or ectoderm, and the invaginated layer which lines the new cavity is the entoderm. The cavity represents the primitive gut and its opening at the site of invagination is the *blastopore* (Fig. 13 B, *Bp.*). This stage of the developing embryo is called the *gastrula*, and the process by which it is formed from the blastula is termed *gastrulation*. Between the ectoderm and entoderm there quickly appears a third layer of irregularly arranged cells, the mesoderm or embryonic connective tissue. In this manner, a type of organization is attained which is comparable to that of simple coelenterates in their mature form.

In higher vertebrates, the process of gastrulation becomes considerably modified. In the development of the amphibian egg, for example, the accumulation of yolk at one side of the blastula (the vegetative pole) makes it mechanically impossible for a simple invagination of the structure as a whole to occur. Instead, a small crescent-shaped groove appears on the surface, representing the earliest formation of the blastopore, and there follows a streaming movement on the surface of the blastula which converges towards it. The movement is the result of the actual migration of the cells of the blastula wall, which grow over the lips of the blastopore

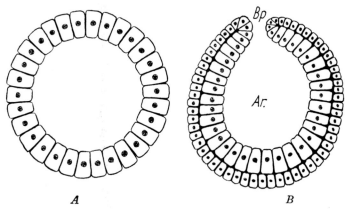

FIG. 13. Diagram illustrating the simplest type of gastrulation as shown, for instance, by *Amphioxus*. *A*. Blastula. *B*. Gastrula. Gastrulation has taken place by a simple invagination of the blastula wall. This has led to the obliteration of the original cavity of the blastula, and the formation of a new cavity, the archenteron (*Ar.*). The latter opens on the surface at the blastopore (*Bp.*).

and tuck themselves inside beneath the surface layer. This process continues until gastrulation is completed, that is to say, until all that part of the blastula wall which is destined to form the entodermal lining of the alimentary tract and the mesoderm has become invaginated (Fig. 14).

The ingrowth of the surface cells over the margins of the blastopore first occurs from the dorsal aspect of the developing embryo, hence the dorsal lip of the blastopore is the first part to become prominent. As we shall see, the tissue which forms the dorsal lip of the blastopore during gastrulation is of extreme importance because of the influence which it exerts on surrounding cells in the initial differentiation of embryonic structures.

Before gastrulation, the cells which will form various structures such as the mesodermal somites, the neural tube, &c., lie on the surface of the blastula. It has been found possible to mark them by vital staining with dyes such as Nile Blue, Neutral Red, and Bismarck Brown, so that they can be followed as they sink in through the blastopore and their fate determined. In this way Vogt[1] originally mapped out on the surface of the

[1] W. Vogt, 'Gestaltungsanalyse am Amphibienkeim mit örtlicher Vitalfärbung', *Arch. Entwicklungsmech.* **120**, 1929.

amphibian blastula areas of presumptive skin, presumptive neural plate, presumptive mesoderm, and presumptive entoderm, &c., that is to say, areas of cells which are normally destined to form these particular tissues after the completion of gastrulation.

In the blastula stage, and also at the beginning of gastrulation, the fate of the presumptive areas is not irrevocably determined—the cells are still plastic in regard to their developmental potentialities. This has been shown by removing, for example, a piece of presumptive skin and grafting it into the area of presumptive neural plate and vice versa, when the transplanted cells become differentiated to form the tissues corresponding to the area in

NB ·

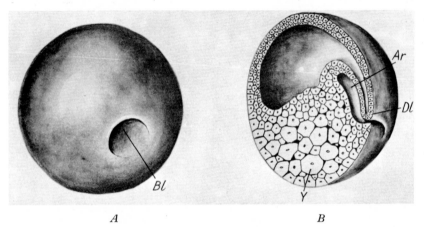

A   B

FIG. 14. Diagram showing a developing frog's ovum during the process of gastrulation. A. Surface view showing the blastopore (*Bl.*). B. In section, showing the dorsal lip of the blastopore (*Dl.*), the formation of the archenteric cavity (*Ar.*), and the large yolk cells (*Y.*) at the vegetative pole of the ovum.

which they now find themselves. Moreover, by this experimental method, presumptive ectoderm can become transformed into mesoderm, or mesoderm into ectoderm, so that at this stage even the germ layers are quite plastic. After the commencement of gastrulation, however, the cells lose their primary plasticity and their immediate developmental destiny becomes fixed. In other words, if now a piece of presumptive skin is grafted into the area of presumptive neural plate, it continues to differentiate into skin only. It appears, therefore, that during gastrulation a change occurs affecting the embryonic cells even though at this stage they show no histological modification. This change must presumably be of a physico-chemical nature, and is termed *chemo-differentiation.*

The earliest part of the blastula to undergo chemo-differentiation is the dorsal lip of the blastopore, which is destined, after it becomes invaginated, to form certain tissues in the roof of the primitive gut, i.e. the notochord and axial mesoderm. More important, however, is the property which the dorsal lip acquires of bringing about the differentiation of other tissues from the cells which lie in its immediate proximity. In fact, as first shown by the classical experiments of Spemann, it is primarily responsible for inducing the chemo-differentiation of other parts of the blastula wall whereby the developmental potencies of the various presumptive areas become irrevocably fixed, and it is also responsible for inducing the organization of the adjacent tissues in such a way that they become converted into neural tube and brain, mesodermal somites, nephric tubules, &c. If a portion of the dorsal lip of the blastopore is grafted into another embryo of the same developmental stage, it induces the formation of these structures from otherwise indifferent tissue in its vicinity. A secondary embryo may thus

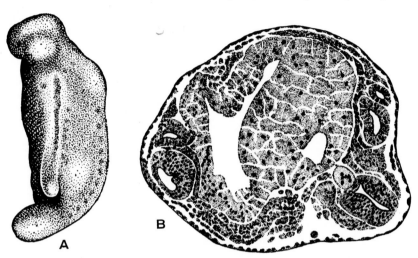

FIG. 15. An experiment in which the organizer from the dorsal lip of the blastopore of one species of larval amphibian (*Triton cristatus*) has been implanted into another species (*Triton taeniatus*). *A* shows a secondary embryo developed on the left side of the host and consisting of medullary tube, somites, otocyst, and tail; *B*, the same specimen in section: on the left the axial organs of the host are seen; on the right the medullary tube and otocyst of the secondary embryo. (From H. H. Woollard, after Spemann.)

develop in relation to the grafted tissue as an appendage of the main embryo (which is itself developed under the influence of its own blastoporic lip). Experiments such as this have led to the conclusion that the dorsal lip of the blastopore contains some substance which directly initiates tissue differentiation. This localized zone in the developing embryo has therefore been called the *organizer*.[1]

A considerable amount of work has been accomplished during recent years on the nature of the organizer. As the result of this, it is known that the organizing influence spreads by direct continuity to surrounding parts, for if the dorsal lip of the blastopore is separated from adjacent tissues by a simple cut, organization does not occur. It is known also that its activity does not entirely depend on the properties of living cells, for portions of the dorsal blastoporic lip retain their capacity to play a part in

[1] For an account of the role of organizers in embryonic development, see H. Spemann, *Embryonic Development and Induction*, Oxf. Univ. Press, 1938; J. S. Huxley and G. R. de Beer, *The Elements of Experimental Embryology*, Camb. Univ. Press, 1934; and J. Needham, *Biochemistry and Morphogenesis*, Camb. Univ. Press, 1942.

initiating tissue differentiation even after coagulation by high temperatures or by treatment with alcohol. Lastly, it is known that the chemical basis of the organizer is present elsewhere in the early embryo (though not in an active form), and is also to be found in a variety of tissues in the adult.

If the dorsal lip of the blastopore is excised experimentally and replaced by a small portion of the blastula wall taken from an indifferent area, the grafted tissue becomes itself endowed with organizing properties. In other words, the chemo-differentiation of the cells of the dorsal blastoporic lip which leads to the development of the organizer depends not on the intrinsic character of these cells, but rather on the position in which they find themselves in the early stages of gastrulation, that is, at the dynamic centre of the movement involved in this process.

The action of the organizer appears to be quite complex, for it involves at least two processes which are probably separable—*evocation* and *individuation.* Its active chemical basis has been termed the *evocator*, and this is responsible for starting off the histological differentiation of surrounding tissues. The morphogenetic processes whereby these tissues are organized or individuated into a definite system (e.g. a neural tube and brain, or a series of mesodermal somites) depend on physiological gradients in the surrounding tissues as well as on a regional differentiation within the organizer. On the basis of these factors an 'individuation field' is built up in relation to which the main embryonic structures are developed in a harmonious manner to form a complete organism.

There are two further points concerning the organizer to which reference should be made, since they throw some light on its mode of action. In the first place, the accumulation of evidence has shown that the chemical basis of the organizer is probably a sterol-like substance, and therefore relatively simple in structure. The second point is that the organizer is non-specific; that is to say, the organizer of one species of animal is still effective even if it is grafted on to the embryo of another species. It is important to note, however, that in such an experiment the developing tissues become organized in a manner which is characteristic of the host, and not of the species from which the grafted organizer was obtained. These facts suggest that the organizer simply provides the necessary stimulus which sets off an organizing activity *already* latent in the cells which are involved in the process. It has been suggested, therefore, that the term 'organizer' is not very appropriate in the sense in which it has been used. However, Spemann emphasized the fact that he originally coined the term as descriptive of a process, without intending to imply an explanation of how this process is brought about. He suggests as an alternative terminology that in any process of induction the inductive tissue may be referred to as the 'acting system', and the tissue whose organization is induced, the 'reacting system', and he makes it quite clear that induction, in its dominant features, is a *releasing* process which is initiated in the reacting system by the inductor.[1] In Needham's terminology, the induction effect of a grafted organizer involves two processes, (1) evocation, which is always performed by the

[1] H. Spemann, op. cit.

graft, and (2) individuation, which is performed by the graft and the host working together.[1]

Tissues which are differentiated under the influence of the primary organizing mechanism of the dorsal lip of the blastopore may themselves become endowed with organizing properties. Thus the induced neural tube can organize tissues in its neighbourhood, and the segmented axial mesoderm has the capacity of inducing organization of the neural tube. Again, the optic cup (i.e. the embryonic primordium of the retina) is developed from the medullary plate under the influence of the underlying mesoderm, and is itself able to induce the differentiation of a lens from the ectoderm which overlies it. The optic cup may therefore be termed an 'organizer of the second order' (Spemann). In the later sections of this book we shall encounter in the various tissues of the body a number of examples of this organizing capacity which is responsible for tissue differentiation and growth in post-embryonic life. Indeed, we may conceive that, during the process of development, the action of the primary organizer gives rise to a number of secondary organizers, and these again to tertiary organizers and so forth, until the whole embryo is made up of a mosaic of morphogenetic fields within which organizing effects can be produced by differentiated inducing agents. Ultimately this same process will be expressed in the mutual action of one cell upon another, which is manifested in the adjustment of growing and mature tissues to the formation and maintenance of every morphological structure.

We have seen that, in the amphibian embryo, the primary organizer is localized in the dorsal lip of the blastopore. When this tissue is invaginated into the roof of the embryonic gut, it becomes differentiated to form the notochord and axial mesoderm. This is also the fate of the tissue of the primitive streak in the embryos of higher vertebrates, and it thus appears that the primitive streak is morphologically homologous with the dorsal lip of the blastopore.

Streaming movements of a characteristic type, similar to those which are directed towards the blastopore in the amphibian embryo, occur in the blastoderm of higher vertebrates in relation to the primitive streak. These movements are directed backwards on each side, and then forwards along the centre of the streak where the tissue sinks in and spreads out laterally between the ectoderm and entoderm. The primitive streak therefore represents the centre of gastrulation in these embryos, a process which is considerably obscured in comparison with lower forms.[2] This homology has been finally demonstrated experimentally, for the primitive streak in chick and rabbit embryos has been observed to possess organizing functions similar to those of the dorsal lip of the amphibian blastopore. By analogy, it is to be presumed that in the human embryo the primary organization centre first develops at about the fifteenth day of development, that is to say, the stage at which the primitive streak appears.

---

[1] J. Needham, op. cit.

[2] The method of gastrulation adopted by mammalian embryos is taken to indicate their evolutionary derivation from an ancestral type in which the ovum was heavily laden with yolk.

It has been noted that, as a result of local chemo-differentiation under the influence of the primary organizer, the early embryo becomes mapped out into various regions whose developmental potencies are fixed. Following on this stage, each region acquires the capacity for self-differentiation independently of extrinsic control. This phenomenon of self-differentiation has been abundantly demonstrated by experimental methods, such as growing tissue fragments *in vitro* in suitable culture media, or (in the case of the chick embryo) grafting them on to the vascular chorio-allantoic membrane. Under these circumstances, an isolated limb-bud rudiment will undergo a remarkable degree of differentiation to form its constituent skeletal elements, the primordium of the otic vesicle will shape itself into a semicircular canal system, or the optic diverticulum of the fore-brain will form the various layers of the retina.

At first each of these self-differentiating regions forms an equipotential system, that is to say, any small portion of it is capable of developing into a whole structure. If, for example, only a fragment of the presumptive limb region of an early amphibian embryo is grafted elsewhere, it will still form a complete limb. Soon, however, each region undergoes a further chemo-differentiation which leads to its division into a mosaic of sub-regions, each with a *fixed* potentiality for development into one part of the mature structure. When this stage is reached in the limb region, a transplanted portion will only form the restricted portion of the limb into which it is normally destined to develop. This mosaic phase of development is progressively elaborated down to the finer details of structure until, in higher vertebrates at least, the capacity for tissue differentiation becomes relatively more and more limited, except in so far as tissues are able to reproduce their own kind in the process of regeneration and repair.

In the whole process leading up to structural differentiation, three main phases may be recognized. The first of these is *chemo-differentiation*, in which no anatomical change is to be discerned in the cells.[1] It is important to realize that the process of chemo-differentiation is an essential preliminary to all types of histological or structural differentiation, for tissues which, either in the embryo or in the mature organism, appear anatomically to be identical, may yet have different properties and potentialities depending on intrinsic physico-chemical differences. This is illustrated by the observation that if connective-tissue cells (fibroblasts) taken from different regions of the same developing embryo are grown in tissue cultures, they are found not always to show precisely the same properties. They may have different growth rates, and show different reactions to chemical agencies.

Chemo-differentiation also underlies the important principle of *tissue specificity* in the adult, which is expressed (*inter alia*) by the tendency of some cells to react to certain stimuli while others, even of the same general type of tissue, remain inactive. This is particularly well shown in the response of tissue growth and differentiation to the influence of the secretory products of endocrine glands. For instance, disturbances of the

---

[1] It remains possible, of course, that a more refined histological technique or improved staining methods may in the future demonstrate anatomical differences in chemo-differentiated cells which otherwise *appear* to be structurally similar.

activity of the pituitary or thyroid gland lead to a modification in skeletal growth, but the response is of a selective nature, some osteogenetic tissues being more affected than others.

Chemo-differentiation is followed by *histo-differentiation*, that is to say, the transformation of anatomically indifferent cells into those characteristic of a particular type of tissue, such as muscle fibres, blood corpuscles, or nervous tissue. It will be realized, however, that histo-differentiation is really an extension of chemo-differentiation to a point where the chemical changes become manifested to the eye by actual structural alteration.

Lastly, the tissues become moulded into characteristic shapes to form definite organs or anatomical systems; this process is termed *organogenesis*.

All the early stages of structural differentiation in the embryo occur before the tissues assume their functions. Thus, the heart and the main blood vessels are laid down before the circulation commences, nervous elements appear in the brain and spinal cord before they are capable of transmitting nervous impulses, and skeletal elements are shaped before they are subjected to the mechanical stresses and strains for which they are constructed.

This 'pre-functional' period of differentiation is followed by a functional period during which the final touches are put to the modelling of the various anatomical structures. The finer details of the shape and surface features of a mature bone are conditioned by the pressure and tension of surrounding structures, by joint movements, by adequacy of blood supply, &c.; the vascular architecture of the smaller blood vessels is related to the haemodynamic factors of the circulation; muscles and ligaments are modified by the work involved in their exercise, and so forth.

The extensive studies in experimental anatomy which have been directed to these problems have rather served to emphasize the capacity for structural differentiation in the pre-functional period. Whereas the older anatomists had sought to explain the processes involved in the development of a particular organ in terms of chemical tropisms, the pressure and tension of adjacent structures, the available space and material, and other factors depending on the interaction of different parts of the growing embryo, methods of tissue culture and transplantation have now shown that isolated tissues, growing quite independently of their normal environment, can undergo self-differentiation to a remarkable degree. It must be recognized, however, that for their full normal development as component parts of a total organism, the environment in which tissues and organs are undergoing differentiation remains an obviously essential factor.

III

CONNECTIVE TISSUE

THE process of preparing a dissection involves the clearing away of a matrix of fibrous and cellular material in which more highly organized structures such as muscles, vessels, nerves, &c., lie embedded. This material varies considerably in its consistency, in some places forming a very delicate reticulum of loose texture, and in other places becoming condensed into a firmly woven feltwork or into tough fibrous sheets. In its various forms it is commonly termed connective tissue or—because of its tendency to form fibrous bands—'fascia'.

Loose connective tissue is a relatively generalized and undifferentiated tissue, preserving to some degree an embryonic potentiality for undergoing differentiation into more highly organized supporting structures. To this extent, it may perhaps be regarded as the persistence of embryonic mesodermal tissue left over with comparatively little change among the highly organized structures which have become differentiated from the same parent tissue. Although, in the adult, it is partly made up of specialized cellular and fibrous elements, it does contain unspecialized cells similar in their morphology to the branching cells of embryonic mesenchyme and retaining the same potentialities for differentiation into other tissues.

Embryonic connective tissue, or *mesenchyme*, is made up of small cells with slender branching processes which often appear to be continuous from one cell to another, forming an exceedingly fine network (Fig. 16). The body of each cell consists of little more than the nucleus surrounded by a minimal amount of cytoplasm. It is from a cellular reticulum of this type that the diverse elements of mature connective tissue are differentiated.

The anatomical composition of connective tissue is most conveniently studied where it forms a loose meshwork as, for instance, in the subcutaneous layer of certain parts of the integument. Here it has an open texture and is usually called *areolar tissue*. If a film preparation of this is made and appropriately stained, it is found to contain fibres and cells embedded in a substratum of a semifluid gelatinous substance. The latter is homogeneous in structure, and can be made visible by treatment with silver nitrate solution which, by reduction, stains it a uniform brown colour. In its composition it is essentially a complex of mucopolysaccharides and scleroproteins. This ground substance (which, according to some authorities, is a product of cell secretion) has been conceived to be of the nature of a fine reticulum the meshes of which contain *tissue fluid*, but it would be more accurate to say that the latter is absorbed in intimate combination with the ground substance. Tissue fluid, which is in part derived from blood plasma, is regarded as an essential medium occupying the intercellular spaces of all tissues, through which nutritive material from the blood stream reaches the actual cellular elements and the waste products of cell metabolism are conveyed into blood and lymphatic capillaries. It has been questioned whether free interstitial fluid exists in normal connective tissue on the grounds that (1) clear fluid cannot be obtained if the skin is punctured with a needle, (2) a bleb of fluid injected into the skin retains its shape, being presumably walled off by a jelly-like connective-

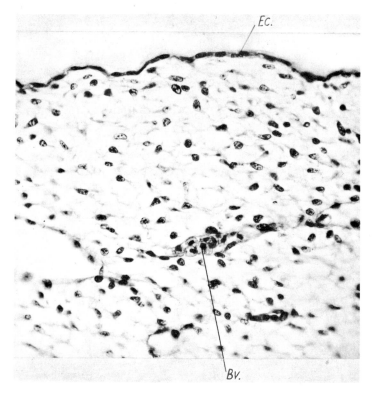

Fig. 16. Microphotograph showing mesenchymal tissue in an early (6 mm.) pig embryo. The surface of the embryo is covered by ectoderm (*Ec.*) which at this stage consists of a single layer of cells. An embryonic capillary vessel, lined by endothelium and containing nucleated red blood corpuscles, is also seen (*Bv.*). × 320.

tissue substance, and (3) Brownian movement of minute particles cannot be observed in the substratum of normal connective tissue. The fact is, however, that free fluid quickly accumulates in tissues subjected to pressure and friction, vascular stasis, irritation, and inflammation (as well as certain other conditions), and it is important to note that this facilitates the rapid spread of any diffusible substances which may be present. On theoretical considerations, it is probably correct to assume that a small amount of free tissue fluid is normally present in the intercellular spaces of some tissues, and that in oedema the amount is greatly increased leading to a separation of the tissue elements. The consistency of the ground substance (which in any case is a complex colloidal state) certainly varies from one type of tissue to another and also with age. It probably varies also in the same tissue under different physiological conditions, affecting what is called 'tissue pressure'. This pressure is an important factor in the filtration of fluid from the blood capillaries to the tissues.[1]

1. THE FIBROUS BASIS OF CONNECTIVE TISSUE

The fibres in areolar tissue are mainly of two kinds, *collagenous* or white, and *elastic* or yellow; and it is upon the relative amount and proportion of these fibres that the consistency of connective tissue depends.

Collagenous fibres (Figs. 17 *e* and 18) are extremely delicate and practically colourless, and in concentration form a tissue which has a dead-white appearance (white fibrous tissue). They are arranged in bundles and, except when under tension, run a characteristically wavy course (Fig. 18). The individual fibres do not branch, neither do they anastomose with each other. Though soft in consistency they are very resistant to tensile stress—hence they provide the basis for such structures as tendons and ligaments which are subjected to a pulling force. It has been estimated that large bundles may have a tensile strength of as much as 18,000 lb. to the square inch.[2] The substance of these fibres, *collagen*, is an albuminoid which can be converted into gelatin by boiling. On treatment with strong acid the fibres are ultimately destroyed, a fact which is made use of in the maceration of tissues in anatomical preparations. The study of collagenous fibres by means of the electron microscope has shown that they have an unexpectedly complex physical structure. They present a segmented or striated appearance, depending on a periodic arrangement of their constituent protein molecules (Fig. 19). The distance between the alternate main bands is about 600 Å (0·06 μ), and each band itself shows subsidiary bands. The functional implications of this rhythmic orientation of the molecular basis of collagenous fibres have yet to be elucidated. Evidently, however, the characteristic banding is an expression of the ultimate molecular configuration of the protein basis of collagen, for it has been demonstrated that if collagenous fibres are dissolved by a dilute acid and precipitated again by the addition of salts, the redeposited fibres may still show the same banded structure.[3]

Elastic fibres are much less frequent in most types of connective tissue. In contrast to collagenous fibres they run singly, branch freely, and anastomose

[1] See S. H. Bensley, 'On the presence, properties and distribution of the intercellular ground substance of loose connective tissue', *Anat. Rec.* **60**, 1934, and J. M. Yoffey and F. C. Courtice, *Lymphatics, Lymph and Lymphoid Tissue* (sections on the permeability of blood and lymphatic capillaries). Reference should also be made to the important work on hyaluronidase, an enzyme believed to be identical with the 'spreading factor' which, by lowering the viscosity of tissue 'fluids', increases tissue permeability and thus increases the invasive capacity of infective agents (see E. Chain and E. S. Duthie, 'Identity of hyaluronidase and the spreading factor', *Brit. Journ. Exp. Path.* **21**, 1940).

[2] It is a matter of considerable practical importance to realize that there is a limit to the resistance which white fibrous tissue in the living body can offer to tensile forces. Ligaments and fibrous bands which are exposed to continued excessive and abnormal tension show a tendency to stretch fairly rapidly. This, however, is due not to an actual elongation of the fibres themselves but presumably to a proliferative activity of fibroblasts and the production of more collagenous tissue, leading to an increase in the length of the structure as a whole.

[3] See *Nature and Structure of Collagen*, ed. by J. T. Randall, London, 1953.

with each other (Fig. 23). They are highly refractile in appearance, and have considerable elasticity; broken fibres are often seen in microscopic preparations of teased tissue, curled up in an elastic recoil (Fig. 17 e). Their

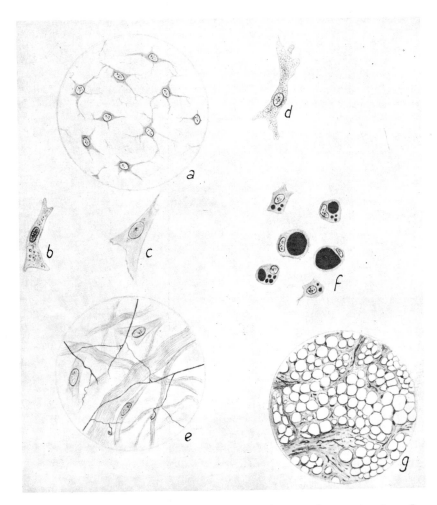

FIG. 17. Schema illustrating some of the component elements of connective tissue. In a is shown the appearance of mesenchyme or embryonic connective tissue, from which the more organized elements of mature connective tissue are derived. Some of the cellular elements are shown in b, histiocyte, c, fibroblast, d, pigment cell, and f, fat cells. In e is shown diagrammatically the appearance of a film preparation of areolar tissue, with fibroblasts and collagenous and elastic fibres. In g is seen the appearance of a section of adipose tissue in which the fat has been removed by solvents in the course of histological preparation.

yellow colour can be recognized macroscopically in tissues in which they are particularly abundant (e.g. in certain ligaments such as the ligamenta flava on the neural arches of the vertebrae). They consist of a protein, or possibly a combination of proteins, called *elastin*, which is highly resistant to treatment by acids or to boiling. As seen under the electron microscope, elastic fibres are sharply contrasted with collagenous fibres by the complete lack of a periodical structure;[1] each fibre appears to consist of a fused mass of fibrillae twisted in rope fashion and embedded in an amorphous matrix. Elastin may also occur in the form of fenestrated membranous sheets, such as those found in the walls of arteries.

It will be seen later that both white and yellow fibres enter into the composition of many structures such as walls of blood vessels, capsules of glands, and sheaths of muscles. In microscopic sections they can be shown up by using certain selective stains. Collagenous fibres, for example, are

FIG. 18. Collagenous fibres in loose areolar tissue. ×470.

stained a bright red with van Gieson's stain[2] (which, however, also stains certain other elements), while elastic fibres are coloured reddish-brown with orcein.

Besides collagenous and elastic fibres, the cellular elements of many tissues are supported in a most delicate basketwork of exceedingly fine fibres called *reticular fibres* (Fig. 21). These are somewhat similar to collagenous fibres, but they also offer a strong contrast in a number of features. For example, they show no tendency to run in bundles, they branch freely and anastomose with each other, and they are uneven in thickness. It has also been stated that the substance of which they are composed, *reticulin*, is not converted into gelatin when boiled in water. Since they can be demonstrated by impregnation with silver salts, they are sometimes called *argentophil fibres*. In spite of their distinctive appearance, there is good reason to suppose that reticular fibres represent an immature form of

[1] J. Gross, 'The structure of elastic tissue as studied with the electron microscope', *Journ. Exp. Med.* **89**, 1949. See also A. I. Lansing *et al.*, 'The structure and chemical characterization of elastic fibers as revealed by elastase and by electron microscopy', *Anat. Rec.* **114**, 1952.   [2] Acid fuchsin and picric acid.

collagenous tissue. For example, electron microscopy shows that they are banded, the period of the striations being typical of collagen fibres. In appropriately stained sections (e.g. of the liver) they can be seen to grade insensibly into ordinary collagenous fibres. Moreover, argentophil fibres have been observed to develop in relation to fibroblasts in tissue cultures, and it is to be noted that even fully developed collagenous fibres reduce silver salts to some extent. Lastly, evidence has been adduced that with increasing age there is a gradual transformation of reticular fibres into collagenous tissue.[1]

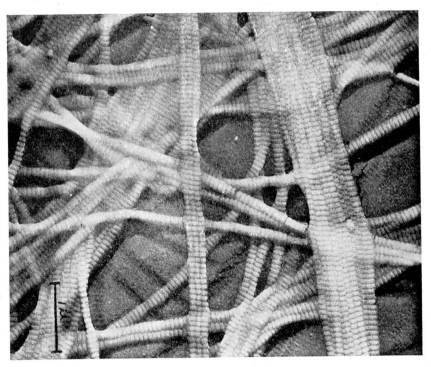

FIG. 19. The appearance of collagen fibrils as seen with the electron microscope (shadowed with chromium). × 19,300. (From Gross and Schmitt, *Journ. Exp. Med.* **88**, 1949.)

It is interesting to conjecture the possible effects of this transformation. Ultimately, of course, it must lead to a loss of pliability and resilience in the tissues generally, and this, again, will be reflected in a corresponding loss of physiological efficiency. But it may also have an important influence on the regenerative and reparative capacity of the tissues. Medawar[2] has pointed out that 'The *tactics* of regeneration, as of embryonic development, is primarily and fundamentally a matter of the movement of cell substance, cells, and cell groups. It is not primarily a matter of cell division nor of

[1] J. M. Wolfe *et al.*, 'The effect of advancing age on the connective tissue', *Amer. Journ. Anat.* **70**, 1942.
[2] P. B. Medawar, 'Biological aspects of the repair process', *Brit. Med. Bull.* **3**, 1945.

synthesis in general, though in due course these are called upon to play their part.' It is important to recognize this kinetic factor, for if the process of natural repair and regeneration means the *movement* of cells, there must be freedom to move. And the freedom to move, whether in the repair of damaged tissues or in the normal repair and replacement of tissue elements that occur in response to the daily wear and tear of physiological activity, is evidently dependent on the nature of the interstitial fibrous framework. As this framework loses its open texture and its resilience (with the progressive conversion of the delicate reticular basketwork into layers of inelastic collagenous fibres), it may be supposed that the movement of cellular elements will certainly become more and more restricted.

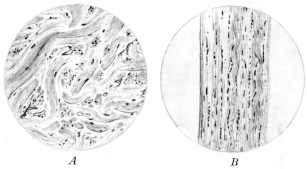

A   *B*

FIG. 20. Drawings showing, *A*, the disposition of interlacing bundles of collagenous fibres in dense connective tissue (e.g. in the dermis of the skin), and *B*, the alignment of collagenous fibres in the formation of a tendon or ligament.

2. THE CELLULAR BASIS OF CONNECTIVE TISSUE

There are various types of cell in connective tissue; of these, two are specially common—*fibroblasts* and *histiocytes*. Superficially they appear somewhat similar in ordinarily stained preparations, but functionally they are very different. Fibroblasts are stationary cells which are concerned with the production of white collagenous fibres, and so play an important part in building up the supporting element of connective tissue. On the other hand, histiocytes have phagocytic properties (i.e. they can ingest foreign material), and, though apparently fixed and stationary under normal conditions, can readily become converted into freely moving amoeboid cells (*macrophages*). They appear thus to play the part of scavengers, removing cell debris and particulate matter from the connective-tissue spaces. In short, of the two main cellular elements of connective tissue, one is concerned with the construction of a passive supporting matrix, and the other with actively protective functions. The contrast between the fixed fibroblasts and the freely moving cells is very striking in cultures of explanted embryonic connective tissue whose growth has been recorded cinematographically. When such pictures are speeded up, the non-motile fibroblasts can be seen steadily proliferating by mitotic division, while at the margin of the explant numbers of motile cells wander actively and energetically among them, sometimes attacking and destroying a damaged cell.

Fibroblasts. Fibroblasts (also called fibrocytes or lamellar cells) vary their appearance according to the point of view from which they are seen (Figs. 17 *c*, 24, 26). They are thin flat cells, so that in profile they appear spindle-shaped. In surface view they tend to be somewhat stellate, with broad-pointed processes. The nucleus is oval and rather large. The cytoplasm of the cells is fairly clear, and, except for fine fat globules, usually contains no obvious particulate inclusions. The outline of the cell is ill

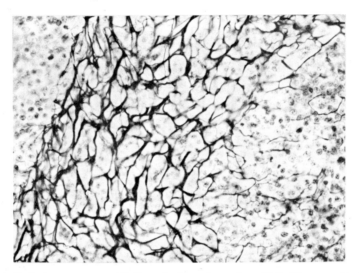

FIG. 21. Reticular or argentophil fibres in the spleen of a rabbit. They provide here a delicate supporting framework for the lymphoid tissue. The section has been impregnated with silver carbonate. × 340.

defined—indeed, it has been supposed by some that adjacent cells (at any rate in young, growing tissue) are in direct protoplasmic connexion to form a continuous syncytium (Fig. 24). They are frequently found in very intimate relation to bundles of collagenous fibres, an association which suggests that these fibres are directly produced by their agency. It is because of this function, indeed, that the cells are called fibroblasts.

The manner in which fibroblasts lead to the deposition of collagenous fibres has for many years been a source of controversy. One view held that the fibres are differentiated within the cytoplasm, to become subsequently extruded; another that the fibres are primarily deposited in the intercellular substance under the influence of some enzymic action by the fibroblasts. Experimental studies during recent years have shown that, in fact, both intracellular and extracellular factors are involved in fibrogenesis. In particular, reference may be made to observations on the behaviour of fibroblasts in transparent chambers in the rabbit's ear.[1] During the process of fibrogenesis, the cells appear to throw off small vesicular masses of cytoplasmic material which become detached and set free in the intercellular

[1] M. L. Stearns, 'Studies on the development of connective tissue in transparent chambers in the rabbit's ear', *Amer. Journ. Anat.* **66** and **67**, 1940.

FIG. 22. Fibroblasts in tissue culture which have been exposed to a regional tension. Along the lines of tensile force the cells have become orientated in parallel formation and their proliferation has been accentuated. (From P. Weiss.)

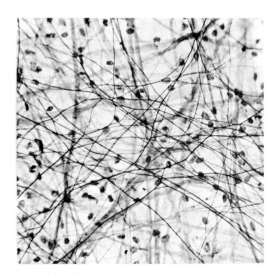

FIG. 23. Microphotograph of a film preparation of areolar tissue, stained with brazilin. The elastic-tissue fibres are well shown, while in the background can be seen the fainter outlines of bundles of collagenous (white) fibres. The nuclei of connective-tissue cells have also been stained. × 225.

substance. It seems that these masses are then directly utilized in the formation of the fibres, the latter becoming spun out from them in a fine web-like formation (Fig. 25). The process is a rapid one, for in three to four hours extensive networks develop round each fibroblast, and in forty-eight hours they may become so dense as to obscure the cells completely. In more recent studies of fibrogenesis as it occurs during the regeneration of a tendon, electron microscopic examination has shown that the primary fibrils which first make their appearance are not only present in bundles

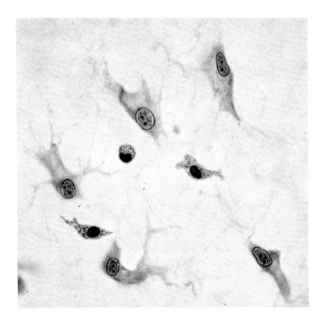

FIG. 24. Section of oedematous connective tissue showing fibroblasts and histiocytes. Three of the latter are seen, with darkly staining nuclei and granular cytoplasm. The fibroblasts have larger nuclei and send out fine cytoplasmic processes which seem to fade into a reticular ground substance. × 570.

outside the fibroblast near its surface, but also apparently within the cytoplasm close to the cell membrane.[1]

The relation of fibroblasts to the production of collagenous fibres is of great importance in providing a basis for the mechanical functions of white fibrous tissue. It has been demonstrated that if a thin film of fibroblasts grown in tissue culture is subjected to a regional tension, the cells exposed to the tensile force multiply more rapidly and orientate themselves in parallel lines in the direction of the stress (Fig. 22).[2] In other words, tension appears to be a mechanical stimulus determining directly or indirectly not only an increased production of fibrous tissue but also the direction in

---

[1] F. Wassermann, 'Fibrillogenesis in the regenerating rat tendon', *Amer. Journ. Anat.* **94**, 1954.
[2] P. Weiss, 'Erzwingung elementarer Strukturverschiedenheiten am *in vitro* wachsenden Gewebe', *Arch. Entwicklungsmech.* **116**, 1929.

which it is deposited.[1] This response has been observed in the transparent chamber experiments referred to above, in the course of which it was found that the rate, amount, and direction of fibre formation are all influenced by tension. It has also been observed in adult animals in subcutaneous tissues which have been exposed to the mechanical stimulation of intermittent tension, and it is particularly well illustrated in the regeneration of cut tendons (see p. 140). It will be appreciated, therefore, that in the body bundles of white fibres will tend to be laid down 'automatically' wherever they are required to resist a tensile force. In this way, tendons and apo-neuroses will be formed by the traction of muscles on indifferently arranged connective tissue, ligaments will develop wherever the tensile force in the

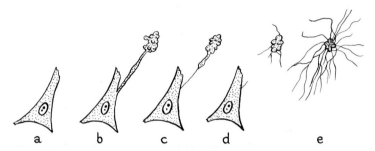

FIG. 25. Diagram showing how collagenous fibres are formed by fibroblasts (the diagram has been constructed on the basis of illustrations in the paper by M. L. Stearns, *Amer. Journ. Anat.* **67**, 1940).

capsule of a joint demands their aid in limiting movements, and retention bands, fibrous pulleys, and other restraining mechanisms will become differentiated in response to similar mechanical requirements.

It should be noted that the deposition of white fibrous tissue in the form of a scar is an essential part of the reparative process by which a wound is healed, and it may be supposed that tensional forces in the inflammatory tissue of the wound exert an important influence on the process of scar formation. But, apart from such local factors, it should be recognized that the formation of collagenous tissue may be profoundly affected by general metabolic disturbances. Of particular importance in this connexion is vita-min C (ascorbic acid). It has long been known that in patients suffering from scurvy due to vitamin C deficiency wounds heal slowly or not at all, and experiments have shown that this is partly due to the fact that the fibroblasts are unable to elaborate collagen. In guinea-pigs deprived en-tirely of vitamin C, no matrix of collagenous fibres is formed at the site of a wound, and in animals suffering from a partial deficiency of the vitamin the tensile strength of healing wounds is less than normal. Lastly, tissue-culture experiments have also demonstrated the part played by vitamin C in the production by fibroblasts of white fibrous tissue.

While the deposition of collagenous fibres is one of the main functions of fibroblasts, the developmental origin of elastic fibres is still obscure.

[1] There is reason to suppose that the tensional forces probably do not act *directly* on the fibroblasts, but that they influence the degree and direction of their proliferation by modifying the ultra-structure of the substratum in which the cells lie.

They appear to be constructed from the alignment of refractile granules laid down in the intercellular matrix, but whether under the agency of fibroblasts or of other cell types is unknown. The fact that their chemical and physical properties contrast so strongly with those of collagenous fibres suggests that they are derived from a correspondingly different source. It is also not known whether (like white fibrous tissue) the deposition of elastic tissue is favoured and directed by the mechanical stimulus of tensional forces. On *a priori* grounds, it might be inferred that its differentiation is likely to be in response to tensional forces of a fairly regular intermittent character.

Some cytologists hold the view that fibroblasts are specialized cells—incapable of becoming transformed into other types of connective-tissue cell, but it is more commonly accepted that they are generalized to the extent that they can be converted into cartilage cells, bone-forming cells or osteoblasts, endothelial cells, and so forth. These contrasting views are probably to be explained by the different potentialities of young and fully mature fibroblasts.

Histiocytes.[1] Histiocytes are irregular in shape, and their cell outline is much better defined than that of the fibroblasts (Fig. 24). Their irregularity is related to their amoeboid properties. The nucleus, also, is smaller, stains more deeply, and often shows a characteristic indentation which gives it a reniform appearance. In accordance with their phagocytic powers, the cytoplasm is usually filled with granules and with vacuoles of various sizes, of which the latter can be stained rather conspicuously with neutral red. The phagocytic activity of histiocytes is particularly well seen in the immediate neighbourhood of a haemorrhage in the tissues. A microscopic section taken from such a region shows masses of histiocytes in the process of clearing away the debris and filled with granular masses of blood pigment.

It seems that normally connective-tissue histiocytes are not always motile—they may be fixed phagocytes (macrophages) of the connective tissue. In this respect they are sometimes contrasted with the freely moving macrophages also found in connective-tissue spaces and which are believed to be derived from the blood stream by emigration from capillaries. However, this criterion of motility is hardly adequate to make such a distinction, for under conditions of inflammation, all these cells appear to be capable of free amoeboid movement. Probably, therefore, such histiocytes are merely wandering macrophages which are in a temporary phase of quiescence. The fact that they are particularly numerous in vascular tissues suggests that they may be derived in part from blood vessels (see p. 231). The general opinion, however, is that they are also formed *in situ* in the connective tissue by the transformation of fibroblasts, or by the differentiation of mesenchyme cells of an embryonic type.

Histiocytes are contrasted with other types of connective-tissue cells by their reactions to vital dyes—that is to say, dyes which, when injected into the

[1] Histiocytes are cells which have suffered much from a plurality of nomenclature, for they have been variously known as clasmatocytes, adventitial cells, rhagiocrine cells, endothelial leucocytes, pyrrhol cells, macrophages, or resting wandering cells.

body, are taken up by the cytoplasm of normal living cells. To demonstrate this phenomenon, trypan blue (an acid aniline dye) and lithium carmine have been commonly used in the form of colloidal solutions. In the tissues they are assimilated by the histiocytes, which store them in the form of coarse granules in their cytoplasm (enclosed in smooth-walled vesicles derived from invaginations of the cell membrane) in the same way that they ingest particulate matter such as carbon particles, fat droplets, or bacteria. This characteristic reaction towards vital dyes is also shown by other types of large phagocytic cells or macrophages in other tissues; it is, in fact, a common property of all cells of what is termed the 'macrophage' or 'reticulo-endothelial' system (*vide infra*, p. 55). The histiocytes are the main

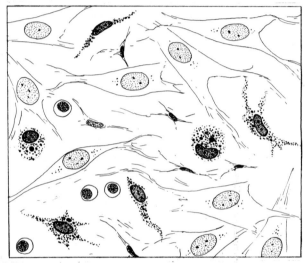

Fig. 26. Drawing of a spread preparation of the mesentery of a rat which had received repeated injections of trypan blue. The histiocytes are contrasted with the fibroblasts by the fact that they contain coarse granules of the dye stored in their cytoplasm. The fibroblasts have taken up the dye only to a slight extent, in the form of a few small and discrete granules close to the nucleus. Note, also, the varying shapes of the histiocytes. To the left and below are seen a few lymphocytes. × 500 (approx.).

representatives of this system, and, under the stimulus of inflammatory conditions, they become extremely active. Fibroblasts, in so far as they react at all to vital dyes in moderate dosage, take them up in very fine granules which are commonly stored in the cytoplasm close to the nucleus (Fig. 26).

Although intermediate types of reaction have been observed, their behaviour towards vital dyes does provide a general criterion for distinguishing between fibroblasts and histiocytes. Nevertheless, the occasional occurrence of intermediate forms suggests the possibility that the one may be capable of being transformed into the other. This question is still disputed. On the one hand, however, prolonged observation of the cells in living tissues has tended to the opinion that, although under varying conditions they may sometimes be difficult to distinguish morphologically, they are specific cell types whose functions are not interchangeable. On the other, it is also held by competent histologists that, at any rate under certain conditions, fibroblasts and histiocytes may become interconvertible. Such a cellular transformation (or *metaplasia*), if it really occurs, would find a parallel among other types of cells—for example, the transformation under changing conditions locally of osteoblasts, which are bone-forming cells, into osteoclasts, which are bone-destroying (see p. 83). It may also be mentioned that typical histiocytes are said, on occasion, to be derived from such specialized elements as the sheath cells of nerve fibres,[1] striated muscle fibres,[2] and even hepatic cells,[3] though the validity of the evidence for such transformations has been questioned.

The growing suspicion that such extreme examples of cellular metaplasia may be a more common phenomenon than was formerly supposed naturally introduces a complication into general conceptions of cell morphology and function. For, on theoretical consideration, it seems less difficult to suppose that different cell types, each with their characteristic function, are the end products of a diversification from a common, embryonic type of cell (i.e. the mesenchymal cell), than to suppose that cells which are evidently highly differentiated for one specific function can be converted (either by direct transformation or by a process of de-differentiation and subsequent re-differentiation) into cells capable of functioning quite otherwise. Perhaps one of the most interesting (and also surprising) examples of this kind of metaplasia demonstrated in recent years is the ease with which, in tissue-culture experiments, a squamous type of epithelium can be replaced by a mucus-secreting and ciliated type by altering the concentration of vitamin A in the culture medium.[4]

Besides fibroblasts and histiocytes, the following cellular elements are also to be found in connective tissue—mesenchyme cells, mast cells, plasma cells, fat cells, and pigment cells. Of these, mesenchyme cells are perhaps the least obtrusive because of their inconspicuous appearance. They pervade connective tissue everywhere and are of great importance since they provide stem cells for the differentiation of other more specialized elements. As we have noted, they are small cells with very scanty cytoplasm and slender branching processes. In lower vertebrates they have been observed to be amoeboid, but it is uncertain whether they have this property in the human body. They are commonly regarded as essentially an embryonic type of cell which persists in an undifferentiated state in the adult tissues, but it has to be admitted that they are not certainly distinguishable in their morphological appearance from small fibroblasts.

Mast cells.[5] Mast cells are rather characteristic cells of connective tissue, though their numbers appear to show considerable variation in

[1] P. Weiss and H. Wang, 'Transformation of adult Schwann cells into macrophages', *Proc. Soc. Exp. Biol. Med.* **58**, 1945.

[2] M. Chèvremont, 'Transformation d'éléments musculaires en macrophages', *Arch. Biol.* **51**, 1940.

[3] J. Frederic, 'La transformation histiocytaire des cellules hépatiques', *Rev. d'Hématologie*, **6**, 1951.

[4] H. B. Fell and E. Mellanby, 'Metaplasia produced in cultures of chick ectoderm by high vitamin A', *Journ. Physiol.* **119**, 1953; H. B. Fell, 'The effect of excess vitamin A on cultures of embryonic chicken-skin', *Proc. Roy. Soc.* H, **146**, 1957.

[5] The term 'mast' is of German origin, implying that these cells have a nutritive function.

different species. Round or polygonal in shape, they are distinguished by the fact that their cytoplasm is filled with granules which stain deeply with basic dyes (e.g. neutral red or methylene blue). They are relatively very numerous in the connective tissue of certain lower mammals, e.g. rodents, and they are more frequent in the immediate neighbourhood of blood vessels. The significance of these cells is not fully known, but there is good evidence that they are concerned in the formation of an anticoagulant substance called *heparin*. This evidence depends partly on the staining reaction of the cell granules, and partly on the observation that active heparin can be isolated from tissues rich in mast cells. The fact that these cells have an intimate relationship to the endothelial lining of the heart and blood vessels (particularly certain of the larger arteries and veins), and also to the cellular lining of the peritoneal cavity, has suggested that, perhaps by the liberation of a 'spreading factor', they may be concerned in alterations of the permeability of these membranes.[1]

Plasma cells. Plasma cells are rarely found in typical connective tissue. They appear in considerable numbers around inflammatory foci associated with certain chronic infections such as tuberculosis, and they are normally common in the subperitoneal tissue of the omentum. The cells are round or oval and their nuclei show a characteristically regular arrangement of chromatin material which is disposed in a radial direction like the spokes of a wheel. It has been generally accepted that plasma cells are derived directly from lymphocytes which find their way into the connective tissue from blood vessels, but there is now some evidence that they are independently differentiated from the reticular cells of the spleen and lymph nodes, and possibly from lymphoblasts in the latter. Their association with chronic infections suggests that they may have a protective function of some kind, and there is now little doubt from a variety of experimental procedures that they are concerned with the production of antibodies for the neutralization of bacterial toxins, &c.[2] The electron microscope shows that cytoplasm of plasma cells is crowded with massed formations of endoplasmic reticula studded with ribosomes. Such an appearance indicates very active synthesizing properties.

Pigment cells. Branched cells containing melanin granules may be present in the subcutaneous connective tissue (Fig. 17 *d*). They are comparatively infrequent in the white races of mankind, except beneath the skin of the eyelids, the nipple area, and the external genitals. In the choroid coat of the eyeball and in the iris, however, they are very numerous. These cells are probably to be divided into two groups, those that are specific cells concerned with the elaboration of the melanin in their cytoplasm, and those that are connective-tissue cells (? histiocytes) which have taken up melanin granules from some other source. Further reference to these cells is made in the section dealing with the skin (p. 300).

---

[1] V. J. McGovern, 'Mast cells and their relationship to endothelial surfaces', *Journ. Path. Bact.* **71**, 1956.
[2] A. Fagraeus, 'Antibody production in relation to the development of plasma cells,' *Acta med. Scand.* **204**, 1948.

3. ADIPOSE TISSUE

Fat cells are numerous and widespread in normal healthy connective tissue, sometimes in such large quantities as to form a definite adipose tissue. A typical fat cell is little more than a thin envelope of cytoplasm covering a relatively large globule of fat. The latter is formed by the collection in the cytoplasm of small particles of fat which, as they accumulate and increase in size, run together in a single drop (Fig. 17 *f*). As a result, the cell body is distended and the nucleus pushed to one side.[1] Increased adiposity, at any rate as it has been studied in fattened geese, is the result of an increase in size of the fat cells rather than an increase in their number.[2]

It is still a matter of some doubt whether fat cells are specific in the sense that they are differentiated cells whose sole function is that of storing fat (and perhaps also of synthesizing it *in situ* by enzyme action). In many various types of cell, fine fat granules are normally present in the cytoplasm. This is the case, for example, with fibroblasts, and it has been supposed by some cytologists that fat cells are simply fibroblasts in which fat has accumulated to a conspicuous extent. Fibroblasts cultivated in a medium containing olive oil develop intracellular fat in considerable quantity,[3] but it remains to be shown that this process of storage is strictly comparable to fat storage in normal adipose tissue.

From the direct observation of fat formation in the living tissues of the rabbit's ear,[4] it appears that fat cells are derived from connective-tissue cells indistinguishable morphologically from fibroblasts. The cells change from an elongated to a rounded shape as droplets of fat begin to accumulate in their cytoplasm, and the droplets increase in size and number until they coalesce to form a single large globule.

Although the pre-adipose cell has thus been shown to be morphologically indistinguishable from a fibroblast, there is some evidence (not altogether conclusive) in favour of the conception that it is actually a specific cell type from the developmental and functional point of view. In the embryo, during the fifth month of foetal life, fat cells first appear in the subcutaneous tissue as derivatives of undifferentiated mesenchyme cells, and it may be supposed that they have a similar origin in the adult. When, as the result of inadequate nutrition, fat disappears from the cells, the latter revert again to what appear to be mesenchyme cells of an embryonic type. These remain quiescent until, with improved nutrition, they again become filled up with a new store of fat. Adipose tissue is characteristically organized into well-defined lobes and lobules, each being encapsuled in a delicate

[1] It should be realized that, in ordinary microscopical preparations of tissues, the fat-content of the cells has usually been entirely dissolved away by the preliminary treatment with alcohol or xylol. In order to demonstrate fat histologically, therefore, it is necessary to avoid using these fat solvents, and the tissues must be mounted in a medium such as glycerine. In a preparation of this kind, fat may be stained selectively with osmic acid, or with dyes known as Scharlach red and Sudan III.

[2] M. Clara, 'Bau und Entwicklung des sogenannten Fettgewebes beim Vogel', *Zeitschr. f. mikr. Anat.* **19**, 1929.

[3] D. E. Dorley and B. A. Hewell, 'The reactions of fibroblasts in tissue cultures to olive and mineral oils', *Amer. Journ. Anat.* **45**, 1930.

[4] E. R. and E. L. Clark, 'Microscopic studies of the new formation of fat in rabbits', ibid. **67**, 1940.

sheath of collagenous fibres and surrounded by a perilobular network of blood capillaries, and in the embryo this vascular arrangement is defined in regions where fat will later accumulate even before it is deposited. It is also to be noted that, when fat is deposited in the adult, it tends to be laid down first in the neighbourhood of blood vessels. Moreover, the study of fat formation in living tissues has shown that it develops primarily in regions in which the circulation is moderate or sluggish. On the other hand, it tends to diminish rapidly when the circulation in adjacent blood vessels becomes over-active, an observation which affords some evidence for the effectiveness of massage, exercise, and the local application of heat for decreasing subcutaneous adipose tissue.

In a well-nourished individual fat is not laid down in all connective tissues where fibroblasts are present, but selects out particularly certain regions—such as the subcutaneous tissues, the omentum, and the mesenteries of the peritoneum. This selective deposition of fat may perhaps also be taken to indicate that it accumulates in specific cells which have a definite regional distribution, rather than in ordinary fibroblasts which form a constituent of connective tissue everywhere. Such a conclusion is corroborated by experimental observations. For example, if a piece of *apparently* indifferent connective tissue in an immature animal, normally destined to form a local deposit of fat in a mature animal, is transplanted to some other part of the body where similar accumulations of fat do not occur, it still becomes differentiated into adipose tissue.[1] The specificity of fat cells is further suggested by the occasional occurrence of fat tumours (*lipomata*) which develop as circumscribed benign growths irrespective of the development of adipose tissue generally in the body.

There is a common tendency to regard fat as a somewhat passive and inert tissue. On the contrary, it is the site of considerable metabolic activity (as evidenced by its oxygen consumption), it is unusually well supplied with blood vessels, and it is rich in enzyme systems. Correlated with its specialized function is the fact that mature adipose tissue does not appear to have any proliferative capacity, for, if a localized pad of fat is partially resected, there is no compensatory hypertrophy of the remaining portion.[2]

Normal accumulations of fat represent essentially a storage of nutritive material which can be drawn upon in response to the needs of other tissues. Their size and extent, therefore, are largely dependent on the maintenance of an adequate nutrition, and a poor diet leads very rapidly to the depletion of these natural food stores in the body. In some degree, however, this depletion is selective. For example, in parts of the body where fat serves a mechanical function—as in the soles of the feet and the palms of the hands —it remains relatively unaltered in quantity even when fat in other parts has largely disappeared.

The removal of fat from each individual cell apparently involves in the first place a breaking up of the larger fat globules into separate discrete

[1] F. X. Hausberger, 'Über die Wachstums- und Entwicklungsfähigkeit transplantierter Fettgewebskeimlager von Ratten', *Virch. Arch. f. path. Anat. u. Physiol.* **302**, 1938.
[2] C. R. Cameron and R. D. Seneviratne, 'Growth and repair in adipose tissue', *Journ. Path. Bact.* **59**, 1947.

particles. When these disappear the cytoplasm of the cell is left with vacuoles empty of fat, and subsequently shrinkage of the cell body leads to the formation of a small mesenchymatous type of cell with branching processes. It has been shown that—at least in some circumstances such as inflammation—the fat may actually be removed from the cell by the activity of histiocytes and other types of phagocyte.

Besides the general adipose tissue of the body, local accumulations of fat which serve a supporting function occur in certain regions. The kidney, for example, is normally surrounded by an adipose capsule which probably has a mechanical function in helping to support the organ in its normal position. The contents of the orbit are likewise embedded in a mass of adipose tissue and, since during life fat is in a fluid or semi-fluid state, it forms a very suitable medium for allowing free movement of the eye and its muscles, at the same time providing a supporting matrix.

The experimental induction of fatty deposits has recently been demonstrated to follow the formation of an artificial cavity within the body by means of a perspex frame.[1] Such a cavity (in the subcutaneous tissue of rabbits) becomes at first filled with a clear fluid of high protein content, but the latter is replaced in a few weeks by a circumscribed mass of fat. This observation may have an important bearing on certain adipose formations of a pathological nature as, for example, in cases of exophthalmic goitre in which an accumulation of fat serves to fill the abnormal space in the orbit occasioned by the protrusion of the eyeball.

As is well known, some mammals accumulate local stores of fat which can be drawn upon as reserves in case of need. Examples are seen in the camel's hump and in the fat tails of a certain breed of sheep. Similar deposits are built up by hibernating mammals in the form of conspicuous masses of a dark-coloured fat at the base of the neck and in the scapular region, and because of their appearance these are sometimes called 'hibernating glands'. They differ from ordinary adipose tissue by the fact that the fat is stored in separate granules in each fat cell (and not in a single large globule), and when the fat disappears the somewhat regular arrangement of the emptied cells suggests the structure of an endocrine gland.

One of the most striking instances of the local storage of fat in man is found in certain African peoples (particularly the Bushmen and Hottentots) in whom adipose tissue accumulates to a remarkable degree in the buttocks. This condition, which is termed *steatopygy*, often gives the individual quite a grotesque appearance. While it occurs in both sexes, it is manifested to a much greater degree in women. It appears thus to be in part a secondary sexual character, and its development takes place at puberty.

4. THE MACROPHAGE SYSTEM

It has been noted that histiocytes are large phagocytic cells in the connective tissue which are able to ingest bacteria and particulate material,

[1] E. E. Pochin, 'Local deposition of adipose tissue experimentally induced', *Clin. Sc.* **8**, 1949.

and which behave in a characteristic way towards vital dyes by storing them in their cytoplasm in the form of coarse granules (p. 49). Throughout the body in various tissues there are found cells which show the same reactions. Together they comprise the *reticulo-endothelial* or *macrophage system*. The former term was originally used by Aschoff[1] (who defined in systematic detail the common features of these cellular elements), but although it has reference to the association of certain elements with connective-tissue fibres of a special type (reticular fibres), it has the disadvantage of suggesting that the system includes all types of endothelium. Actually this is not the case, and the term 'macrophage system' is more frequently employed. There is some doubt whether all its constituent elements are morphologically identical cell types. However, even if the macrophage system, as usually defined, includes cells of heterogeneous origin, it is none the less of great practical value to include in the same category all those cells which show a common property in their functional reactions.

Besides the histiocytes of connective tissue, the macrophage system includes the following elements: the cells which form the reticular matrix of the spleen and lymph nodes, the endothelium lining the sinusoidal blood vessels of red bone marrow, certain cells lining the blood sinuses in the liver (*Kupffer cells*), endothelial cells lining the sinusoids of the suprarenal glands and the anterior lobe of the pituitary gland, and the large phagocytic mononuclear cells which circulate in the blood stream. Those cells of the system which are normally stationary are capable of being detached and converted into freely moving elements.

The reticular cells of lymph nodes line the lymph sinuses which pervade these organs, and play an important part as a protective mechanism by removing from the lymph stream deleterious material (see p. 254). In the spleen reticular cells have a corresponding relation to some of the venous blood sinuses, and here they are sometimes seen to contain effete red blood corpuscles which are in the process of destruction by them. In the lymph nodes and the spleen the reticular cells appear to represent continuous extensions of the endothelium lining the lymphatics and blood vessels which enter and leave them. Nevertheless, they are distinguished from endothelial cells generally by their phagocytic properties.

Kupffer cells ('stellate' cells) are found scattered among the non-phagocytic endothelial cells lining the venous sinuses of the liver, and it has been observed that under certain conditions they become pedunculated and eventually set free in the blood stream as circulating phagocytes. In the liver they take part in the destruction of red corpuscles with the formation of one of the bile pigments—bilirubin. Contrary to former views, it is now realized that bile pigments are not only produced in the liver but probably also by other cells of the macrophage system (including histiocytes) all over the body, for experimental evidence has demonstrated that, in an animal in which the liver has been removed, bilirubin can still be produced in the tissues, particularly in regions where effete red blood corpuscles are in process of disintegration. Kupffer cells, therefore, are

[1] L. Aschoff, *Lectures on Pathology*, New York, 1924. See also K. Kiyono, *Die vitale Karminspeicherung*, Jena, 1914.

simply one group of the widespread system of large phagocytic cells concerned in this process.

In the macrophage system should perhaps be included the flattened mesothelial cells which form the fine membranes (*arachnoid* and *pia mater*) investing the spinal cord and brain. Although these cells in normal conditions are not conspicuously phagocytic, under the influence of irritants they become rounded up and set free in the cerebrospinal fluid as mobile phagocytes capable of ingesting particulate material and assimilating vital dyes.[1] In other words, these cells are still generalized to the extent that they can become converted into macrophages. Herein they contrast with the cells of serous membranes such as the peritoneum, which are apparently too differentiated to be capable of taking on other than their specialized normal functions.[2]

In the substance of the central nervous system are somewhat specialized cellular elements, called *microglial cells*, which are probably of mesodermal origin and which are amoeboid and phagocytic. There is reason to suppose that these cells should also be included in the macrophage system (see p. 397).

In summary, it may be emphasized again that the cells of the macrophage system form a protective and scavenging mechanism which pervades most of the tissues of the body and which can be called into vigorous action when the body is invaded by infective micro-organisms. This is not to say, however, that other types of cells may not have phagocytic properties. Carleton has shown, for example, that the vaginal epithelial cells of the rabbit are phagocytic, and can take up carmine particles, hydrocollag (a colloidal suspension of graphite), and bacteria. The same author has also demonstrated that the epithelial cells lining the alveoli of the lung can be set free and phagocytose particulate or colloidal suspensions introduced into the alveoli by the respiratory tract.[3] These cells, whose origin had previously been in doubt, are sometimes known as 'dust cells'. Their possible histiocytic origin is discounted by the fact that they do not show the characteristic reaction towards vital dyes. Such examples of epithelial phagocytosis can usually be fairly sharply distinguished on morphological grounds from the amoeboid and phagocytic properties of connective-tissue elements.

5. CONNECTIVE TISSUE AS A SUPPORTING MATRIX

It has been noted that connective tissue is an all-pervading matrix in which are embedded more highly organized tissues such as muscles, nerves, vessels, &c. To this extent it obviously provides a mechanical support for these structures. Around muscles it becomes condensed to form epimysial sheaths and intermuscular septa, penetrating also into the muscle substance between individual muscle fasciculi. Around vessels and nerves it forms

[1] H. H. Woollard, 'Vital staining of the leptomeninges', *Journ. Anat.* **58**, 1924.
[2] R. S. Cunningham, 'On the origin of the free cells of the serous exudates', *Amer. Journ. Phys.* **59**, 1922.
[3] H. M. Carleton, 'Studies in epithelial phagocytosis', *Proc. Roy. Soc. London*, **108**, 1931, and **114**, 1934.

perivascular and perineurial sheaths, while it also provides fibrous capsules for glands and other structures.

Where movement occurs between one structure and another, this is facilitated by a loosening in the texture of the intervening connective tissue. The free mobility of many parts of the skin over underlying structures is due to this. For example, the scalp is separated from the surface of the skull by what has sometimes been termed a 'lymph space', but which is really a layer of extremely loose areolar tissue. This same modification of connective tissue is shown in the formation of tendon sheaths which allow tendons to move freely during contraction and relaxation of muscles.

Here and there in the body where a tendon lies in contact with a bony surface or with another tendon, the connective tissue may become so loose and open in texture as to be converted into a well-defined and circumscribed sac in which flattened fibroblasts form a smooth mesothelial lining apparently identical with the synovial membrane of a joint cavity. These sacs are called *bursae*. They function as small 'water cushions' in minimizing the effects of pressure and friction, and they contain a minute amount of fluid similar to the synovial fluid of a joint. All stages in the differentiation of bursae may be recognized in various parts of the body, from a slight loosening of the connective tissue associated with an opening up of the tissue spaces, to the formation of a clear-cut cavity. A knowledge of the position of some of these bursal sacs in the human body is of some importance since they are liable to inflammatory or traumatic affections which may require surgical treatment. They are particularly numerous in the neighbourhood of joints, where tendons are commonly found in close apposition to each other and to bony eminences, and in some cases a bursa secondarily establishes a direct communication with a joint cavity. Subcutaneous bursae may develop over bony prominences which lie immediately deep to the skin (e.g. the patella and the olecranon process of the ulna), but it is probable that, as well-defined sacs, these only appear adventitiously in response to some trauma.

6. SUPERFICIAL FASCIA

Besides forming a general matrix throughout the body, connective tissue is also differentiated into sheets of fascia which are often so well defined as to merit detailed description in textbooks of topographical anatomy. Immediately deep to the skin is the superficial fascia. In the human body this is also called the *panniculus adiposus* because in normal health it contains an abundance of fat (Figs. 27 and 28). In some areas, however, fat is practically absent, e.g. in the eyelids, the scrotum, and penis; here the superficial fascia consists simply of loose areolar tissue. The amount of subcutaneous fat shows a sexual difference, for in women it forms a thicker and more even layer, softening the surface contours of the body in a characteristic manner. The panniculus adiposus is not developed in animals with a hairy coat; in them the superficial fascia is loose areolar tissue which, incidentally, allows the skin to be stripped off quite readily from underlying tissues. The relative absence of hair in man demands the provision of a subcutaneous layer of fat to take its place for the conservation of body heat.[1]

The superficial fascia is not everywhere homogeneous, for it may be differentiated into more than one layer of fat by intervening layers of fibrous tissue. Over the lower part of the abdominal wall it contains elastic

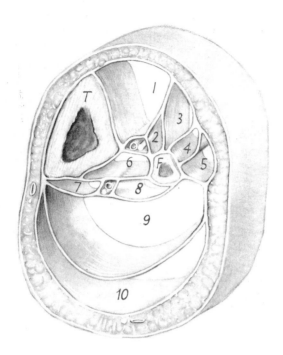

FIG. 27. Diagram of a section through the leg at the middle of the calf, in which the fascial planes have been preserved after removal of the muscles. Deep to the skin is seen the superficial fascia, and deep to this the ensheathing layer of deep fascia. Processes of the latter extend in among the muscles to form intermuscular septa, and also surround vessels and nerves. *T.* Tibia. *F.* Fibula. The muscular compartments are indicated as follows. 1. Tibialis anterior. 2. Extensor hallucis longus. 3. Extensor digitorum longus. 4. Peroneus brevis. 5. Peroneus longus. 6. Tibialis posterior. 7. Flexor digitorum longus. 8. Flexor hallucis longus. 9. Soleus. 10. Gastrocnemius.

tissue in sufficient abundance to comprise a separate stratum—the 'fascia of Scarpa'. This seems to play the part of a natural elastic belt, aiding in the support of the abdominal wall and, through it, of the abdominal viscera.

The superficial fascia provides a medium through which run the cutaneous nerves, blood vessels, and lymphatics which supply the skin. It also contains the more deeply situated sweat glands, the mammary glands, and, here and there, localized groups of lymph nodes. Among its other occasional contents may be mentioned sheets of cutaneous musculature, whether

---

[1] The panniculus adiposus is particularly well developed in whales, where, as a thick layer of blubber, it is an important factor in maintaining body temperature.

this is striped muscle (e.g. in the face and neck) or unstriped muscle (e.g. in the scrotum and at the base of the nipple).

Superficially, the superficial fascia passes rather abruptly into the connective tissue of the integument (the corium or dermis). On its deep aspect, however, it is sharply demarcated from the deep fascia except where this is but feebly differentiated.

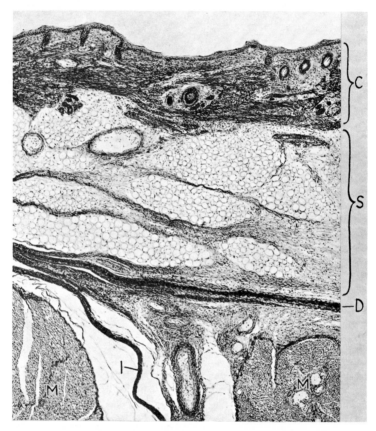

Fig. 28. Section through the integument of the leg of a human foetus aged approximately 7 months, showing the various fascial layers. Above is the skin, consisting of the epidermis and the dense corium (C). In the latter are some hair follicles. Beneath the corium is the superficial fascia (S), consisting of fatty tissue supported by delicate layers of collagenous tissue. Two 'cutaneous' blood vessels are seen in the superficial fascia. The deep fascia (D) is formed by a compact layer of collagenous tissue. Deep to it are seen portions of two muscles (M) and between them an intermuscular septum (I). × 54.

7. DEEP FASCIA

The consistency of the deep fascia varies quite considerably. Over the limbs it is extremely well defined as a tough sheet of white fibrous tissue. This forms a non-elastic and tightly fitting sleeve, keeping underlying structures in position and preserving the characteristic surface contour of the limbs. Where bony ridges or eminences approach the surface, it often blends with the periosteum and is bound down to the bone. Muscles frequently gain an extensive attachment to the deep aspect of the ensheathing fascia, and the latter is further strengthened in these places by the deposition of parallel bundles of collagenous fibres so as to form a tough, tendinous sheet, or aponeurosis. Moreover, where the deep fascia is subjected to tension it is reinforced in the same way. For example, where it covers the tendons at the wrist or ankle, transverse bands are differentiated which serve the purpose of natural wrist and ankle straps by holding the subjacent tendons in position when, during muscular contraction, they tend to start out of place. In the palm of the hand and the sole of the foot it is unusually thick, forming a mechanically protective layer—the palmar and plantar fasciae.[1]

From the ensheathing layer of deep fascia processes extend down among subjacent structures to form fascial septa which, in the case of the limbs, mark out a series of compartments containing separate muscle groups, bundles of vessels and nerves, &c. (Figs. 27 and 28). Similar processes have been described in other parts of the body (e.g. the pelvis and the neck) in very complicated detail. Many of them, however, are not so well defined as to deserve the descriptive attention which they have sometimes received, for there is little doubt that, to the bewilderment of the student of anatomy, accounts have been given in textbooks of fascial planes which do not exist as such in the undissected body. In the process of manual dissection, as one structure after another is 'cleaned' with forceps and scalpel, the investing and relatively homogeneous matrix of connective tissue in which they are embedded collapses into sheets of fascia which are really quite artificial in their definition. Apart from these manufactured planes, however, there are certain well-defined fascial septa the existence of which it is important to recognize since they play a part in directing the spread of pathological effusions of fluid or of new growths. Fluid tends to track along such fascial planes and also to be limited by them. Moreover, lymphatic drainage routes also tend to follow them, and these provide paths by which infections can spread from one part of the body to another. Hence a knowledge of fascial planes is of considerable practical importance.

Among their other functions, differentiated sheets of deep fascia which form intermuscular septa and interosseous membranes serve as additional surface areas for the attachment of muscles. The muscles of the forearm, for example, take their origin very largely from the ensheathing layer of deep fascia, the intermuscular septa between them, and the interosseous membrane connecting the radius with the ulna.

Circulatory functions of the deep fascia. One of the incidental functions which may be ascribed to the ensheathing layer of deep fascia is that of promoting the circulation in lymphatic and venous channels in association with muscular activity. The limb musculature is contained within the deep fascia under some tension—as is shown by the bulging out of muscle through

---

[1] The palmar and plantar fasciae actually represent the degenerated remains of the expanded tendons of muscles (the *palmaris longus* and *plantaris*) which in man have undergone considerable evolutionary atrophy.

a puncture wound. When the muscles contract against the resistant fascial sheath of the limb, the compression of the soft-walled veins and lymphatic vessels which lie among them necessarily accelerates the flow of their contents in the direction determined by their valves (see p. 201). The veins in the superficial fascia are not exposed to this mechanical influence and support, and in the lower limb they are for this reason particularly liable to varicosities in association with a poor circulation.

Retention bands, &c. It has already been noted that, when connective tissue is exposed to any tensile force, deposition of collagenous fibres by the activity of fibroblasts leads to the formation of structures which have a purely mechanical restraining function, such as retention bands, fibrous pulleys, check ligaments, and so forth. Many examples of these mechanisms could be given in the human body. Tendons occupying bony grooves are commonly bound down by retention bands. One of the best-defined pulleys is that through which the superior oblique muscle of the eye turns outwards and downwards to its insertion, and as a check ligament we may instance the fibrous expansion by which the anterior end of the lateral rectus muscle of the eye is hitched to the malar bone, and which serves to put a brake on its contraction. In the more elaborate fibrous structures— such as the fibrous pericardial sac and the processes of the dura mater in the skull—the collagenous fibres which form their basis are found on close analysis to be orientated in well-defined bundles along the lines of tension to which they are subjected during life. The various mechanical factors which determine these arrangements (in many cases quite complicated) have been worked out in some detail.[1]

8. SUMMARY OF FUNCTIONS OF CONNECTIVE TISSUE

The preceding account of the anatomy of connective tissue has shown it to serve a multiplicity of functions in the body. It is convenient to summarize them in tabular form.

1. In virtue of the mesenchymatous cells of an embryonic type which it contains, connective tissue provides a generalized tissue capable of giving rise under certain circumstances to more specialized elements.
2. As a packing material, connective tissue provides a supporting matrix for more highly organized structures.
3. Incidentally, it provides, by its fascial planes, pathways for nerves, blood vessels, and lymphatic vessels.
4. Where it is loose in texture it facilitates movement between adjacent structures, and, by the formation of bursal sacs, it minimizes locally the effects of pressure and friction.
5. It supplies restraining mechanisms by the differentiation of retention bands, fibrous pulleys, check ligaments, &c.
6. The ensheathing layer of deep fascia preserves the characteristic contour of the limbs, and aids in the promotion of the circulation in veins and lymphatic vessels.

[1] See, for example, G. T. Popa and E. Lucinescu, 'The mechano-structure of the pericardium', *Journ. Anat.* **67**, 1932–3.

7. The ensheathing layer of deep fascia, together with intermuscular septa and interosseous membranes, provides additional surface area for the attachment of muscles.

8. The superficial fascia, which forms the panniculus adiposus, allows for the storage of fat and also provides a surface covering which helps to conserve body heat.

9. In virtue of its fibroblastic activity, connective tissue aids in the repair of injuries by the deposition of collagenous fibres in the form of scar tissue.

10. The histiocytes of connective tissue comprise part of an important defence mechanism against bacterial invasion by their phagocytic activity, as well as playing the part of scavengers in removing cell debris and foreign material.

11. The meshes of loose connective tissue contain the 'tissue fluid' which provides an essential medium through which the cellular elements of other tissues are brought into functional relation with blood and lymph. Connective tissue thus plays a part in nutritive processes.

Besides these general functions of connective tissue, it also contributes to the special functions of the more highly organized structures and organs into the composition of which it enters. Some of these functions will be noted in succeeding chapters.

From the account which has been given of connective tissue, it will now be realized that this is by no means merely an inert packing material which occupies the interstices between highly organized structures such as muscles, vessels, nerves, glands, &c. It is a tissue of manifold functions of great importance, and therefore requires the closest attention of the student of anatomy.

9. BODY CAVITIES

Mention has been made of bursal cavities which are formed in connective tissue in order to permit of free movement between adjacent structures. Very similar to these are the cavities of synovial joints, which will be dealt with in a later chapter. A third type of cavity which occurs in connective tissue is represented by the serous sacs of the body, which include the pericardial, pleural, and peritoneal cavities. Although in their mode of formation and in the nature of their lining cells these are rather special formations, it is convenient to refer to them in this section.

The several serous sacs of the adult body are derived in the embryo from one continuous body cavity, the coelom (see p. 26). At a very early stage of human development (third week) this is first indicated by clefts appearing in the primary mesoderm which fills the embryonic blastocyst. Subsequently, further isolated spaces appear in the embryonic mesoderm and, by coalescence, contribute to the formation of a cavity separating the body wall from the developing viscera (Fig. 11).

Anteriorly, the coelom surrounds the heart to form the pericardial sac. This is linked up behind with the peritoneal sac by two pericardio-peritoneal canals, one on either side of the midline. The peritoneal sac becomes occupied by the abdominal viscera, while the pericardio-peritoneal canals accommodate the developing lungs and form the pleural sacs. These different compartments of the coelom are soon separated off from each other. The large ducts of Cuvier carrying blood to the heart encircle the openings from the pericardial sac to the pleural sacs and ultimately occlude them entirely. The pleural sacs become in their turn cut off from the peritoneal cavity by the development of the diaphragm. In the male, a fourth sub-division of the coelom is formed by the herniation of a part of the peri-toneal sac which surrounds the testicle as it descends into the scrotum. This becomes separated off as a closed sac, usually just before birth. Although these serous sacs are in common speech said to 'contain' the viscera with which they are associated, it should be realized that normally they are in effect empty potential cavities the lining membrane of which is everywhere

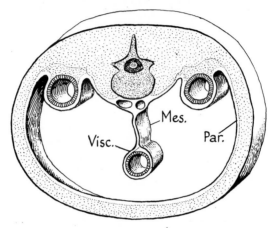

FIG. 29. Schematic diagram representing a transverse section through the abdominal cavity. The peritoneum is shown lining the cavity. Where it lines the body wall it is called the parietal peritoneum (*Par.*), and where it is reflected over the viscera it is called the visceral peritoneum (*Visc.*). The small intestine, which is an 'intraperitoneal' viscus, is shown almost completely invested by peritoneum and suspended from the posterior abdominal wall by a mesentery (*Mes.*). On either side are seen the ascending and descend-ing portions of the colon which are only partially covered with peritoneum, and are thus said to be 'retroperitoneal'.

reflected over the viscera. The latter are thus covered by a visceral layer of the serous membrane which is continuous, often through folds, or *mesen-teries*, with the parietal layer lining the body wall.

The nature of a mesentery can readily be appreciated by reference to the diagram in Fig. 29, which represents a transverse section through the abdominal cavity. In the midline is shown the mesentery of the small intestine, a double fold of peritoneum suspending it from the ventral aspect of the vertebral column. It will be realized that the mesentery allows considerable mobility to the small intestine, and at the same time provides a route by which vessels and nerves reach it from the posterior abdominal wall. On either side are seen in section parts of the large intestine—the ascending and descending colon. These usually have no mesentery, and are thus much less mobile than the small intestine. Being in direct contact with the posterior abdominal wall with no intervening peritoneum, they are said to be 'retroperitoneal'. It may be noted, however, that the whole length of the colon, and also certain other viscera which are normally retroperitoneal such as the duodenum and pancreas, are provided with a mesentery in the embryo. During development they become secondarily retroperitoneal by the apposition and fusion of one of the peritoneal layers of their mesentery with the parietal peritoneum, and the subsequent absorption of the fused membranes. It is of some practical importance to be acquainted with this embryological process, for it may occur as the result of mal-development that an embryonic mesentery persists, leading to an abnormal mobility of

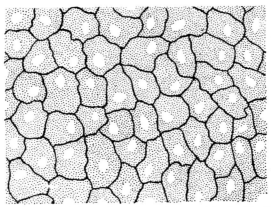

FIG. 30. Spread preparation of the peritoneal membrane, stained with silver nitrate. The intercellular cement substance has become stained by reduction of the silver salt, outlining the serosal cells. × 340.

the viscera concerned, and this introduces a liability to certain pathological conditions such as strangulation of the intestine. It is unfortunate that some mesenteric processes still retain the old designation of 'ligaments', for this suggests that they play a significant part in holding the viscera in their normal position. In fact, however, they are not concerned with this function, and surgical operations which in the past were devised to correct misplacements simply by shortening peritoneal attachments proved unsuccessful.

It is probable that a coelom was developed primarily in animal evolution as a sac into which the sex glands could discharge their reproductive cells. The further elaboration of the body cavity, however, was directly related to the development of a circulatory system with a rhythmically contracting heart, of a pulmonary mechanism of respiration, and of a mobile alimentary tract.[1] In order to permit of free movement in any visceral organ, the latter must of necessity be surrounded by a potential space. The existence of the coelom allows the smooth surface of the visceral layer of serous membrane to glide freely over the opposed parietal layer. The heart can thus alter its shape in its rhythmic contractions, the lungs can expand and collapse in respiration, and the free portions of the intestinal tract can undergo peristaltic movements with minimal resistance from surrounding structures.

[1] F. Wood Jones, 'The functional history of the coelom and the diaphragm', *Journ. Anat.* **47**, 1912–13.

The essential structure of a serous membrane is a lining layer of flat pavement cells, cemented edge to edge at their intercellular margins (Fig. 30). This mesothelial[1] or serosal layer is supported by an underlying layer of connective tissue. For many years it was generally accepted that the mesothelial lining is not complete—that minute gaps or 'stomata' occur between adjacent cells through which fluids or suspensions of particulate matter can escape from the cavity. Critical studies with carefully controlled technique have now led many anatomists to the conclusion that, at least in the case of the peritoneum, these stomata are merely artefacts in histological preparations and do not exist in normal living tissue. Absorption from the peritoneal cavity, therefore, must take place through the mesothelial cells or through their intercellular junctions.[2]

By using the technique of staining with vital dyes, Cunningham[3] has shown that serosal cells are of a specialized type, incapable of being transformed into macrophages or fibroblasts. In acute inflammatory conditions they may be shed into the cavity, but they undergo rapid disintegration there. A destructive lesion of part of the mesothelial lining of a serous sac is repaired by proliferation of surrounding mesothelial cells. If the denuded area is too large, however, fibrous adhesions are formed by the activity of fibroblasts in the connective-tissue substratum, but these cells are apparently incapable of forming new serosal cells. The practical significance of this is that, following extensive destruction of the mesothelial lining of the serous membrane by inflammatory processes, gross adhesions may be formed which seriously limit the normal mobility of the viscera.

[1] The term 'mesothelium', originally applied to the lining membrane of the coelomic cavities, is often applied also to the lining membrane of connective-tissue spaces such as joint cavities and bursae. In spite of the difficulty of defining the term with precision on embryological grounds, it is used conveniently as a distinction from 'endothelium' which is customarily applied to the lining membrane of the heart and blood vessels and the lymphatic vessels.

[2] However, there may be an exception to this generalization in the case of the peritoneum covering the diaphragm, for Allen (*Anat. Rec.* **67**, 1936) has described peritoneal stomata as normal openings in this region, overlying lymphatic vessels. Through these stomata fluids within the peritoneal cavity are believed to gain direct access to the lymphatic endothelium. See also H. Florey, 'Reactions of, and absorption by, lymphatics, with special reference to those of the diaphragm', *Brit. Journ. Exp. Path.* **8**, 1927.

[3] R. S. Cunningham, 'On the origin of the free cells of serous exudates', *Amer. Journ. Phys.* **59**, 1922.

IV

CARTILAGE

CARTILAGE is a form of supporting tissue in which a firm, resilient matrix is deposited intercellularly by the agency of cells called *chondroblasts*. The matrix incorporates connective-tissue fibres to a variable degree, and upon this depends the recognition of three main types of cartilage: hyaline, elastic, and fibro-cartilage. In contrast to bone, cartilage as a supporting structure combines a certain amount of rigidity with considerable flexibility and resilience.

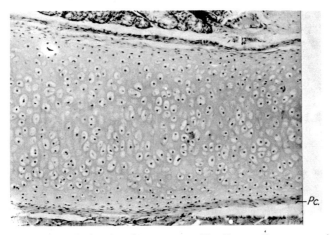

FIG. 31. Microphotograph showing the structure of hyaline cartilage as seen in the wall of the larynx. The cartilage is covered by a layer of fibrous tissue, the perichondrium (*Pc.*). × 110 (approx.).

Hyaline cartilage. Hyaline cartilage is a widespread tissue during embryonic life, for it provides the basis for the ossification of the greater part of the skeleton. In the adult it persists in the cartilages of the larynx, trachea, and bronchi, in the cartilage covering the articular surfaces of bones, and in the costal cartilages. It is freely sprinkled with cells which are usually scattered fairly evenly in the matrix, completely enclosed in small spaces or lacunae (Fig. 31). These cells are rounded in the central region of a piece of cartilage, but become flattened towards the periphery. Except over articular surfaces, hyaline cartilage is enclosed in a connective-tissue sheath, the *perichondrium*, and here can be observed a gradual transition from typical cartilage cells to connective-tissue cells which are not distinguishable morphologically from fibroblasts. In the matrix the cells are commonly disposed in discrete groups of two, three, four, or more which have arisen from the division of a single chondroblast. Each group is enclosed in a 'capsule' of more recently deposited matrix which shows up by its deeper staining. Mitosis is the normal mechanism of cell division in immature cartilage, but it appears that in the mature tissue cell division is uncommon except in articular cartilage.

The interstitial matrix of hyaline cartilage consists of a translucent substance, a tough gel whose main constituent is a sulphated mucopolysaccharide adsorbed to a protein; it is particularly resistant to pressure forces and at the same time possesses considerable elasticity. Embedded in it is an extremely fine meshwork of collagenous fibrils which can only be detected by special histological techniques. They are arranged in a characteristic pattern with reference to the groups of cartilage cells and to the tensional requirements of the cartilaginous mass as a whole.[1] In its structure hyaline cartilage has been compared mechanically to corded motor-tyres in which the rubber (equivalent to the interstitial hyaline substance of cartilage) serves to resist compression, and the cord (equivalent to the collagenous fibrils) to resist tension.

The problem of the nutrition of cartilage cells embedded in the middle of the matrix has attracted considerable attention. Some anatomists have described a system of fine canaliculi permeating the cartilage and linking up the lacunae in which the cells are confined. It is generally agreed, however, that these are artefacts, and it seems clear, therefore, that nutriment can only reach the cells by diffusion through the matrix itself. Vital dyes brought into contact with hyaline cartilage during life appear quite rapidly to diffuse through the matrix.

In relatively large masses of cartilage, vascular canals are sometimes found penetrating into the cartilaginous substance from the perichondrium, often to quite an elaborate extent. These have been described in some detail by Haines.[2] In human development they first appear in the third month, and by the seventh month all the larger masses of cartilage in the foetus are richly permeated with them. They contain blood vessels surrounded by loose cellular tissue (Fig. 32).

There is little doubt that cartilage canals are formed to meet the nutritional requirements of cartilaginous masses which exceed a certain bulk. While they are a prominent feature of the epiphysial cartilages (at the end of long bones) in man, they are absent in those of small mammals such as the rat. On the other hand, they are absent in the relatively slender laryngeal cartilages of man but well developed in the much larger laryngeal cartilages of the ox. It has been supposed that, in the case of those elements of the skeleton which are ossified in cartilage, they are formed as a direct preliminary to the process of ossification; but this is certainly not the case for they appear in the foetus quite a considerable time before ossification commences.[3] Having been formed, however, the contents of the canals certainly provide the osteoblastic tissue and the vascularization for ossification when this ultimately begins, and also determine by their distribution the position where the centre of ossification appears.

[1] A. Benninghoff, 'Der funktionelle Bau des Hyalinknorpels', *Ergebnisse Anat. Entwickl.* **26**, 1925.   [2] R. W. Haines, 'Cartilage canals', *Journ. Anat.* **68**, 1933-4.
[3] D. J. Hurrell, 'The vascularization of cartilage', ibid. **69**, 1934-5.

There is some doubt as to the mode of development of cartilage canals. One view is that they are the result of an active invasion of perichondrial tissue which erodes the cartilage by the chondroclastic action of vascular endothelium, but there is some evidence to suggest that they are formed by a process of inclusion of perichondrial vessels as the cartilage expands by a surface accretion.

Histogenesis, growth, and repair of hyaline cartilage. In the embryo cartilage appears as a differentiation of mesenchymatous tissue. The mesenchyme cells retract their processes and become more rounded, while

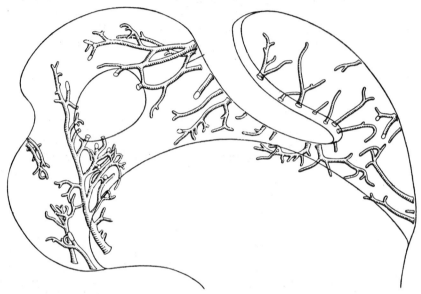

FIG. 32. Drawing of a reconstruction of a thick section showing the cartilage canals of the upper end of the femur of a child of eighteen months. (From R. W. Haines.)

an intercellular matrix of clear mucinoid substance is deposited between them. This transitional stage is termed *precartilage*. Then, round each individual cell (which is now termed a *chondroblast*) true cartilage is deposited as a thin film of hyaline material which has a basophil reaction. The tissue thus takes on a honeycomb appearance, and this rapidly changes to the appearance of more mature cartilage by a thickening of the intercellular septa, leading to a wider separation of the enclosed cells. In the adult, new cartilage can be formed in the same way by a metaplasia of connective-tissue cells.

In young, growing cartilage expansion takes place mainly by the progressive differentiation of mesenchyme cells in the surface perichondrium to form chondroblasts, that is to say, by surface accretion. It also involves some degree of interstitial growth, for in a growing cartilaginous epiphysis which has begun to ossify, a zone of proliferating cartilage cells separates older cells at the surface from older cells immediately surrounding the centre of ossification.

The regeneration of cartilage following injury has been studied experimentally in animals, and it appears to result mainly from the activity of the perichondrial cells. It has also been observed from tissue-culture experiments that perichondrial cells have a specific potentiality for forming cartilage, though these cells are at first indistinguishable morphologically from ordinary fibroblasts. The formation of new cartilage has further been studied by the transparent-chamber technique in rabbits' ears.[1] Here the cells which later form cartilage are similar to fibroblasts, though it has not been possible to determine whether they are derived from indifferent connective-tissue cells or from perichondrial cells of the adjacent ear cartilage. It is also uncertain whether the intercellular cartilaginous matrix is a direct product of cell secretion by the chondroblasts, or an extracellular formation effected as the result of some chemical change induced by the chondroblasts. The fact that localized pressure and friction may lead to the production of cartilage by the metaplasia of ordinary connective-tissue elements is strong evidence that cartilage formation is not necessarily dependent on specific chondroblasts. The same inference may be drawn from the experiments of Lacroix, as the result of which he reports that cartilaginous tissue may be formed at the site of the intramuscular injection of alcoholic extracts of epiphysial cartilage.[2]

Structures which are normally composed of hyaline cartilage have a tendency to calcify and ossify in old age. For example, the cartilages of the larynx usually show a partial conversion into bone in later life, commencing with the deposition of calcified plaques on the surface. The costal cartilages undergo similar changes.

Fibro-cartilage. This type of cartilage consists usually of little more than dense, white fibrous tissue with small scattered islands of cartilage cells. The latter, which are commonly disposed in rows, are by no means conspicuous. Fibro-cartilage is found in symphysial joints, and covering the articular surfaces in diarthrodial joints between membrane bones (see p. 167). It is also a usual component of tendons where they are exposed to frictional pressure, or at the point where they are inserted into bone (see p. 77). It has already been noted (*vide supra*) that hyaline cartilage contains collagenous fibrils embedded in its matrix. With advancing age this collagenous tissue becomes increasingly predominant so that the hyaline cartilage of youth may become the fibro-cartilage of senility. Indeed, it has been said that the transformation of hyaline cartilage into fibro-cartilage is one of the earliest signs of ageing in the body.[3]

Elastic cartilage. In elastic cartilage the matrix is permeated with a rich network of elastic fibres, which provides the tissue with considerable resilience. This type of cartilage is only found in a few isolated regions in man, e.g. the cartilage of the external ear, and certain of the laryngeal cartilages, such as the epiglottis. In contrast to hyaline cartilage, it normally shows much less tendency to calcify or ossify with advancing age.

[1] E. R. and E. L. Clark, 'Microscopic observations on new formation of cartilage and bone in the living mammal', *Amer. Journ. Anat.* **70**, 1942.
[2] P. Lacroix, *L'Organisation des os*, Paris, 1949.
[3] E. V. Cowdry, *Problems of Ageing*, London, 1939.

V

BONE

T H E main supporting structure of the vertebrate body is an endoskeleton of bony tissue. This is in strong contrast to the chitinous exoskeleton which serves similar functions in invertebrates. The functional and growth adaptations of an exoskeleton are limited and, in their adoption of this alternative method of bodily support, the invertebrates are at a disadvantage in evolutionary possibilities. Many vertebrates supplement their endoskeleton with a partial exoskeleton. Such are the bony dermal plates in certain fishes, the carapace of the tortoise, and the armour of the armadillo.

The human skeleton can be divided into somatic and visceral categories. The somatic skeleton is developed in the body wall and comprises the bones of the limbs, the trunk, and the greater part of the skull. In addition to this major part of the skeleton, which in its composition is morphologically very stable throughout the higher vertebrates, variable osseous elements may be developed in tendons and ligaments forming 'sesamoid' bones. The visceral skeleton comprises the branchial arch derivatives such as the mandible, the hyoid bone, the ossicles of the middle ear, and the cartilages of the larynx. In certain mammals, also, bony structures develop in other splanchnic tissues. Examples of these are the os cordis, found in the interventricular septum of the heart of large ungulates, and the os penis which is very frequently present in lower mammals.

True bone consists essentially of calcareous material in a complex crystalline form deposited in a fine matrix of collagenous fibres, and it varies considerably in its density. It is important to recognize the living and plastic nature of bone. The dry bones which are studied in an anatomy department give an impression of stability and immutability which is very illusory. They tend to be regarded as a completely rigid inorganic framework which persists in the mature individual uninfluenced by the factors which so readily modify, mould, and distort the softer structures around. This erroneous conception has not infrequently led anatomists and others to attribute to the smaller details of bone form an intrinsic morphological value which is by no means justified. Bone is essentially a living tissue supplied with blood vessels and nerves, and not only its external form but also its internal architecture can change in response to the stresses and strains to which it is subjected during life. Indeed, it has been remarked that, next to blood, bone is the most plastic tissue in the body. Even if this is an exaggerated statement, it serves to emphasize a highly important conception.

The primary function of bone as a supporting framework for the soft tissues of the body is self-evident. The skeleton not only provides for rigidity of the body as a whole, it also forms a series of mechanical levers to which are attached muscles and ligaments, allowing of free mobility. Certain parts of the skeleton, again, have equally obvious protective functions, e.g. the bony thorax and the skull. Bone also serves as an important reservoir for calcium which can be drawn upon when required for special metabolic activities; indeed the skeleton contains approximately 97 per cent. of the total calcium in the body. Lastly, bony tissue serves as a nidus for the haemopoietic activites of red marrow (see p. 235).

1. THE STRUCTURE OF BONE

If a limb bone such as the femur is sectioned longitudinally, the shaft is seen to be a tubular structure with walls of dense, compact bone and a central cavity—the medullary or marrow cavity. The latter contains, in life, yellow marrow which is mainly composed of fatty tissue. The extremities of the bone are filled with a sponge-work of bony trabeculae in the interstices of which is yellow marrow mixed with a variable amount of marrow of a different type, red marrow. The term 'cancellous bone' is given to this open trabecular type of bony tissue (Fig. 47). There is every gradation between widely meshed cancellous bone and compact bone, and even the latter is really porous in its minute structure. The cancellous tissue at the extremity of a long bone is covered on the surface by a relatively thin shell of compact bone, and where this takes part in the formation of a movable joint it is again overlaid with articular cartilage. As we shall see, the bony trabeculae which compose cancellous tissue are not disposed in a haphazard manner; their arrangement is determined by mechanical and growth factors.

Up to 60 per cent. of the weight of compact bone is formed by inorganic material, which consists mainly of calcium phosphate.[1] This salt is deposited in and impregnates a fibrillary basis of collagenous tissue, and the 'fibrous' character of bone can be readily appreciated by scraping the surface of a decalcified bone. It is owing to its fibrous basis that a completely decalcified bone retains its shape and even its general histological appearance. The calcified matrix forms the interstitial substance of bone. This substance is arranged in thin lamellae, is perforated by fine canals containing blood vessels and nerves, and is also richly permeated with bone cells or *osteocytes*. The lability of bone during life, even in the mature individual, is emphasized by the observation that, if salts of radioactive calcium, or phosphates containing radioactive phosphorus, are administered, the radioactive element can be detected in the bones usually within a few hours. By this method, also, it has been found that about 30 per cent. of the phosphorus deposited in the skeleton of an adult rat has been removed again in three weeks. It appears, indeed, that there is a continous physiological turnover of the inorganic constituents of bony tissue; they are employed

---

[1] In addition to calcium phosphate, the inorganic material of bone also contains some calcium carbonate and magnesium carbonate, as well as small quantities of fluoride and citrate. The composition of bone salts is far more complex than this simple statement suggests. For reference to this subject see F. C. McLean and M. R. Urist, *Bone, an Introduction to the Physiology of Skeletal Tissue*, University of Chicago Press, 1955; and *The Biochemistry and Physiology of Bone*, ed. by G. H. Bourne, New York, 1956.

as a reserve to be drawn upon to meet metabolic requirements elsewhere in the body, and are as continuously replaced by the deposition of new material. In other words, the inorganic basis of a bone is to a considerable extent undergoing repeated renewal during life.[1]

In Fig. 34 is shown a section of typical compact bone, taken transversely across the shaft of a long bone. On the surface, the lamellae are disposed in flat layers immediately beneath the periosteal tissues. In the substance

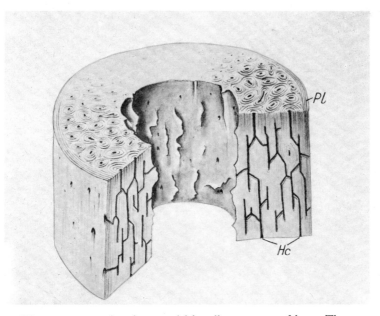

FIG. 33. Diagram representing the essential lamellar structure of bone. The concentric lamellae of the Haversian systems are indicated, and also the anastomosing system of Haversian canals (Hc.). The surface of the bone is covered by peripheral lamellae (Pl.). It should be emphasized that in the diagram the Haversian systems are represented entirely out of proportionate scale in order to indicate schematically their arrangement in a long bone.

of the shaft they are arranged in concentric rings round the fine canals which penetrate the bone everywhere, usually in a longitudinal direction. These are the *Haversian canals*. They range from 20 μ to 100 μ in diameter, and each contains minute blood vessels accompanied by fine nerve fibres and loosely enmeshed in reticular tissue. The canals branch and anastomose with each other to some extent, and provide essential channels for carrying blood into the bone substance. They open at the surface of the bone, and they also communicate with the marrow cavity (Fig. 33). Each canal and its

[1] It is a matter of supreme importance to realize that some radioactive elements which become incorporated in the inorganic matrix of bone may be retained for a long time. This is the case with radioactive strontium, the most dangerous of the isotopes which are produced by nuclear explosions. The possibility that the gradual accumulation of strontium-90 in bony tissue (particularly in growing children) may eventually lead to serious pathological lesions is recognized to be a very real hazard (see J. M. Vaughan, 'The effects of radiation on bone', in *The Biochemistry and Physiology of Bone*, ed. by G. H. Bourne, New York, 1956).

series of encircling lamellae comprise what is called an *Haversian system*. In the angular spaces between adjacent concentric systems the lamellae are disposed more irregularly, and here are termed *interstitial lamellae*. Between the lamellae are minute oval cavities—*lacunae*—from which extend on all sides fine, branching canaliculi. The latter connect up adjacent lacunae with each other, perforating the bony lamellae at right angles. They ultimately lead into communication with the central Haversian canal. There

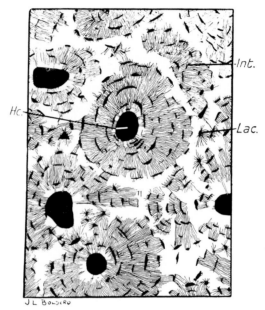

FIG. 34. Section of adult bone (human femur) showing the Haversian systems. *Hc.* Haversian canal. *Int.* Interstitial lamellae. *Lac.* Lacunae. × 150 (approx.).

seems, however, to be little or no anastomosing connexion between the canaliculi of one Haversian system and those of an adjacent system.

Each lacuna is completely occupied by an osteocyte. These bone cells send off fine protoplasmic processes from their cell body, which penetrate through the canaliculi and effect direct continuity with similar processes of neighbouring cells. In other words, in each Haversian system there is a true cellular syncytium.[1] Osteocytes are derived from bone-forming cells (osteoblasts), and retain the capacity to take on osteoblastic functions if the need should arise for bone repair. It may be presumed, also, that they subserve the modified osteoblastic processes which must be involved in the continuous physiological turnover of bony tissue which occurs during life. Under normal circumstances they appear histologically to be relatively inert cells, receiving nutrient material from the blood circulating in the vessels of the Haversian canals by imbibition along the communicating system of canaliculi.

[1] At least, this is the case in young mammalian bone, but there is some doubt whether continuity of the cell processes persists into adult life.

As with bony tissue in general, Haversian systems are by no means im-
mutable structures once they have been laid down. There is evidence that
their pattern and arrangement can shift and change from time to time in
response to physiological requirements.

2. THE PERIOSTEUM

Each bone is ensheathed by a relatively tough membrane called the
periosteum. This covers the bone shaft everywhere up to the margin of

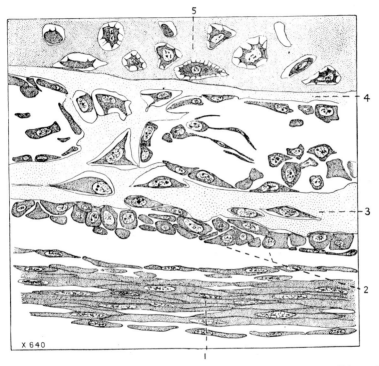

FIG. 35. High-power view of subperiosteal bone of a developing femur. 1. Fibrous layer
of periosteum containing fibroblasts. 2. Osteogenetic layer of periosteum containing
osteoblasts. 3. Subperiosteal bone, outer layer. 4. Subperisoteal bone, inner layer. 5. Cartilage. (From P. D. F. Murray and J. S. Huxley.)

articular surfaces, and while it is firmly adherent at the extremities and also
at the sites of attachment of muscles, tendons, and ligaments, it can otherwise be readily stripped off. In structure the periosteum consists mainly
of white fibrous tissue with a small element of elastic fibres. Beneath this
layer is looser connective tissue which is highly vascular and contains cells
capable of becoming active osteoblasts. The deeper stratum of the periosteum may therefore be called the *osteogenetic layer*. At the articular ends
of long bones the fibrous layer of the periosteum becomes continuous with
the fibrous capsule of the joints, so that in this way the periosteal sheaths
of adjacent bones are indirectly continuous with each other.

The function of the periosteum was at one time the subject of considerable controversy. Obviously it provides a medium for the attachment of muscles, tendons, and ligaments, and, in so far as it contains the blood vessels which communicate with the capillary network in the Haversian canals, it has a nutritive function. The dispute to which reference has just been made concerned its bone-forming properties.

In the middle of the last century, an Edinburgh surgeon, James Syme, carried out experiments which convinced him that the periosteum has osteogenetic functions. If, for instance, a segment of the shaft of a long bone is excised by subperiosteal resection, leaving intact the periosteal sheath, repair ensues and the two ends of the bone become united. If the periosteal covering is also removed, he found that there was no regeneration. He also carried out the experiment of raising the periosteum and inserting between it and the bone shaft a sheath of tinfoil. In this case, new bone was deposited on the surface of the tinfoil beneath the periosteum. These and other observations suggested that the periosteum is essential for regeneration of bone. Many years later, Sir William Macewen, a Glasgow surgeon, repeated several of these experiments with contradictory results. He found, also, that bone grafts denuded of their periosteum could survive and grow, and that if a ring of silver wire is placed round a bone shaft stripped of its periosteum, the wire becomes in the course of time enclosed by new bone formation. Macewen concluded that the periosteum is not only not essential for regeneration of bone, but it has no osteogenetic capacity. On the contrary, he believed that it is a limiting membrane, playing an important part in confining the osteoblastic tissue *inside* bone to its proper sphere. If the periosteal sheath of a bone is torn, osteoblasts may escape into the surrounding tissue and form bony excrescences. Furthermore, union of a fracture may be prevented by a layer of periosteum becoming interposed between the broken ends.

The argument regarding the osteogenetic capacity of periosteum was really based on a verbal quibble—the precise definition of 'periosteum'. If the term is confined to the superficial fibrous layer—its most conspicuous component—it is a limiting membrane. If it includes the deeper layer which is now known to contain cells capable of forming osteoblasts, it is also osteogenetic. Both these properties are important to recognize for practical purposes. Especially is this the case with the limiting function of the fibrous layer. In order to emphasize this function, it is useful to compare a bone to a very viscous fluid enclosed under pressure in a membranous tube. If the latter is torn, the viscous fluid slowly extrudes itself through the opening. Similarly, if the periosteal sheath of a bone is lacerated, osteoblasts from the deeper layers of the periosteum and from the bone itself may escape with the formation of exostoses—the bone gives an appearance of 'flowing' through the gap into the surrounding tissues. For this reason, it is important in the operative treatment of fractures to preserve the periosteal sheath intact if possible, for, if exostoses occur in the neighbourhood of joints, they may lead to considerable disability by limiting movement.

The conception of the periosteum as a limiting membrane makes it easy to understand how ridges and tubercles may be produced on the surface of a bone by the attachments of muscles and other structures. A tendon, for instance, at the point of its attachment to the periosteum will tend by its traction to pull up the membrane from the underlying bone. The latter will then extend into this area with the formation of a raised eminence.[1] In this way, the sites of the attachment of individual muscles, tendons, ligaments, and intermuscular septa are usually marked out clearly on the surface of a bone and can be delineated with little difficulty on close inspection. Where a muscle has a fleshy attachment over a wide area, its traction is correspondingly diffused. In this case its area of attachment is marked by a smooth surface usually bounded by faint ridges caused by the attachment to the periosteum of the intermuscular fibrous septa between which the muscle lies.

Grooves and depressions on the surface of a bone may be produced by the pressure of adjacent structures such as nerves, vessels, glands, or the convolutions of the brain in the case of the skull bones. It may be supposed that local pressure on the vascular periosteum disturbs the blood supply to the underlying bone, leading to a loss of vitality and subsequent absorption. It is evident, therefore, that hard bone can be quite extensively moulded in the smaller details of its surface features by the soft tissues with which it is in contact. This circumstance makes it possible for the palaeontologist, in his study of the fossil bones of extinct creatures, to draw important inferences concerning the relative development and disposition of many of the soft structures during life, and hence to learn something of the habits and movements which characterized the living animal.

From the deep surface of the periosteum, fine, pointed bundles of fibres run straight into the underlying bone at fairly regular intervals. These are the *perforating fibres of Sharpey*. Inside the bone they are calcified, and seem to have a supporting function by 'nailing' together the superficial lamellae of the bone. Where a tendon is attached to a bone, the perforating fibres are massed together so that they form a direct continuation of the tendon into the substance of the bone. Thus, although a tendon is primarily attached to the bone through the medium of the periosteum, it may subsequently assume the appearance of penetrating the bone substance directly.

In the adult there may be a gradual transition in structure at the insertion of a tendon. The latter, as it terminates, shows cartilage cells arranged in columns between the tendon fibres, forming a zone of fibro-cartilage. This is followed by a zone of calcified cartilage which, in turn, passes over into a zone of true bone.

---

[1] That this is not a completely adequate explanation is shown by the fact that the attachment of a tendon may sometimes be marked by a bony pit or depression rather than by an eminence. The reason for this is not clear. It is evident, however, that both a pit and an eminence serve the same purpose of increasing the surface area available for the attachment of the tendon.

3. OSSIFICATION

The ossification of bone is commonly described as taking place by two methods: in membrane (*intramembranous ossification*) or in cartilage (*endochondral ossification*), but the fundamental process is the same in either case. In endochondral ossification the bone is preceded by a cartilaginous model. However, there is no question here of a direct transformation of cartilage into bone, for the former has first to be removed by a process of tissue erosion before its place is taken by the deposition of true bone. In intramembranous ossification, bone is laid down directly in connective tissue.

The bones of the appendicular skeleton (with the exception of the clavicle), the trunk, and the base of the skull are cartilage bones. Among the membrane bones are the bones of the skull vault (all those that lie in direct contact with the scalp), the bones of the face, and the clavicle.[1] Some membrane bones may be developed *around* a cartilaginous precursor without the occurrence of endochondral ossification. Such are the mandible, which appears as an ossification in the membranous sheath of the cartilage of the first visceral arch (Meckel's cartilage), and the vomer, which is a membrane bone formed on the surface of the cartilaginous nasal septum. In these cases, the cartilaginous framework is simply absorbed and disappears when its place is taken by bone.

The distinction of cartilage and membrane bones is sometimes assumed to have a morphological basis in the sense that a cartilaginous skeleton is more primitive and that it preceded a bony skeleton in phylogeny just as it does in ontogeny. However, the palaeontological evidence is against such a supposition, since bony fishes are known to have existed in palaeozoic times before the appearance of cartilaginous fishes of the shark type. Cartilage is to be regarded rather as an embryonic adaptation, and not as representing an ancestral skeletal material.[2] On this interpretation, the skeleton of cartilaginous fishes represents a specialized retention in this group of an embryonic phase of development and is not a primitive feature. In some primitive fishes, the head and other parts of the body are protected by a surface covering of dermal denticles. At the base of these denticles, in the underlying dermis, bony plates are developed to form a superficial armour of *dermal bones*. The latter provide the basis for the evolutionary development of membrane bones in higher vertebrates. From the morphological point of view, therefore, the membrane bones of the mammalian skeleton, since they were originally derived from dermal plates, might well be regarded as elements of what was primarily an exoskeleton.

The question whether cartilage bones and membrane bones are really distinct morphological entities has been raised from time to time by comparative anatomists, and there is certainly some evidence that homologous elements may, on rare occasions, show different methods of ossification

[1] The ossification of the clavicle is said to take place in 'precartilage'. That is to say, the membranous *anlage* shows histological changes which suggest imminent chondrification. No cartilage, however, is formed except for isolated patches at the two extremities.
[2] A. S. Romer, 'Cartilage an embryonic adaptation', *Amer. Nat.* **76**, 1942.

in different vertebrate types. For example, it has been found that the caudal vertebrae in some teleostean fishes may ossify directly in membrane, although vertebrae are typically cartilage bones. In this case it must be assumed that the usual process of chondrification has been suppressed for some reason. Again, membrane bones, as they sink in from the surface during evolutionary development, may come secondarily into relation with underlying cartilaginous elements. As an example of this, it may be noted that comparative studies in lower vertebrates suggest the derivation of the ethmoid (a cartilage bone) from the supra-ethmoid (a membrane bone).

The problem is further complicated by the fact that *secondary cartilage* may be laid down in tissues where true membrane bones are undergoing ossification. This occurs in the coronoid and condyloid portions of the mandible, and also (in certain mammals) in the malar and frontal bones. Secondary cartilage, however, is to be distinguished from the primary cartilage of cartilage bones in its structure—the component cells are larger and the intercellular matrix is very sparse. Probably it is a secondary adaptation for resisting stresses to which the membrane bones may be subjected in their early development.[1]

Although the same fundamental processes are involved in intramembranous and endochondral ossification, in certain pathological conditions these two types of bone formation may be independently disturbed. In achondroplasia the growth of cartilage bones is everywhere retarded, leading to deformities of the limbs and the base of the skull.[2] In a rare disease called cleido-cranial dysostosis, on the other hand, the ossification of membrane bones is affected. The pathological basis of these conditions is not known.

Intramembranous ossification. Ossification in membrane may be conveniently described in a simple skull bone such as the parietal. Here there is usually a double centre of ossification which appears in the region corresponding to the middle of the bone at the eighth week of intra-uterine life in the human embryo, and from it the process spreads peripherally. Before this time, the parietal area of the skull is formed by a thin membrane of embryonic connective tissue.

The onset of ossification is preceded by an increased vascularity of the tissue. Collagenous fibres (the so-called *osteogenic fibres*) are collected into bundles in a gelatinous ground substance in which are also congregated numerous branching cells. The latter are osteoblasts, and through their fine processes they appear to effect direct protoplasmic continuity with each other. Calcium salts are then deposited in the interstitial substance and on the osteogenic fibres, apparently through the agency of the osteoblasts (*vide infra*). Many of the latter become imprisoned in the osseous matrix to form osteocytes, their branching processes occupying canaliculi. The bone is laid down at first as fine spicules which interlace to form a spongy texture, and at the margin of the expanding centre of ossification they radiate out in

---

[1] See the discussion on cartilage and membrane bones in G. R. de Beer, *The Development of the Vertebrate Skull*, Oxford, 1937.

[2] Many of the dwarfs which used to be exhibited at fairs were affected by the condition of achondroplasia.

a characteristic way. Later on, with increasing growth, the bone becomes thicker and more dense, and it extends superficially until it meets with the margins of the adjacent bones of the skull along a sutural line.

Endochondral ossification. This process will be described by reference to a comparatively simple long bone, such as the tibia. The shaft of

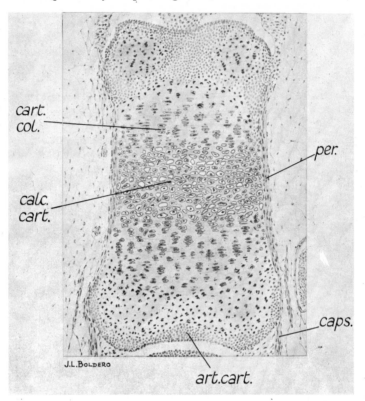

FIG. 36. The initial stage in the ossification of a long bone, as seen in the metatarsal of a foetal kitten. At this stage, deposition of actual bone has not yet commenced. Enlarged cartilage cells are seen in the zone of calcified cartilage (*calc. cart.*), and on either side of the latter the cells are becoming arranged in characteristic columns (*cart. col.*). The cartaliginous metatarsal is enclosed in a sheath of perichondrium (*per.*) which is continuous with the capsule (*caps.*) of the developing joint. The cartilage at the extremities of the metatarsal is taking on the appearance of articular cartilage (*art. cart.*). × 55.

this bone is ossified from one primary centre. Before the eighth week in the human embryo it is composed entirely of hyaline cartilage, and is enclosed in a membranous sheath which at this stage is called the perichondrium. The site of commencing ossification is indicated by changes in the cartilage cells. In the middle of the shaft these become large and vacuolated, and they tend to arrange themselves in rows which radiate up and down the shaft from the centre. Coincidentally, calcium salts are deposited in an amorphous form in the cartilaginous matrix. This process of calcification apparently leads to the cutting off of nutrient supply to many of the enclosed cartilage cells, and the latter undergo rapid degeneration and die. The enlarged and empty spaces which they occupied are called the *primary areolae* (Fig. 36).

Meanwhile the perichondrium at the middle of the shaft becomes active, and true bone is laid down as a thin shell on the surface by the process of

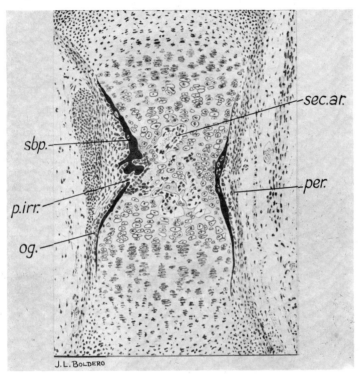

FIG. 37. Ossification in a phalanx of a foetal kitten. At this stage subperiosteal bone (*sbp.*) has already been laid down by the osteogenetic layer of the periosteum (*og.*). The shaft of the phalanx has been invaded by an irruption of osteogenetic tissue (*p. irr.*) which has led to an erosion of the calcified cartilage and the production of secondary areolae (*sec. ar.*). The perichondrium has now become the periosteum (*per.*). × 55.

intramembranous ossification. This is the subperiosteal bone of the shaft (for the perichondrium must now be called periosteum). From the deeper osteogenetic layer of the periosteum, an active ingrowth, richly supplied with blood vessels, pushes its way into the centre of the shaft, eroding the calcified cartilaginous matrix to do so. The centre of the shaft thus becomes vascularized. The irruption of subperiosteal tissue carries in cells which play a fundamental part in the construction of bone. These cells are of two types, *osteoblasts* and *osteoclasts* (Fig. 39). The former are concerned with the deposition of the calcareous basis of bone, and are rounded or oval cells with a large nucleus which stains rather deeply. They can be seen arranged in rows on the surface of bone which is being newly deposited, while they are not apparent in regions where there is no active osteogenesis or where bone is being absorbed. Osteoblasts appear to be derived in large part from undifferentiated connective-tissue cells, and also from some of the surrounding cartilage cells.

Osteoclasts are very large multinucleated cells of irregular shape. Histological study of stained sections suggests very strongly that they are concerned with the absorption of bone. Wherever local absorption is taking

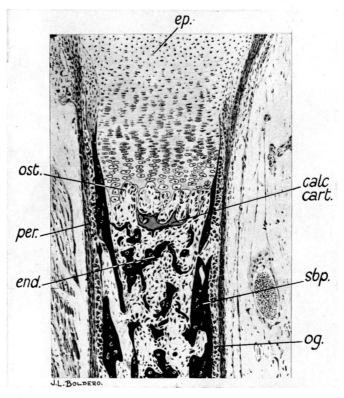

FIG. 38. Ossification in a metatarsal bone of a foetal kitten. The shaft is almost completely ossified, the cartilage having been replaced by endochondral bone (*end.*) and subperiosteal bone (*sbp.*). Osteoblasts (*ost.*) can be seen in close relation to the calcified cartilage at the advancing margin of ossification and elsewhere. In the non-ossified part of the shaft the cartilage cells are arranged in characteristic rows which extend as far as the epiphysial cartilage (*ep.*) × 55.

place these cells are characteristically seen directly applied to the bony tissue, sending out blunt-pointed processes which appear to be eroding it. They are commonly thought to be derived from the same connective-tissue cells which give rise to osteoblasts and also by a direct metaplastic transformation of the latter, though it is not known what factors determine the differentiation of such functionally contrasted types of cell.

The invasion by vascular osteogenic tissue is accompanied by the absorption in this area of the calcified cartilage. In this way large spaces are opened up in the centre of the shaft (*secondary areolae*). On the walls of these spaces osteoblasts can be seen in large numbers, in intimate relation to the deposition of bone on the surface of the cartilage, and they may be so closely packed in regular rows as to simulate an epithelium. It will now be apparent that the calcified cartilage is a temporary scaffolding which is removed bit by bit as it is replaced by true bone.

In the process of ossification, the bony architecture requires continual rearrangement in adaptation to the increasing size of the bone. For instance, the interior of the shaft of a long bone consists at first of a closely meshed core of osseous trabeculae. Later on this sponge-work is removed in the formation of the marrow cavity. It is therefore clear that bone deposition must proceed in harmonious association with a continual process of bone absorption. The latter is apparently carried out by osteoclasts. These cells also play a part in removing the calcified cartilage in the early stages of ossification.[1] The osseous tissue first laid down is lamellar in structure, but without showing the regularity of lamellae characteristic of compact bone in the adult. The Haversian system of lamellae is not distinctly evident till after birth.

True endochondral ossification in the centre of the shaft proceeds *pari passu* with the deposition of subperiosteal bone on the surface. These two processes extend up and down the shaft till the latter is completely ossified. The extremities of most long bones remain cartilaginous until after birth, when secondary centres of ossification appear in one or both of them to form the *epiphyses*. In the small bones of the carpus and tarsus, subperiosteal bone is not laid down—or only in a negligible amount. They are ossified practically entirely from the endochondral centres.

4. THE ROLE OF THE OSTEOBLAST AND OSTEOCLAST[2]

In the preceding account, the description of the process of ossification has been reduced to its simplest terms. We may now consider some of the details of this process.

It has been mentioned that the osteoblast is a bone-forming cell. This interpretation of its function rests primarily on the direct histological observation that osteoblasts are found in large numbers at the site of the deposition of new bone. Such an association, however, is clearly not a final proof of the bone-forming properties of these cells. Indeed, it has been held by some that osteoblasts are merely hypertrophied fibroblasts undergoing degeneration and destruction as the result of the precipitation in the sur-rounding matrix of calcium salts.[3] On this view the deposition of calcium salts is a biochemical phenomenon not directly determined by any local cellular elements, but controlled by extrinsic chemical factors related to the general metabolism of the body, and local variations in the blood supply.

---

[1] G. S. Dodds, 'Osteoclasts and cartilage removal in endochondral ossification', *Amer. Journ. Anat.* **50**, 1932.

[2] For critical essays on the osteoblast and osteoclast, see the chapters by J. J. Pritchard and N. Hancox in *The Biochemistry and Physiology of Bone*, op. cit.

[3] R. Leriche and A. Policard, *The Normal and Pathological Physiology of Bone*, London, 1928.

Tissue-culture experiments with explanted osteogenic tissues, however, have thrown very considerable doubt on this conception.

Fell[1] has shown that bone is developed *in vitro* from fragments of periosteum taken from chick embryos, while other tissues under the same conditions show no tendency to ossification. In other words, the capacity to form bone depends on the intrinsic properties of periosteal tissue and not

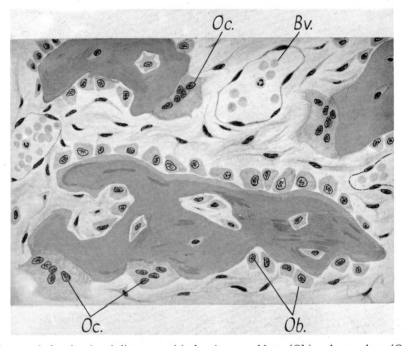

FIG. 39. A drawing (semi-diagrammatic) showing osteoblasts (*Ob.*) and osteoclasts (*Oc.*) in ossifying bone. Note that some of the osteoblasts become enclosed in newly deposited bone to form bone cells or osteocytes. The tissue is richly vascular, and some thin-walled blood vessels (*Bv.*) can be seen in section. × 500.

on chemical variations in the tissue fluids with which it is in contact in the living body. Furthermore, the same investigator found that the periosteal tissue of six-day chick embryos which forms bone in a culture medium contains recognizable osteoblasts. The important conclusion from this and other experimental studies is that, whereas *fibroblasts* from normally non-osteogenic connective tissue apparently do not form bone under the usual conditions of the tissue-culture technique, under the same conditions differentiated osteoblasts do so rather consistently. These observations strongly support the inference of the older histologists that osteoblasts are cells of specific type directly concerned with the deposition of bone, and this interpretation may be now accepted as probably quite correct. It may be observed that, in any case, the regular 'epithelioid' arrangement

[1] H. B. Fell, 'The osteogenic capacity *in vitro* of periosteum and endosteum isolated from the limb skeleton of fowl embryos and young chicks', *Journ. Anat.* **66**, 1931–2.

which is assumed by osteoblasts at the site of bone deposition is hardly consonant with the suggestion that they are merely degenerating fibro-blasts.

The origin of osteoblastic cells remains in doubt. There is evidence that they can be differentiated from mesenchymal cells, and perhaps also by the metaplastic transformation of fibroblasts. That such a process may occur in postnatal life is indicated by the occasional appearance of true ossification in tissues quite unrelated to bone, e.g. in the fibrous tissue of old scars, in muscles, or in the walls of arteries. Such heterotopic ossification is, of course, a pathological phenomenon. It has been suggested that, in association with the reparative processes following on local injuries or infections, the connective-tissue elements in any part of the body can assume general developmental potentialities similar to those of embryonic tissue. Under these conditions, osteoblasts may be differentiated locally and may even become active in bone deposition. In any case, however, the nature of the immediate stimulus which leads to the differentiation of osteoblasts re-mains obscure. In the case of ossification in tendons and muscle (*myositis ossificans*) which occurs under certain pathological conditions, the process may be partly due to the laceration of the periosteum at the site of the muscle attachment as the result of trauma, leading to the liberation of subperiosteal osteoblasts into the muscle tissue.

A number of interesting observations have been recorded on the experi-mental induction of heterotopic ossification. For example, it has been demonstrated that if the bladder epithelium of a dog is transplanted into the rectus abdominis muscle the surrounding connective tissue after a few days may become converted into bone, and that this process of ossification is associated with the appearance locally of an enzyme called phosphatase.[1] Similar experiments were later reported with the same results, though the authors note that bone was not always formed in spite of the presence of phosphatase in the epithelium.[2] Chondrogenesis and osteo-genesis have been reported after the injection into muscle of an alcoholic extract of normal bone, or of osseous tissue from the callus newly formed at the site of a fracture.[3] On the basis of more detailed experiments, Lacroix[4] has confirmed these results and has also found that an alcoholic extract of epiphysial cartilage injected into muscular tissue likewise leads to hetero-topic ossification. He has postulated the existence in the cartilage of a specific substance, 'osteogenin', which he believes is capable of causing ossification in surrounding tissues by a process of induction. However, it has been demonstrated that heterotopic ossification can also be produced in some cases by the local injection of alcoholic extracts of other tissues (such as muscle and tendon), or by the injection of chemical substances such as formic acid. The precise nature of this experimental osteogenesis remains obscure, but it does demonstrate that in certain circumstances osteoblasts

---

[1] C.B. Huggins, 'Influence of urinary tract mucosa on the experimental formation of bone', *Proc. Soc. Exp. Biol. and Med.* **27**, 1930.

[2] F. R. Johnson and R. M. H. McMinn, 'Transitional epithelium and osteogenesis', *Journ. Anat.* **90**, 1956.

[3] G. Levander, 'A study of bone regeneration', *Journ. of Surg. Gyn. and Obst.* **67**, 1938.

[4] P. Lacroix, *L'Organisation des os*, Paris, 1949.

may become activated in indifferent connective tissue, whether by the transformation of fibroblasts proper or by the local differentiation of mesen-chymatous cells of an embryonic type.

Experiments *in vitro* indicate that there is no fundamental difference in their functional potentialities between young cartilage cells (chondroblasts) and osteoblasts. The one may become transformed into the other in either direction. It is probable, indeed, that in the process of endochondral ossifi-cation many cartilage cells do not die but become active osteoblasts.[1] The experiments of Selye (see p. 90) suggest that osteoblasts can become con-verted into chondroblasts in the regeneration of the epiphysial cartilage after amputation of the growing end of a long bone. There is good evidence also, that the first appearance of osteoblasts in a developing long bone occurs round the zone of cartilage activity which *precedes* the irruption of sub-periosteal tissues into the shaft.

The question now arises, how does the osteoblast form bone? It was at one time supposed that this is the direct result of a secretory activity on the part of the cell body and its processes. It was shown by Robison and Fell,[2] however, that in developing cartilage-bones osteoblasts (or their precursors, the enlarged cartilage cells) elaborate an enzyme, *alkaline phosphatase*. These observations led to the hypothesis that the enzyme hydrolyses a soluble calcium phosphoric ester[3] which is circulating in the blood stream, with the consequent formation of a free salt in the form of calcium phosphate; the latter, it was supposed, is then precipitated locally. In the embryo, phosphatase is found to be concentrated at those points where ossification is imminent. Further, it has been demonstrated in tissue-culture experi-ments that there is a correlation between the quantitative production of phosphatase and the degree of bone formation. In other words, it appears that the enzyme is actually synthesized by osteogenic tissue. It is interesting to note, also, that the action of phosphatase is accelerated by magnesium which, as already noted, is a constituent of bone in the form of a carbonate. It is generally agreed now that the phosphatase mechanism postulated by Robison does not by any means provide a complete explanation of the immediate process of osteogenesis. For one thing, a local concentration of alkaline phosphatase is not necessarily associated with ossification (although it is involved in many other metabolic processes of cells) and, for another thing, the complex crystalline nature of bone salt (as demonstrated by X-ray diffraction methods) makes it clear that it must be formed indirectly by crystallization, and not by simple precipitation. There is probably a local factor such as an increase in pH in the region of bone deposition, and some other enzymatic mechanisms are certainly involved in the process.

The role of the osteoclast. The supposition that osteoclasts are directly concerned with the absorption of bone has been called into question. They are regarded by some authorities as degenerative products—the result of

---

[1] Others possibly help to form the reticular tissue of bone marrow.

[2] R. Robison, 'The possible significance of hexophosphoric esters in ossification', *Biochem. Journ.* **20**, 1926; H. B. Fell and R. Robison, 'The growth, development and phosphatase activity of embryonic avian femora and limb buds cultivated *in vitro*', ibid. **23**, 1929.

[3] Calcium hexose monophosphate.

incomplete cell division, by mitosis or amitosis, of worn-out osteoblasts. According to Arey,[1] for example, osteoclasts can frequently be observed in broad protoplasmic continuity with undoubted osteoblasts. Their presence in areas of localized bone absorption, however, is so consistent, and the histological picture they present is so persuasive, that there can be little doubt of their participation in the removal of bone (possibly by the secretion of an osteolytic enzyme, or perhaps by effecting a local change in the reaction of the tissue fluid) and in the remodelling process which is fundamental to the development of a bone. Sections of certain growing bones are specially convincing in this respect. For example, the parietal bone of the skull in late foetal life expands mainly by the continual deposition of bone on its outer surface and a corresponding absorption from its inner surface, and during this time the former is seen to be in contact with rows of osteoblasts, while the latter gives the appearance of being eroded by numerous multinucleated osteoclasts. The visual demonstration of osteoclasts over the inner surface of skull bones during growth has been made possible by a method of supravital staining with neutral red. With this technique, Barnicot has shown (in young mice) that quantitative estimates of the density of the cells in different areas may be obtained by directly counting them under low-power magnification, and these observations seem to make it clear that osteoclasts are concentrated over areas where (on other evidence) bone absorption is proceeding actively.[2]

In a developing long bone such as the femur, osteoclasts are present in great numbers on the inside of the tubular shaft where erosion is taking place with the progressive enlargement of the marrow cavity, while on the outside they may be found in the regions of the extremities which are involved in the remodelling process (see p. 91). Sections through the growing mandible provide further evidence of the activity of osteoclasts, for they are here seen lining the walls of the developing tooth sockets, apparently excavating the latter in preparation for the reception of the tooth germs as they sink down from the surface of the gums. That osteoclasts are essentially decalcifying agents is suggested by Dodds's observation on the removal of cartilage in endochondral ossification.[3] It has already been noted that in this type of ossification the cartilage undergoes calcification. This process, however, is not complete, for portions of the original cartilaginous matrix undergo absorption without preliminary calcification. Dodds found that while the uncalcified matrix is absorbed without the intervention of osteoclasts, these cells are always abundant in regions where the calcified matrix is being removed. All this evidence for the decalcifying property of osteoclasts is admittedly indirect. More direct evidence, however, is available from the observations of Kirby-Smith on a small piece of bone embedded in a transparent chamber in a rabbit's ear.[4] He kept this fragment under

[1] L. B. Arey, 'The origin, growth and fate of osteoclasts and their relation to bone resorption', *Amer. Journ. Anat.* 26, 1920.

[2] N. A. Barnicot, 'The supravital staining of osteoclasts with neutral red', *Proc. Roy. Soc.* B, 134, 1947.

[3] G. S. Dodds, 'Osteoclasts and cartilage removal in endochondral ossification', *Amer. Journ. Anat.* 50, 1932.

[4] H. T. Kirby-Smith, 'Bone growth studies', *Amer. Journ. Anat.* 53, 1933.

observation for many weeks and watched the process of absorption taking place, apparently by the agency of large granular cells similar to the osteoclasts seen in fixed preparations. In one instance these cells were found to be present before the actual process of absorption began. Moreover, growth and absorption were seen to occur simultaneously in different parts of the same bone fragment without any noticeable change in the local circulation or any difference in vascular pattern.

Some histologists have suggested that osteoclasts are true phagocytes, for they claim to have observed fine bone spicules in their cytoplasm in pathological foci of bone absorption, as well as the ingestion of dead bone corpuscles. But this is not in accord with the generally accepted view that the absorptive process is rather one of dissolution of the bony substance, and not of fragmentation accompanied by the phagocytosis of discrete particles.[1] It has also been argued that osteolysis can occur without the presence of osteoclasts, the result of physico-chemical changes in the tissue fluids. This factor may explain the diffuse bone resorption which occurs on a large scale in certain pathological conditions. It can hardly be the basis of very localized changes, as for instance when, in a small spicule of bone only a few microns in thickness, bone deposition is found to be occurring on one side and bone absorption on the other. Undoubtedly of decisive importance for settling the question of osteoclastic activity is the evidence which has now accrued from electron microscopy that, where they lie in contact with bony substance undergoing resorption, osteoclasts have a finely 'ruffled' border through which bone crystals are seen to pass and to become actually enclosed in cytoplasmic vacuoles.[2]

Available evidence shows that osteoclasts are probably produced by the fusion of a number of individual cells (and not by the aborted divisions of one cell). They are believed to be derived either from osteoblasts or from the same undifferentiated mesenchyme cells which also produce osteoblasts. When their work is completed, they either revert to osteoblasts or to the reticular cells of bone marrow, or they undergo degeneration and die.

5. THE GROWTH OF BONE

The mode of growth of a long bone is a somewhat complicated process. At birth, a bone such as the tibia has an osseous shaft (derived from the primary centre of ossification) with an incipient medullary cavity. Its extremities are still cartilaginous. Growth in thickness occurs as the result of the continuous deposition of layers of subperiosteal bone on the surface of the shaft. Coincidentally, the marrow cavity becomes enlarged as the bone grows in size owing to the continuous absorption of the medullary wall of the shaft by the activity of osteoclasts. In the adult bone, therefore, the main part of the shaft must have been derived entirely from the intramembranous

[1] It may be noted, incidentally, that osteoclasts do not exhibit the reaction to vital dyes which is characteristic of cells of the macrophage system (see p. 49).
[2] B. L. Scott and D. C. Pease, 'Electron microscopy of the epiphyseal apparatus', Anat. Rec. 126, 1957.

ossification of subperiosteal bone, the site of the original bony shaft (ossified in cartilage) having been now replaced by the marrow cavity.

Growth in length of a typical long bone is the result of the gradual extension of endochondral ossification into the cartilaginous extremities. The latter, as they become encroached upon in this way, grow superficially by the proliferative activity of the cartilage cells and so maintain their integrity. In the first few years after birth, secondary centres of ossification appear in the middle of the cartilaginous extremities. These centres

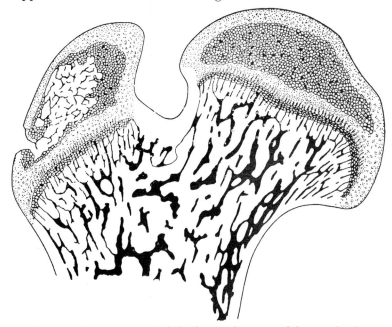

FIG. 40. Section of the upper end of the femur of a rat aged four weeks showing the ossification in the epiphyses of the head and great trochanter of the bone, and the zone of epiphysial cartilage separating them from the diaphysis. (From R. W. Haines.)

rapidly extend until the extremities are almost completely ossified, forming bony *epiphyses* (Fig. 40). The latter remain separated from the shaft (or *diaphysis*) by a thin plate of cartilage, the *epiphysial cartilage*, and where they enter into the composition of movable joints their surface retains a thin covering of articular hyaline cartilage. In a macerated immature bone, the epiphysis can be readily detached from the diaphysis. Their opposed surfaces are then seen to be irregularly nodular, the irregularities interlocking with each other. Such an arrangement gives mechanical strength to the epiphysio-diaphysial junction.

The epiphysial plates of cartilage are of extreme importance for the continued growth in length of long bones until maturity is reached. The growth is maintained by the extension of ossification from the diaphysis into the diaphysial surface of the epiphysial cartilage, while the opposite surface of the cartilage continuously proliferates at a corresponding rate. In this way, the epiphysial cartilage is retained, and continues to separate the bony epiphysis from the diaphysis. This process goes on until the bone has reached its maximal length, when the proliferation of the epiphysial cartilage ceases. As a result, the extension of ossification from the diaphysis soon involves the whole thickness of the cartilage, leading to its obliteration. In other words, the epiphysis becomes fused with the diaphysis, and no further growth in length of the bone takes place.

Some interesting experiments by Selye[1] have shown that, in the *early* stages of development, growth in length of a bone is not necessarily dependent on the integrity of the epiphysial cartilage which has already been differentiated. In rats 12–15 days old, the leg was amputated *above* the level of the epiphysial cartilage at the lower end of the femur. The stump of the femur became rapidly closed up by a mass of proliferating osteoblasts which were transformed into chondroblasts. The latter produced islets of cartilage and, in a few days, the end of the bone was closed by an even layer of cartilage showing a cellular arrangement typical of epiphysial cartilage. In other words, the epiphysial cartilage was regenerated and growth in length of the femur proceeded normally. No regeneration occurred if the operation was performed on rats over five months old. It is interesting to note that, if the femur is sectioned obliquely, the cut end becomes straightened up again before the layer of epiphysial cartilage is re-formed. In this way the original axis of growth is maintained. Similar experiments were later carried out by Nunnemacher[2] with corresponding results, but he attributed the regeneration of the epiphysial plate to cartilage cells in the diaphysial extensions of the epiphysial cartilage not completely removed by amputation.

The line of fusion between an epiphysis and diaphysis is usually quite evident in the mature bone. It is marked on the surface by a slight heaping up of bone into a low ridge, while the cancellous tissue inside the bone shows a line of condensation of trabeculae in the same position. This line of condensation, which is sometimes termed an 'epiphysial scar', is shown in Fig. 47 at the junction of the head of the femur with the shaft.

The part of the diaphysis which is immediately adjacent to the epiphysis is sometimes called the *metaphysis*. This term is convenient for differentiating the region of the shaft where growth in length is actually taking place. It is particularly well supplied with blood, and it is the site of attachment of many of the muscles and ligaments related to the adjacent joint. It is also metabolically more active than other parts of the bone since (as far as long bones are concerned) the calcium storage function of the skeleton is mainly served by the cancellous tissue in the metaphysis. It is for this reason, possibly, that it is more particularly prone to certain pathological affections. The vascularization of the metaphysis undoubtedly has an important relation to the proliferative activity of the epiphysial cartilage during growth, and experimental evidence has shown that disturbance of the metaphysial circulation may lead to the arrest of normal growth.[2]

Although in the large limb bones epiphyses are formed at both extremities,

[1] H. Selye, 'On the mechanism controlling the growth in length of the long bones', *Journ. Anat.* 68, 1933–4.
[2] R. F. Nunnemacher, 'Experimental studies on the cartilage plates in the long bones of the rat', *Amer. Journ. Anat.* 65, 1939.

growth takes place mainly at one end, the so-called *growing end* of the bone. In the case of the upper limb, this is at the end further away from the elbow, and in the lower limb nearer the knee. For instance, the humerus grows in length mainly from its head end, and the femur from its condylar end. Until maturity is reached, the epiphysial cartilages at the growing ends of the bones are the seat of great proliferative activity. Injury or disease affecting them may seriously disturb this activity, leading to irregular growth and subsequent deformity, or even to the complete cessation of growth. As the result of accidents, the epiphysis may be dislocated from the end of the shaft at the level of the epiphysial cartilage (separation of the epiphysis).

The growth of a bone involves not only a continuous accretion on the surface of the shaft and at the extremities, but also a continuous process of 'remodelling' whereby the characteristic shape of the bone is retained. This phenomenon can best be illustrated by a diagram showing the superimposition of outlines of the lower end of the femur and the upper end of the tibia at different phases of growth (Fig. 41). It is clear from this diagram that as the shaft grows in length at the epiphysial plate, bone must at the same time be removed from the surface of the metaphysis by an absorption mechanism in order to preserve the contour of the shaft in this region. This phenomenon of remodelling is particularly well de-

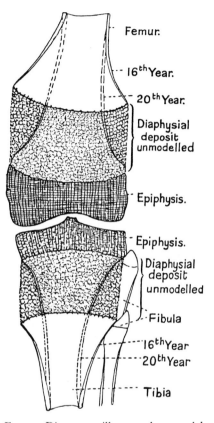

FIG. 41. Diagram to illustrate the remodelling process which occurs at the lower end of the femur and the upper end of the tibia during growth. The extent of bony tissue which is removed between the 16th and 20th year in order to preserve the characteristic contour of each bone is indicated. (From A. Keith.)

monstrated in such bones as the mandible by the use of madder staining (*vide infra*, p. 96). It is the result of the absorptive activity of osteoclasts working in complete harmony with the bone-forming activity of osteoblasts. The two processes, however, appear to have an independent basis, for under certain circumstances they may become dissociated. One process can occur without the other. This is found in a pathological affection of bone growth described by Keith[1] under the name of *diaphysial aclasis*, in

[1] A. Keith, 'Studies on the anatomical changes which accompany certain growth disorders of the human body', *Journ. Anat.* **54**, 1919–20.

which the long bones grow in length at the normal rate, but the remodelling process at the metaphysis is interrupted or irregular. The extremities of the bones as a result become thickened and ungainly in appearance. This condition is evidently due to a disturbance of the activity of the osteoclasts, and it may also be associated with exuberant outgrowths of bone near the joint surfaces forming multiple exostoses. Evidence has been brought forward suggesting that the remodelling process depends upon a separate genetic factor, for Grüneberg[1] has described a very curious mutation in mice which is characterized by the absence of this remodelling factor in the growth of bone. It affects not only the limb bones, but also the jaws, leading to difficulties in tooth eruption.

It should be emphasized that remodelling not only occurs on the surface of bones during growth, but also in the cancellous tissue inside. Indeed, the pattern of the bony trabeculae of cancellous tissue is continually being rearranged. Some of the skull bones become excavated during growth by air sinuses which are in open communication with the nasal cavities or the nasopharynx. Such excavations are also brought about by osteoclastic activity, with the subsequent ingrowth of an epithelial lining.

6. EPIPHYSES

Bony epiphyses are formed from secondary centres of ossification. Those which occur at the articular extremities of long bones are sometimes termed *pressure epiphyses*. They are presumably developed to protect the delicate and highly important epiphysial cartilage from the mechanical stresses which are associated with the free mobility of joints. Growth is thus allowed to continue, unimpeded by the exercising of the limbs. Epiphyses may also play a part in facilitating the modelling by osteoclastic action of the articular ends of the bones. Hence, in some lower vertebrates in which bony epiphyses are not developed, the extremities of the long bones are bulky and cumbrous in appearance as compared with the more delicately modelled bones of higher forms.

Epiphyses are in some cases formed at the site of attachment of tendons and ligaments, usually at the extremities of the long bones. For example, the trochanters of the femur, the tuberosities of the humerus, and the tip of the olecranon process of the ulna are ossified from secondary centres. They have been termed *traction epiphyses*, on the assumption that their development is in some way conditioned by the traction of the attached structures. It is possible, of course, that the initial appearance of these bony elements in evolutionary history may have been determined by, or associated with, such mechanical forces. If this is the case, however, they seem to have acquired a morphological individuality independent of these extrinsic factors. For example, it has been shown experimentally that if the gluteal muscles of a young rabbit are completely removed, the secondary centre for

[1] H. Grüneberg, 'A new sub-lethal colour mutation in the house mouse', *Proc. Roy. Soc. London*, **118**, 1935.

the greater trochanter of the femur (into which they are inserted) appears at the usual time, and ossification proceeds in quite a normal manner.[1]

A third type of epiphysis has been termed *atavistic*, for the reason that such structures represent skeletal elements which have undergone retrogression and have become secondarily fused with bones in the neighbourhood. An example of this is found in the subcoracoid centre of the scapula. This centre represents the coracoid bone which in lower vertebrates (such as the Reptilia) is a relatively large and independent skeletal element of the pectoral girdle. With the transformation of the mechanism and musculature of the shoulder region which took place in the transition from a reptilian to a mammalian phase of evolution, the coracoid bone dwindled to an insignificant epiphysis which in the adult becomes completely merged in the scapula.

Ossification in a cartilaginous epiphysis may commence in two ways— from the vascular tissue in the cartilage canals (see p. 68), or from an invasion of the cartilage by vascular tissue from the perichondrium on the surface. The former is the usual method in human development, while the latter is characteristic of small animals in which no cartilage canals are present.

It may be noted that the calcification or ossification of epiphyses does not occur in all vertebrates. In fishes, the epiphyses remain permanently cartilaginous. Palaeontological evidence suggests that this primitive feature was also common to all land vertebrates in early geological times, at least up to the Triassic period,[2] and it has persisted in certain groups of modern reptiles (crocodiles and chelonians).

7. TIMES OF OSSIFICATION

The onset of ossification and its progress in the early stages of development have been studied mainly by the examination of serial microscopic sections of embryos. In the study of centres of ossification in late embryos or in young animals, staining with alizarin is commonly employed. Alizarin is a vegetable dyestuff prepared from madder root. It forms a red lake with calcium oxide, and in its staining selects out centres of ossification probably because of the concentration of calcium ions in newly deposited bone. If the whole specimen is thus stained and cleared, the centres are well displayed and can be studied with a binocular microscope (Fig. 42). In a young animal a similar picture can be obtained by injecting trypan blue intraperitoneally and killing it forty-eight hours later. The developing bones are stained deep blue and stand out very conspicuously.[3]

In general, the primary centres of ossification of all the larger bones of the human body appear at the end of the second month of foetal life. The earliest bone to start ossifying is the clavicle (during the sixth week). The mandible and femur follow very closely (during the seventh week). In

[1] A. B. Appleton, 'Influence of mechanical factors on epiphysial ossification', *Journ. Anat.* **62**, 1922.
[2] R. W. Haines, 'The primitive form of epiphysis in the long bones of tetrapods', ibid. **72**, 1937–8.
[3] P. G. Shipley and C. C. Macklin, 'The demonstration of centres of osteoblastic activity by use of vital dyes of the benzidine series', *Anat. Rec.* **10**, 1915–16.

the skull the primary centres for the bones of the olfactory capsule and the otic capsule (e.g. the ethmoid, turbinate, and petrosal bones) appear rather late, about the fourth or fifth month. The carpal bones only begin to ossify after birth, the smallest element—the pisiform—not till the twelfth year. In the

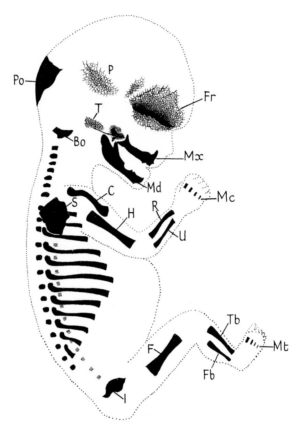

FIG. 42. Drawing of a thirteen-weeks' human embryo in which the centres of ossification have been stained with alizarin and the whole preparation cleared in glycerin. The centres of the following bones can be recognized. Frontal (*Fr*), parietal (*P*), post-occipital (*Po*), basi-occipital (*Bo*), squama of temporal (*T*), maxilla (*Mx*), mandible (*Md*), clavicle (*C*), humerus (*H*), radius (*R*), ulna (*U*), metacarpals (*Mc*), scapula (*S*), ilium (*I*), femur (*F*), tibia (*Tb*), fibula (*Fb*), metatarsals (*Mt*), ribs and vertebrae (centra and neural arches). × 2½ (approx.).

tarsus, the centres for the two largest elements, the calcaneum and the talus, appear at the sixth and seventh months of intra-uterine life, and the remainder after birth.

Secondary centres of ossification almost all appear after birth.[1] The dates of their appearance vary considerably from one bone to another, and to

---

[1] It is a matter of practical interest to note that the centres for the lower end of the femur and for the upper end of the tibia appear in the ninth month of intra-uterine life, at the time when birth normally occurs.

some extent also in different individuals. They are somewhat more precocious in females. A knowledge of the dates of ossification, and more particularly of the dates at which epiphyses become fused to the diaphyses, permits an estimation of the age of an unknown individual whose skeletal remains are available for study. For archaeological or medico-legal purposes, the importance of this will be obvious. An example of particular interest in relation to historical research was provided when, in 1933, the urn containing the bones of the Princes in the Tower was opened for an anatomical study of the remains.[1] There has been some doubt among historians about who was actually responsible for the murder of the two princes, Edward V and his brother Richard, Duke of York. One view attributed the crime to their uncle Richard III, and another to Henry VII after his accession to the throne in 1485. An examination of the skeletal remains attributed to Edward V showed that the stage of epiphysial growth and development (as well as the dentition) corresponded to an age of approximately twelve or thirteen years. In other words, since Edward V was born in 1470, it was inferred on anatomical evidence that he died before the Battle of Bosworth, that is, when Richard III was still in power. Although this inference has been controverted on the grounds of the known variability of times of ossification and dental eruption in modern populations, it seems likely (as a matter of probability) to be correct.

It may be noted that epiphyses at the growing ends of bones are the last to fuse with the diaphysis. In general, also, the epiphyses which join last are the first to commence ossification. An exception to the latter rule is found in the lower extremity of the fibula. Here the epiphysis appears earlier than that at the growing end of the bone, and also fuses earlier with the shaft. The precocious ossification of the lower end of the fibula (during the second year) perhaps has some relation to the strain which it needs to withstand when the infant begins to walk.

In the female sex, epiphyses join with the shaft somewhat earlier than in the male. It is also the case that, contrary to what might be expected, the dates of fusion are usually delayed in normal individuals of short stature, and accelerated in tall individuals.

The factors which determine the time of ossification are obscure. A comparison of the tarsal and carpal bones and to some extent the vertebrae (each series of which form components of a common mechanism) suggests that size is a factor, the larger bones commencing to ossify earlier. This, however, will not account for the early ossification of the clavicle, or the different times of ossification of (say) the bones of the skull. Functional and morphological factors may be involved in such cases.

As an individual anomaly, the union of an epiphysis may be delayed beyond its normal time, or it may never occur at all. The possibility of such aberrations should be recognized since they provide pitfalls for the inexperienced radiographer. An unfused epiphysis may be mistaken for a fracture. Other sources of fallacy in the interpretation of radiographic appearances are the occasional occurrence of accessory epiphysial centres

[1] L. E. Tanner and W. Wright, 'Recent investigations regarding the fate of the Princes in the Tower', *Archaeologia*, **84**, 1935.

(e.g. in the upper end of the ulna) or of accessory ossicles in the carpus or tarsus. In cases of doubt, an X-ray photograph of the opposite side may help to determine the issue, for such anomalies are frequently bilateral.

8. ANATOMICAL METHODS OF STUDYING BONE GROWTH

Reference has already been made to the experimental methods used by the older anatomists in their studies of bone growth, such as the removal and transplantation of the periosteum, encircling a growing bone with silver wire or tinfoil, and removing or grafting portions of bone. In this section, similar methods of critical importance will be described.

One of the earliest facts to be established in regard to bone growth is the absence of all interstitial growth. In other words, a mass of bone never increases its size by internal expansion but always by surface accretion. This can be demonstrated by a simple experiment made many years ago by John Hunter and other investigators. At two points in the length of the shaft of a limb bone in a young animal marks are made or metal pellets inserted, and the distance between them measured. When the bone is examined after it has grown considerably in length, the distance is found to be precisely the same. Similar observations have been made by the use of the madder method.[1] This was also employed by Hunter, and in later years provided the basis for a series of careful studies of bone growth by Brash and other workers.[2] It is essentially a process of vital staining (Figs. 43 and 44).

When certain young animals (e.g. pigs) are fed with madder root (*Rubia tinctorum*) newly deposited bone becomes stained a pink colour.[3] If a very young pig is fed with madder for some weeks, and is then given a madder-free diet for, say, three weeks and killed, the macerated bones show that the osseous tissue laid down in the earlier period is stained, while that deposited during the madderless period is unstained. The most accurate observations can be made by alternating several times madder feeding with a madder-free diet. The shaft and the extremities of the long bones under these circumstances show alternating bands of stained and unstained bone corresponding to the different dietetic periods. Not only is it possible by this method to detect precisely where bone deposition and bone absorption have been occurring during the process of growth, but the actual amount of bone laid down or removed in any particular area may be measured with a fair degree of accuracy. The method has been criticized on the ground that madder merely stains the more vascular part of the bone, and that it

[1] The observation that the bones become stained pink when an animal is fed on madder was first made by a Mr. Belchier in 1735, an observation for which he was awarded the Copley Medal of the Royal Society in 1737. But it does not appear that he applied it to the study of bone growth.

[2] J. C. Brash, 'Some problems in the growth and developmental mechanisms of bone', *Edinburgh Med. Journ.* **41**, 1934; *Journ. Anat.* **66**, 1932.

[3] It has been shown that the colouring substance in madder root which stains newly formed bone is *purpurin-3-carboxylic acid*; this is related to alizarin, and like it forms a red lake with calcium salts (D. Richter, 'Vital staining of bones with madder', *Biochem. Journ.* **31**, 1937).

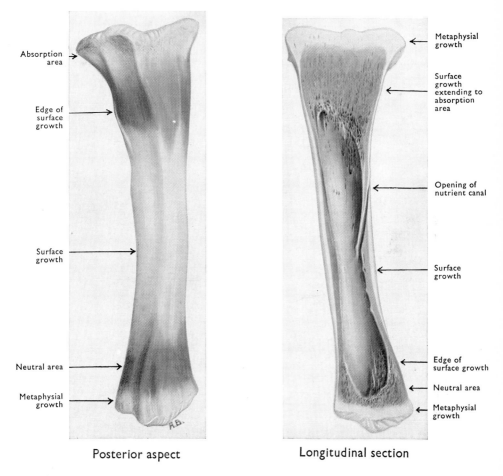

Absorption area

Edge of surface growth

Surface growth

Neutral area

Metaphysial growth

Posterior aspect

Metaphysial growth

Surface growth extending to absorption area

Opening of nutrient canal

Surface growth

Edge of surface growth

Neutral area

Metaphysial growth

Longitudinal section

FIG. 43. Diaphysis of tibia of madder-fed pig to illustrate the mode of growth of a long bone. (After Brash, 1934.)

The animal was $40\frac{2}{7}$ weeks old, and the madder had been omitted from the food for 30 days. The new bone added during that period is white, in contrast to the old, red bone.

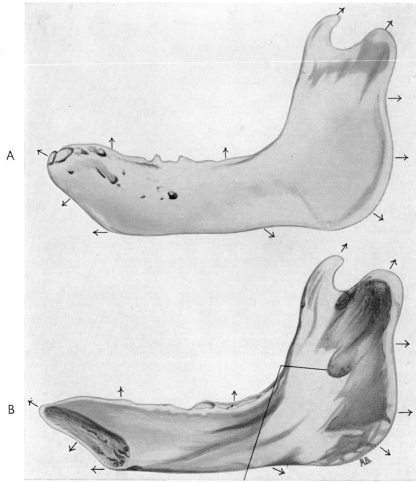

Growing edge of mandibular foramen

FIG. 44. Mandible of madder-fed pig to show the sites of growth. (After Brash, 1934.)

The animal was $20\frac{1}{7}$ weeks old and the madder had been omitted from the food for 19 days. The new bone added during that period is white in contrast to the old, red bone; and the main directions of growth are indicated by arrows.

A. Lateral surface of left half to show the distribution of growth, neutral, and absorption areas.

B. Medial surface of right half. Note particularly the indication of growth at the alveolar and posterior borders of the condyle and coronoid process, and of the upward and backward growth-movement of the mandibular foramen.

is not a true indicator of osteogenesis. These criticisms, however, have been adequately answered.

The method, of course, requires careful interpretation, but if employed with discretion and by the technique of alternate variations in madder feeding, it certainly provides extremely useful information. It has given the final proof that a long bone grows by accretion, accompanied by the absorption mechanism of remodelling. The diaphysis grows in length at the

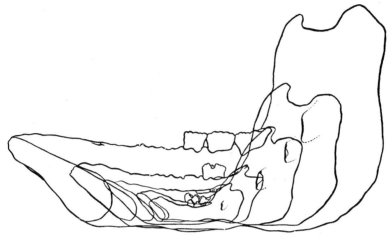

FIG. 45. Superimposed outlines of the medial aspects of the right halves of five pigs' mandibles. The ages of the pigs were as follows: newborn, 4 weeks, 11 weeks, 32 weeks, and 92 weeks. The position of the fourth milk molar is indicated in the first two, and the permanent molars in the others, to show the upward and forward movements of the teeth during growth of the mandible. (From J. C. Brash.)

epiphysial ends only. The bony epiphyses themselves grow by the deposition of bone on their free articular surface. No new bone is added to the diaphysial side of the epiphysis; indeed, some degree of absorption seems to take place here.[1]

The kind of detail which is provided by the madder method may be illustrated by reference to Brash's observations on the growth of the mandible in pigs (Fig. 45). The mandible increases in width as a whole by deposition of bone on its lateral surface and corresponding absorption from its inner surface. At the same time the ascending ramus grows in antero-posterior width by accretion at its posterior border, and to some extent its anterior border undergoes absorption. Thus the whole ascending ramus is shifted back relatively, allowing an increase in the backward extension of the alveolar border. The horizontal ramus of the jaw grows in length mainly by deposition of bone on the anterior surface of the symphysial region. The height of the horizontal ramus is increased at the alveolar border. This last observation necessarily means that the teeth contained within the alveolar border are involved in a continual shift in order to maintain their relative position. Indeed, it is now known that the teeth

[1] C. G. Payton, 'The growth of the epiphyses of the long bones in the madder-fed pig', *Journ. Anat.* **67**, 1933.

move upwards and forwards gradually, partly in order to keep pace with the upward growth of the alveolar border and partly to make room for the eruption of the molar series behind.

The madder technique shows that during growth the bone forming the anterior wall of the tooth sockets is continually being absorbed, while new bone is laid down at an equivalent rate on the posterior wall. This leads to a relative forward migration of the teeth, and when it is realized that a similar process must also occur in the upper jaw in perfect co-ordination with the lower jaw in order to maintain the normal precision of tooth occlusion, the complexity of the mechanism becomes very evident. It will be realized, also, that this tooth movement is a factor of some importance in the study of irregular eruption of the teeth and mal-occlusion in the adult.

This example of bone growth emphasizes the great plasticity of bony tissue during development. It may be deposited and removed again and again, so that in the final model of the mature structure nothing is left of the bony substance which was first laid down by the process of ossification. Yet the characteristic shape of the bone and its relation to neighbouring structures are maintained relatively unchanged.

The bones at the base of the skull grow in much the same way as long bones, at the site of cartilaginous plates which lie between them or between their component elements. The flat membrane bones in the skull vault expand to a considerable extent by accretion on the exposed surface and absorption on the intracranial surface. The importance of the suture lines as zones of growth has been the subject of much discussion, but from recent studies with the madder technique it seems now to be satisfactorily established that, at any rate during the early stages of cranial expansion, growth at the sutures also plays a significant role.[1]

The application of radiography to the study of bone growth. The growth of a long bone in a single individual can be studied continuously over a period of years by radiography. In this way the lengthening of the shaft, the appearance and enlargement of the epiphyses, and also the trabecular pattern of the cancellous tissue can be closely watched. The dates of appearance of ossific centres in the epiphyses have been found to show a slight degree of individual variation and, as has already been mentioned, are earlier in the female sex. However, they are sufficiently constant in some cases to provide a basis for the study of factors which retard or accelerate ossification. They may therefore be used for the nutritional assessment of growing children, for it is well known that ossification is considerably retarded if nutrition is inadequate. The possibilities of this application have been studied by a number of investigators, and for this purpose radiographic atlases of skeletal maturity standards in normal growing children have been compiled to provide a basis for comparison. On X-ray films, the absolute size of a growing epiphysis in a child can be measured with a planimeter, or the relation of its width to that of the lower end of the diaphysis may be expressed as a percentage ratio. The extent of the growth of the epiphyses in relation to the age group of the child is thus recorded.

---

[1] L. W. Mednick and S. L. Washburn, 'The role of the sutures in the growth of the brain-case of the infant pig', *Amer. Journ. Phys. Anthr.* **14**, 1956.

In the case of the carpal bones and the tarsal bones, the time of appearance of the primary centres of ossification and their size at different ages may be used for similar comparisons.[1]

The rate of bone growth and maturation is influenced not only by age and sex, but by economic status, the individual's total body weight, and also, possibly, by function. Racial differences also require to be taken into account, but it is very difficult to obtain reliable evidence on this point. For example, in a study of West African children the authors demonstrated that by comparison with American white children the skeletal age was delayed by an average of sixteen months, but, as they emphasize, this may not be a racial difference properly speaking. It may be related to nutritional, climatic, or other factors.[2]

Reference should be made to the lines of arrested growth which can occasionally be seen radiographically in the metaphyses of long bones. These transverse lines are local condensations of the trabeculae of cancellous tissue, similar in appearance to the line which normally appears at the site of fusion of the epiphysis with the diaphysis in a mature bone. They can often be related to definite illnesses which have occurred in early life, and which have led to a temporary disturbance of ossification in the epiphysial cartilage. Such lines of arrested growth when once formed are fixed, and they can be employed for making accurate measurements in X-ray films of the subsequent rate of growth of the diaphysis. They have provided corroborative evidence of the mechanism of growth in length of long bones.

9. THE MECHANICS OF BONE

In virtue of its inorganic framework of calcareous material, bone has lent itself more readily than other tissues to the study of mechanical adaptation in the body. Close study has revealed that in its internal structure a bone is adapted in a very remarkable way to resist the stresses to which it is subjected during life. This subject was considered by D'Arcy Thompson in his classic book on *Growth and Form*.[3]

Bone is required to have strength of two kinds, to resist compression or crushing forces, and to resist tensile or disrupting forces. The tensile and compression strength of bone are of an equivalent order, and herein bone shows considerable advantage over many constructional materials (such as cast and wrought iron and wood) which are strong in one direction but weak in the other. An engineer may require to use two different kinds of material as stays and struts (to resist tensile and compression forces). Bone is almost equally adapted for both purposes. The tensile strength of bone has been demonstrated by experimental studies[4] on the femur of rats, which showed the average breaking stress to bending to be 35,000 lb./sq. in.

[1] For further information on this subject, see J. M. Tanner, *Growth and Adolescence*, Blackwell Scientific Publ., 1955.

[2] J. S. Weiner and V. Thambipillai, 'Skeletal maturation of West African Negroes', *Amer. Journ. Phys. Anthr.* **10**, 1952.

[3] D'Arcy W. Thompson, *Growth and Form*, Cambridge, 1942.

[4] G. H. Bell, D. P. Cuthbertson, and J. Orr, 'Strength and size of bone in relation to calcium intake', *Journ. Physiol.* **100**, 1941.

as compared with 40,000 for cast iron. On the other hand, bone is about three times lighter than cast iron, and very much more flexible. It was also found that bone material is more than three times as strong as timber and half as strong as mild steel. The interesting observation was made that, even if the axial load on the femur is increased five times by jumping from a height, the strength of the bone is 140 times greater than is necessary for such an action, and the conclusion is reached that 'bones are "made" not as simple props but rather to withstand the severer actions of bending and twisting'. A mechanical analysis of the human femur as a structural element of a machine by Koch has given similar results.[1] According to his determination, the load on the head of the femur in a standing position is 0·3 of the body weight. In running, the dynamic effect of the sudden application of the body weight produces stresses which are more than five times as great—i.e. 1·6 times the body weight. Further, while the femur has a factor of safety of 5·68 for stresses due to running, the factor of safety for the standing position is as much as 30·3.

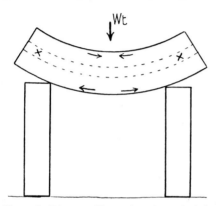

FIG. 46. Diagram illustrating the principle of the loaded beam. A heavy beam supported on a pillar at either end and carrying a weight in the centre tends to sag. The material of the beam at its upper surface is subjected to a crushing force, while at the lower surface it is subjected to a tensile or disrupting force. Along a central zone, X–X, these opposing forces are neutralized. In other words, the material along this zone is not necessary for resisting the stresses to which the beam as a whole is subjected.

The tubular character of long bones is readily explained on mechanical principles. It is the best construction for resisting bending forces in any direction. Such forces concentrate their effect on the surface of the cylindrical shaft so that, as far as strength is concerned, there is no need for a solid core. This can most easily be demonstrated by reference to the principle of a loaded beam.

If a horizontal beam is supported at either end by vertical pillars, it will tend to sag in the middle (Fig. 46). It is now clear that on the lower surface of the beam the material of which it is constructed is exposed to a tensile force and will tend to be torn apart. On the other hand, at the upper surface the material is exposed to an opposite or crushing force. It is equally clear that in the central axis of the beam there will be a zone where the tensile and compression forces are neutralized. In other words, the removal of the material in this zone will not appreciably affect the strength of the beam as a whole.

In the skeleton, the tubular character of the long bones, while not lessening their strength, makes for lightness and economy of material and also, at the same time, provides accommodation for the marrow tissue. Precisely the same mechanical explanation applies, of course, to structures such as

[1] J. C. Koch, 'The laws of bone architecture', *Anat. Rec.* **11**, 1916–17.

hollow reeds, quills of feathers, &c. If a long bone such as the femur is subjected to a bending force, the maximal stress is taken by the middle of the
shaft. Hence the bony wall of the tubular shaft here is thickest, and gradually thins out towards each extremity. In the skull, some of the bones are
excavated by extensive air sinuses. The body of the maxilla, for instance,
is a hollow shell with relatively thin walls. In this case, again, lightness is
gained without loss of strength. The maxilla must be sufficiently large to
accommodate the teeth, provide for the attachment of facial musculature,
and take its part in the construction of the orbit and nasal cavity. If it were
a solid structure, the weight of the facial part of the skull would be seriously
increased. Incidentally, the air sinuses also serve a function as resonating
chambers for the voice.

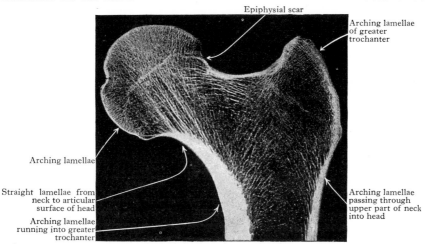

FIG. 47. Radiograph of a section of the upper end of the femur, showing the systems of
lamellae in the cancellous bone. (From Cunningham.)

For many years it has been recognized that the disposition of the
bony trabeculae in cancellous tissue bears a close relation to the lines of
stress to which the bone is subjected. In the middle of the last century,
Meyer[1] compared the trabecular arrangement in the upper end of the
femur with the trajectory lines of a crane-head. As in cancellous tissue elsewhere, the bony lamellae can be divided into two sets, pressure lamellae
which are related to compression forces, and tension lamellae which are
related to tensile forces. These two sets of lamellae should theoretically
form an orthogonal system, crossing each other at right angles. They have
been described picturesquely as the 'crystallization of lines of force', a phrase
which is expressive even if, as a critic has noted, it is inherently meaningless. Meyer's original concept was later elaborated by Wolff, and the
conformation of lamellar patterns in cancellous tissue to lines of stress
is sometimes referred to as 'Wolff's law'.[2] In many bones the lamellae

[1] H. Meyer, 'Die Architektur der Spongiosa', *Arch. f. Anat. u. Phys.* **47**, 1867.
[2] It is interesting to note that the observations of Meyer and Wolff were actually anticipated by a little-known Englishman, F. O. Ward, in a small book, *Outlines of Human
Osteology*, published in 1838.

form fairly clear-cut systems. For instance, in the calcaneum the pressure lamellae are resolved into two main components directed from the upper articular surface downwards and backwards to the point of the heel, and downwards and forwards to the under surface of the arch of the foot, while a system of tension lamellae curves in an antero-posterior direction through the lower part of the bone.

It seems apparent that these lamellae are developed in direct response to mechanical stimuli, though the precise nature of this response is not understood. It must be remembered that in an immature bone cancellous tissue is a closely meshed network of trabeculae. The differentiation of well-defined lamellae therefore comes about largely by a process of thinning out—presumably the result of the absorptive activity of osteoclasts. It is well known that bones which are not subjected to their normal mechanical stimuli, as in a paralysed limb, undergo atrophy in the course of time. It may be supposed that a similar process of atrophy leads to a thinning out of superfluous cancellous tissue in a normal bone, leaving behind the mechanically functional bony lamellae which are necessarily disposed along trajectory lines.

The direct adaptive response of cancellous tissue to functional requirements is shown by the fact that the lamellar architecture in the bones of the lower limb in man only acquires its characteristic arrangement when the child begins to walk. It is further demonstrated, in a still more striking manner, by the complete reconstruction and rearrangement of the trabecular pattern which takes place in bones which are exposed to an abnormal complex of stresses as the result of a badly set fracture or some deformity. In these cases, there is no doubt that the new pattern is more efficient mechanically.

The concept of pressure and tension lamellae has been vigorously attacked by some authorities.[1] It is objected that they do not always intersect precisely at right angles, and that they do not always conform accurately to the theoretically calculated trajectory lines. However, the varying stresses to which any single bone is subjected during life are so complex that it is hardly possible at present to express them with mathematical exactitude. Apart from the major forces which are related to the weight of the body and general movements, the pull exerted by numerous muscles and ligaments in different directions, and even the pressure of adjacent structures such as tendons, vessels, and nerves all exert some influence.

Considering these variable factors, it is perhaps remarkable that the coincidence of the lamellar pattern with the main estimated trajectory lines is as marked as it is. The attempts which have been made to explain the pattern of cancellous tissue as secondary to the arrangement of the cartilage columns which are replaced during ossification, or to the disposition of the blood vessels in the bone, or as simply conforming with the external contour of the bone, are not convincing. There can be no reasonable doubt that, at least

[1] See the discussion on this subject by P. D. F. Murray in his book *Bones*, Cambridge, 1936, and also the chapter on 'Bone as a mechanical engineering problem', by G. H. Bell in *The Biochemistry and Physiology of Bone*, ed. by G. H. Bourne, New York, 1956.

in great part, the disposition of trabeculae in cancellous tissue is directly conditioned by the mechanical requirements of the bone. Experiments *in vitro* have provided important information bearing on this problem, for it has been demonstrated[1] that if a fragment of ossifying cartilage is directly exposed to tensional forces, bone formation is increased, while reducing the tension leads to a corresponding retardation of ossification. Moreover, in osteogenetic tissue subjected to pressure and tension, osteo-blasts and osteogenetic fibres become orientated along the lines of stress. It seems clear, therefore, that mechanical forces do play an important part in determining the pattern of osseous architecture.

A study of the stresses and strains to which skeletal elements are exposed in normal or abnormal activities during life is of considerable interest for the understanding of the characteristic architecture of individual bones, and it is also of practical importance in studying the factors responsible for different types of fracture. The 'stresscoat' technique which has been developed allows an analysis of these forces in some detail. The method consists essentially of coating the surface of the isolated bone with a brittle lacquer (such as is used industrially to detect weak spots in machine parts), and then applying a known force in a particular direction. The position and direction of the resulting tensile forces are then indicated by linear cracks which appear in the lacquer at right angles to the lines of stress. The site of maximum tensile stress in any bone subjected to a force can thus be determined with accuracy.[2]

It is to be noted that many of the experiments designed to test the tensile or crushing strength of bone have been applied to isolated skeletal elements, often in a comparatively dry state. But in the living body a considerable proportion of the energy of a blow is absorbed by the soft tissues before it reaches the bone itself. For example, the empty skull cleaned of its over-lying tissues may be fractured by as little as 40 lb./sq. in. of energy, while the intact head of a cadaver may require as much as 600 lb./sq. in. to produce a fracture. In the living head an additional resistance will be introduced by the greater fluidity of the tissue matrix and the buffering effect of circulating fluids in the blood vessels.

10. THE SELF-DIFFERENTIATION OF BONE

The remarkable adaptability of bones to mechanical requirements sug-gests that, in the embryonic development of their characteristic form, extrinsic factors may exert an important influence. For instance, it was at one time supposed that the shape and curvature of the shaft of a long bone are conditioned by the developing musculature and other soft structures immediately related to it, that the contour of articular surfaces is deter-mined by the position of muscle attachments and the pressure between adjacent bones, and that tubercles and tuberosities are formed in direct response to the pull of tendons or ligaments.

[1] A. Glücksmann, 'Studies of bone mechanics *in vitro*', *Anat. Rec.* **72**, 1938; 'The role of mechanical stresses in bone formation *in vitro*', *Journ. Anat.* **76**, 1942.
[2] F. G. Evans, *Stress and Strain in Bones*, Illinois, 1957.

The application of experimental embryological methods to these problems has shown, on the contrary, that the development of the main anatomical features of a bone depends on intrinsic growth factors to an astonishing degree. As an example of this, we may quote early experiments by Murray and Huxley.[1] These authors grafted a fragment of the posterior limb bud of a four-day chick embryo on the chorio-allantoic membrane of a chick of seven days' incubation. The hind-limb of a four-day chick is only just beginning to appear and at this stage consists of an undifferentiated mass of mesodermal cells with no apparent indication of the future skeleton. After five days, the grafted fragment had developed into a recognizable femur with ossification in the middle part of the shaft (Fig. 48). The femur showed a typical curvature, a head and trochanter of avian form, and an incipient development of condyles at its lower extremity. A separate patella was also formed. This observation shows that a bone can be developed with all its main anatomical characters from a piece of morphologically undifferentiated tissue which is completely isolated from its normal environment. It appears, therefore, that this development is independent of the influence of extrinsic mechanical or chemical stimuli of a local or regional character—the potentiality resides in the tissue itself.

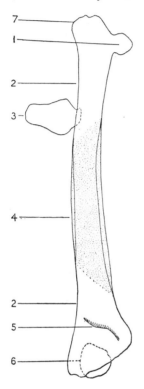

FIG. 48. Reconstruction of a grafted chick femur, viewed from behind. 1. Head. 2. Shaft of femur. 3. Fragment of opposite pelvis. 4. Region, at middle of femur, ensheathed in subperiosteal bone. 5. Line of attachment of muscles. 6. Patella. 7. Trochanter. (From P. D. F. Murray and J. S. Huxley.)

In summarizing the results of the experiments, Murray and Huxley concluded that the limb bud of a four-days' chick is a mosaic structure and not an harmonious equipotential system—'the limb bud at this stage consists of a number of different regions, each predestined to give rise to one segment of the limb only; but each of these regions is totipotent as regards all the parts of the segment to which it is destined to give rise'. Other experiments by Murray led him to suspect that, even before a limb bud appears in the embryo, there already exists in the somatic mesoderm a growth pattern representing the shafts of the limb bones.

The self-differentiating properties of mammalian bone, also, have been demonstrated by Willis,[2] though not in quite so striking a manner because

[1] P. D. F. Murray and J. S. Huxley, 'Self-differentiation in the grafted limb bud of the chick', *Journ. Anat.* **59**, 1924–5.

[2] R. A. Willis, 'The growth of embryo bones transplanted whole in the rat's brain', *Proc. Roy. Soc. London*, **120**, 1936.

he used cartilaginous primordia as grafts. Rudiments of the femur about
1 mm. long, removed from 15-mm. rat embryos, were implanted into
brains of six-weeks-old rats. After nine weeks in this curious situation,
they had grown into fully formed long bones reaching a length of 19 mm.
The rate of growth was approximately the same as that of the corresponding
bones in the intact animal. Femora with shafts and epiphyses of typical
shape and structure were formed.

The bony articular surfaces of joints can similarly be formed by intrinsic
differentiation with no relation to the attachment of muscles, or to the initiation of movements in the foetus. That movement is not a factor in the
development of joint surfaces had been inferred from the embryological
study of fixed and stained sections of normal material. Thus it has been
pointed out that the ossicles of the middle ear become differentiated with
the establishment of their articular extremities while they are still embedded in a mass of connective tissue and apparently immobile. Fell and
Canti[1] have observed the development of joints *in vitro* from explants of
limb buds. Under such conditions, articular surfaces become differentiated in their normal positions at the extremities of the bones. It is also possible for joints to be formed by tissue which is normally some distance from
the articular surface, but in this case they do not develop their characteristic
shape.

It is interesting to note that the epiphyses of the limb bones can develop
independently of the shaft. In a remarkable case of a congenital deformity
of this type, seen by the writer, radiography showed that the diaphysis of
the femur was completely absent, while the osseous epiphyses at the articular extremities appeared almost normal in size and shape. It seems,
therefore, that the normal ossification of the epiphysis is not conditioned by
the normal development of the diaphysis.

11. THE INFLUENCE OF MECHANICAL FACTORS ON THE SHAPE OF A BONE

The preceding account has emphasized the power of self-differentiation
in bone development. Remarkable as this is, it does not account for the
finer details and markings which are formed on a normal bone. For instance, in experiments on the development of isolated femora in chorio-
allantoic grafts, Murray and Selby[2] found that a groove normally present
in the head of the bone, and directly related to the acetabular ligament,
does not appear. It seems, therefore, that the direct contact of this ligament is necessary for the production of the groove. There is no doubt that,
in the later developmental stages at least, the refinements and details of
bone contour are conditioned to a large extent by extrinsic factors and
depend upon normal functional activity. The *accurate* adaptation of joint
surfaces is related to movement. The less conspicuous ridges and depressions on the surface of the bone, or, in other words, the finishing touches

[1] H. B. Fell and R. G. Canti, 'Experiments on the development *in vitro* of the avian
knee joint', *Proc. Roy. Soc. London*, **116**, 1934.
[2] P. D. F. Murray and D. Selby, *Intrinsic and Extrinsic Factors in the Primary Development of the Skeleton*, Roux's Archives, **122**, 1930.

in its final modelling, owe their origin to the tension of muscles and ligaments, or to the pressure of adjacent structures.

The shape of a mature bone can be markedly disturbed by abnormal activity during growth. Thus a dislocated femur may, as the result of pressure, excavate for itself a new acetabular socket in the ilium. In deformities of the feet, the tarsal bones acquire an abnormal shape and their internal architecture is altered or does not become differentiated.[1] A grafted strip of bone, inserted by the surgeon between the ends of a badly fractured bone, may unite and thicken so as to assume the general shape of the bone which it replaces. Removal of one of the forearm bones in an experimental animal is followed by a considerable hypertrophy of the other bone which now takes on the supporting function previously shared with its companion.[2] All these examples emphasize the plasticity of bone and its ability to respond to the mechanical requirements of new functions which are imposed upon it. This phenomenon was studied experimentally in some detail by Appleton,[3] with particular reference to modification in the shape of bones induced by the adoption of an unusual posture.

In a series of young rabbits the common obturator tendon was sectioned at its insertion into the upper end of the femur. This results in a medial rotation of the thigh which may be progressive until the popliteal surface faces almost directly laterally, and there is a consequent displacement of the foot which comes to rest on its inner border. In two-months-old rabbits, structural modifications are recognizable in a fortnight. These are shown in a retroversion of the lateral condyle of the tibia, and alterations in the torsional angles of the femur and the tibio-fibula. In another series of experiments, splints were applied which compelled the animal to displace its foot laterally in walking by means of a medial rotation of the thigh. After three weeks, the torsional divergence between the femur of the splinted limb and that of the opposite side reached 8°, and after six weeks 24°. Similar experiments made on mature rabbits led to no recognizable modification of the bones. In other words, these postural adaptations are the result of an alteration in normal growth, probably due to the changed distribution of pressure on the epiphysial plates of cartilage at the growing ends of the bones. Observations of this type point the way to the orthopaedic treatment of skeletal deformities in children by a proper appreciation of their postural basis.

The influence of muscular activity on the shape of the skull has been studied experimentally by Washburn.[4] He found that removal of the temporal muscles and the long neck muscles in new-born rats resulted in the failure to develop of certain features such as the temporal lines and the nuchal and mastoid crests. He also made the interesting observation, following separation of the neck muscles, that the shape of the interparietal bone

[1] F. Weidenreich, *Über formbestimmende Ursachen am Skelett und die Erblichkeit der Knochenform*, Roux's Archives, **51**, 1922.

[2] See numerous papers by J. Wermel in *Morphol. Jahrbuch*, **75**, 1935.

[3] A. B. Appleton, 'Postural deformities and bone growth', *Lancet*, **1**, 1934.

[4] S. L. Washburn, 'The relation of the temporal muscle to the form of the skull', *Anat. Rec.* **99**, 1947. For experimental evidence of the influence of mechanical factors on the shape of the scapula, see D. M. Wolffson, 'Scapula shape and muscle function', *Amer. Journ. Phys. Anthrop.* **8**, 1950.

is altered by retardation of growth, and since this bony element of the skull is situated anterior to the site of insertion of the neck musculature, it seems evident that its mal-development is due to the absence of mechanical stresses normally transmitted to it through the occipital bone. Washburn found practically no change in the internal form of the brain case in his experimental material, and thus was unable to confirm the suggestion (which had been entertained by some anatomists in the past) that the growth of the brain may be restricted by the compressing action of large temporal muscles, or that the contrasting shapes of head in different races of mankind (e.g. dolichocephaly and brachycephaly) may be related to corresponding differences in the temporal musculature.

Certain racial differences in the skeleton are apparently related to posture. Such, for instance, are retroversion of the head of the tibia, flattening of the shaft of the femur (*platymeria*) or of the tibia (*platycnemia*), and the so-called 'squatting-facets' on the lower end of the tibia and the neck of the talus. It has been presumed that these modifications are related (at least in part) to the habitual adoption of a squatting posture in rest, or to excessive activity of certain kinds such as hill-climbing,[1] but the evidence for such inferences is not always clear. The ease with which abnormal pressure can distort bony tissue in the adult is well shown in certain pathological conditions. A growing tumour, such as an aneurysm, can produce quite extensive erosion in adjacent bone as the result of mechanical pressure.

12. SOME POSSIBLE FACTORS INITIATING AND CONTROLLING THE NORMAL PROCESS OF OSSIFICATION[2]

Various theories have from time to time been put forward to account for the initial appearance of centres of ossification. Some of these will now be considered.

In old age, certain tissues which are badly nourished may undergo calcification. A similar process not uncommonly occurs in the middle of poorly vascularized or avascular tumours. It has even been suggested that the deposition of calcareous salts in the first stages of endochondral ossification is a degenerative change of a similar kind. On this view, there was presumed to be a definite limit to the size which can be attained by avascular tissue such as cartilage which depends for its nutriment on the imbibition of fluid from the surrounding vascular tissues. When this limit is reached the central cells of the cartilage undergo degeneration. This is accompanied by a precipitation of calcareous salts in the cartilaginous matrix, perhaps due to a local increase in the acidity of the tissue fluids. The calcified cartilage is then invaded by the surrounding blood vessels and removed like any foreign body. It can hardly be maintained, however, that the commencement

---

[1] There is some evidence that nutritional factors are also concerned in modifying the shape of the femoral and tibial shafts (see H. L. Dudley Buxton, 'Platymeria and platycnemia', *Journ. Anat.* **73**, 1938).

[2] For a discussion of these factors see P. D. F. Murray, *Bones*, Cambridge, 1936.

of a primary ossification in cartilage is delayed until the latter has increased in size up to its nutritional limits. At the time when they commence to ossify, for instance, the cartilaginous model of a phalanx of the little finger (at the second month of foetal life) is incomparably smaller than the cartilaginous lower extremity of the femur (at birth). On the other hand, the large cartilaginous epiphyses in man are richly permeated with vascular cartilage canals some time before ossification commences, so that there is no reason to suppose that the nutrition of the cartilage is in any way defective.

It is important to recognize that there is a clear distinction between simple calcification and endochondral ossification. The former may be regarded as a degenerative process related to a poor blood supply and a breaking-down of the tissues, but the latter is an active growth process accompanied by a rich vascularization and tissue proliferation. Some cartilages in the human body undergo simple calcification with advancing age without necessarily forming true bone, e.g. the thyroid cartilage.

The onset of ossification has been supposed by some authorities to be dependent on mechanical stimuli, particularly tensile forces and friction. It has already been noted in previous sections that ossification can be accelerated or retarded by tension and pressure. Reference may also be made to the work of Krompecher, who presented evidence that, in the healing of fractures, bone formation is directly promoted by tensional forces; pressure, on the other hand, tends at first to initiate cartilage formation, but with persistent pressure the cartilage also becomes replaced by bone.[1] Many years ago Thoma advanced the theory that practically all bone formation and bone growth, both in embryonic and postnatal life, are directly dependent on the response of osteogenetic tissues to tensile and compression forces. The appearance of centres of ossification in the membrane bones of the vault of the skull, for example, was speculatively related by him to the tensile forces set up by the rapidly expanding brain.[2] It was pointed out, however, that in cases of anencephaly (a developmental anomaly characterized by the almost complete absence of the cerebral hemispheres) the membrane bones are still ossified, though their final shape is distorted by the abnormal conditions of development. It is also the case that the adjacent margins of two skull bones along a sutural line may expand at very different growth rates, though presumably they are here exposed to the same field of tensile forces.[3]

Howell[4] has shown that in the completely denervated limbs of young animals in which the muscles are paralysed, so far as their general shape is concerned the bones grow normally except in thickness and, to a very slight extent, in length. These are therefore the only *major* characters of a bone whose full development may depend to some extent on the stresses

[1] S. Krompecher, *Die Knochenbildung*, Jena, Fischer, 1937.

[2] R. Thoma, 'Untersuchungen über das Schädelwachstum und seine Störungen', *Virch. Arch. f. path. Anat. u. Physiol.* **206**, 1911, and **212**, 1913.

[3] J. J. Pritchard, J. H. Scott, and F. G. Girgis, 'The structure and development of cranial and facial sutures', *Journ. Anat.* **90**, 1956.

[4] J. A. Howell, 'An experimental study of the effect of stress and strain on bone development', *Anat. Rec.* **13**, 1917.

related to normal muscular activity; the general shape and dimensions are determined rather by genetic factors.

The relation of osteogenesis to tensile forces seems to be fairly evident (on superficial consideration) in the case of ossification of tendons. This process is commonly found in the tendons of the leg in some birds, and in the long and powerful tendons of a kangaroo's tail.[1] In these cases, however, the conclusions drawn from direct observation require to be tested experimentally. It is essential to know whether the tendons, relieved of their normal tension by severing their attachments, become ossified in the usual manner.

In the human body, as in other mammals, bony nodules normally develop in certain tendons forming *sesamoid bones*. Examples of these are found in the patella, in the ossicles which develop at the insertion of the short flexor muscles of the thumb and great toe, and (sometimes) in the tendon of the peroneus longus in the sole of the foot. Fibrous or cartilaginous sesamoid bodies may develop without ossification. Sesamoid bones are formed at points of friction, where, for instance, a tendon passes over a bony eminence or ridge and changes its direction slightly. Indeed, the development of a sesamoid bone may help to amplify this change of direction and so increase the leverage power of the tendon. Friction is minimized by the formation of articular surfaces between the sesamoid and the underlying bone, and it has also been suggested that sesamoids serve to maintain the vascular supply of the tendon at a point where pressure might otherwise occlude the blood vessels.

It is commonly assumed that sesamoid bodies are formed under the direct influence of friction, and there is some experimental evidence in support of this conception. For example, it has been found that, when the patella is excised in young puppies, a new bone is formed if free movement at the knee joint is allowed. If, on the other hand, the limb is kept immobile, no regeneration of the patella occurs.[2] The authors of these experiments conclude that these results reveal 'the influence of pressure by the quadriceps extensor femoris muscle upon the ventral aspect of the distal extremity of the femur . . . in the formation of the patella'. It has already been mentioned (*vide supra*, p. 104) that the patella can apparently be differentiated in grafted fragments of the limb bud of chick embryos, when there can be no question of the extrinsic influence of adjacent muscles or tendons. We must conclude from this that while the *regeneration* of the patella after birth depends on appropriate mechanical stimuli, the normal development of the patella in the embryo depends rather on intrinsic growth factors.

There is little doubt that the patella is a true sesamoid bone (and distinct, therefore, from the morphological elements of the typical vertebrate skeleton) for its primary centre of ossification appears extremely late— during the third year of life in man.[3] This late ossification is characteristic

[1] F. Weidenreich, 'Über Sehnenverknöcherungen und Faktoren der Knochenbildung', *Zeitschr. Anat. u. Entwickl.* **69**, 1923.

[2] E. J. Carey, W. Zeit, and B. M. McGrath, 'The regeneration of the patellae of dogs', *Amer. Journ. Anat.* **40**, 1927.

[3] The morphological antiquity of the patella, however, is suggested by its occurrence in some reptiles and birds.

of sesamoid bones in general. It is tempting to suppose that, in its initial evolutionary origin, the formation of the patella was induced by friction and pressure, and that it has now acquired a representation in the gene complex of the germ cell which allows it to develop independently of such mechanical stimuli. A similar interpretation might equally well apply to muscular eminences such as the trochanters of the femur. Although at first sight this theory might seem to involve the acceptance of the principle of the inheritance of acquired characters, this is not necessarily so.

There appears to be a close morphological relationship between sesamoid bones and traction epiphyses. This is indicated particularly clearly by the epiphysis for the olecranon process of the ulna. In most reptiles this element is represented by a sesamoid ossicle in the triceps tendon. In *Sphenodon* (a generalized reptile), the sesamoid fuses in later life with the upper end of the ulna to form an olecranon process as in mammals. Among the latter, bats retain the more primitive condition, for the olecranon centre of ossification remains permanently separate, closely resembling the patella of the lower limb.[1]

In considering the possible influence of mechanical stimuli in the initiation and control of ossification, it is worth noting that in nature a bone may develop quite a complicated form and structure adapted for resisting particular stresses during adult life, without being subjected during its growth to any mechanical stimuli of a comparable kind. This is well shown in the elaborate antlers of deer, which sprout out from the frontal region of the skull quite independently of the pull of muscles and ligaments or the pressure of adjacent structures.

If observation and experiment suggest that in certain types of ossification extraneous stimuli provide the conditioning factors, it remains quite certain that the *normal* appearance of centres of ossification is entirely dependent on intrinsic growth factors. This much has been adequately demonstrated by the self-differentiation of complete bones in explanted tissues entirely removed from their usual environment. It seems evident that the pattern of the adult skeleton is already mapped out in the somatic mesoderm of the early embryo long before ossification, or even chondrification, has occurred. In this embryonic tissue, osteoblasts are destined to become differentiated from mesodermal cells at certain phases of development and to begin the formation of individual bones at a time and in a sequence which are characteristic for each species of animal. In other words, the development of each element of the typical vertebrate skeleton has an intrinsic morphological basis, and it is possible therefore to speak of the morphological individuality of these bones.

Ossification centres may express their morphological individuality with astonishing persistence even though the bones which they represent have long ago ceased to be of any real functional importance. Many of the bones in the human skull, by the multiplicity of their centres of ossification, reveal their phylogenetic derivation from the fusion of bony elements which are separate in lower vertebrates. For example, the adult occipital bone

[1] But see R. W. Haines, 'The evolution of epiphyses, &c.', *Biol. Rev.* **17**, 1942, where a distinction is made between sesamoid bones and 'intratendinous centres of ossification'.

is a composite structure formed by the coalescence of such elements as the basi-occipital, ex-occipital, and post-occipital, which in primitive vertebrate skulls remain distinct. Each of these elements in man is represented by a separate centre of ossification. Again, a small process of the human sphenoid bone (the medial pterygoid process) develops from a separate centre of ossification in relation to a plate of cartilage representing a separate element in lower vertebrates—the palato-quadrate bar. Reference has already been made to similar atavistic centres forming epiphyses. Another striking example of the same phenomenon has been noted in the sheep. In the adult sheep there is no clavicle, and some of the metacarpal and metatarsal bones have completely disappeared. Yet these elements appear in the embryo and undergo ossification, to be later absorbed or indistinguishably fused with adjacent bones.

It sometimes occurs that such rudiments continue to develop to an unusual degree. In man, for instance, the seventh cervical vertebra normally has a separate centre of ossification for its costal element, representing a rib. Instead of fusing with the transverse process of the vertebra (as it normally does), this element occasionally continues to grow and form a cervical rib. The latter, by pressing on the adjacent brachial plexus, may give rise to symptoms which require surgical treatment. It has been remarked that the patient is not to be regarded as a freak who has grown a new cervical rib, but as an unfortunate who has failed to keep the embryonic rib within normal bounds by fusion and suppression.

The conservatism of skeletal elements in evolution is clearly of great significance to the comparative anatomist for the study of phylogenetic history. Ancestral traits may be recorded in the persistence of archaic features which have lost their functional significance. For example, in the metatarsal bone of the great toe of man, the torsion of the shaft and the contour of the proximal articular facet betray its previous mobility as a grasping digit; the remnants of coccygeal (or caudal) vertebrae still bear witness to the tail which was perhaps lost as far back in geological times as the Miocene era (a matter of twenty million years ago, or so); simian features of the skull and jaws persisted for many thousands of years after the brain volume and the limb skeleton had, in the course of evolution, attained to modern human standards.

13. NUTRITIONAL FACTORS IN BONE GROWTH

The growth of skeletal tissue is more conspicuously affected than most other tissues by nutritional defects, leading to retarded growth, deformities, or undue fragility of the bones. Normal bone growth depends in the first place on the availability of its mineral components in an appropriate form. Of these, calcium and phosphorus are the most important, though magnesium and fluorine are also required. Diets grossly deficient in either calcium or phosphorus lead to a rarefaction of the bones (osteoporosis) and rickety deformities.[1] Even moderate calcium deficiencies may affect the mechanical strength of bones to the extent that they are more liable to fracture, as observed in certain groups of people who have had to exist on a restricted diet during war-time. Experiments on young rats have confirmed these observations by showing that the bending .and twisting strengths of the femur are directly related to the percentage of calcium in the diet.[2] It is interesting to note that, within the limits of these experiments, calcium deficiency did not affect the external dimensions of the femur or the quality of the bony material; the strength of the bone as a whole was diminished by a thinning of the wall of the tubular shaft due to increased absorption of the wall of the marrow cavity. It seems, indeed, that in an attempt to maintain its normal growth expansion in the absence of a sufficient calcium intake the bone is compelled to draw on its own calcium inside in order to re-deposit it outside on the surface.

Even if the actual intake of calcium and phosphorus is sufficient to provide all the material required for the growing skeleton, these elements cannot be utilized in the absence of vitamin D. If this vitamin is deficient during the period of growth, ossification is retarded and rickets ensues. In this nutritional disease, ossification at the epiphysial cartilages of long bones is grossly disturbed. The zone of calcifying cartilage is no longer straight and well defined, and the regular columnar arrangement of the cartilage-cells disappears. Instead, the growing end of the bone presents an appearance in section of a confused mixture of calcified and uncalcified cartilage, with patches of imperfectly ossified bone and an abnormal amount of vascular tissue. It is probable that all these changes are secondary to a failure of the cartilage to complete the preliminary calcification which occurs in normal ossification,[3] so that the process of ossification becomes disturbed, and partially arrested, at this stage. In severe cases of rickets, not only is the general growth of the bones retarded, but their deficient calcification also leads to deformities since they are no longer able to resist the stresses to which they are subjected. It appears that vitamin D controls the proportion of calcium and phosphorus compounds in the blood stream, and the availability of calcium phosphate for purposes of ossification depends on the maintenance of the correct relation between these two.

Vitamin C, as already noted (p. 48), plays an important role in stimulating the production of collagenous tissue from fibroblasts. For this reason it is also concerned in the development and repair of bony tissue, since these processes involve the provision of a fibrous ground substance in relation to which calcification occurs. It has been repeatedly shown that the healing of fractures is delayed in animals suffering from mild scurvy as the result of a vitamin C deficiency. It has also been suggested that by stimulating the activity of osteoblasts, or by influencing the phosphatase system,

[1] M. D. D. Boelter and D. M. Greenberg, 'Severe calcium deficiency in young rats', *J. Nutrition*, **21**, 1941; R. H. Follis, H. G. Day, and E. V. McCollum, 'Histological studies of rats on a low phosphorus diet', ibid. **20**, 1940.
[2] G. H. Bell, D. P. Cuthbertson, and J. Orr, 'Strength and size of bone in relation to calcium intake', *J. Physiol.* **100**, 1941.
[3] S. B. Wolbach and O. A. Bessey, 'Tissue changes in vitamin deficiencies', *Physiol. Rev.* **22**, 1942.

the vitamin may be directly concerned with the process of calcification of bone.

The influence of vitamin A on bone growth has been experimentally demonstrated by Mellanby.[1] A deficiency of this vitamin in the diet of growing animals leads to curious abnormalities of skeletal growth so that the bones become thickened and coarse in appearance. This is partly due to the laying down of superfluous bone in certain regions (as the result of excessive osteoblastic activity), but in the main it seems to be the result of an interference with the remodelling process which normally occurs during ossification (see p. 91). Thus, for example, while the vertebrae grow actively by accretion on their outer surface, there is a diminished absorption of the bone lining the vertebral canal, so that the latter becomes relatively narrowed as growth proceeds. Similarly, retarded absorption of the intra-cranial aspect of the expanding skull leads to a gross thickening of the skull bones with a corresponding restriction in the volume of the intracranial cavity. The growing brain may thus become exposed to such abnormal pressure that many of its nerve cells and fibres undergo degeneration. New subperiosteal bone (apparently normal in structure) is also deposited at the margins of foramina for the passage of some of the cranial nerves and, as a result, the latter become compressed and distorted and undergo extensive degeneration. Involvement in this way of the auditory nerve leads to deafness. Histological examination of the affected tissue shows a relative scarcity of osteoclasts in those regions where the normal process of bone absorption is interrupted, while an excessive number of osteoblasts are found in the areas of superfluous bone formation. It is thus inferred that, in some manner, vitamin A controls the activity, distribution, and co-ordination of the osteoblasts and osteoclasts. The addition of vitamin A to the diet of an A-deficient animal during the period of growth is followed by a return of osteoblastic and osteoclastic activity to the sites where it is normally found, and the reaction is often of a very intense nature (suggesting an attempt by the tissues to restore the normal shape of the affected bones).

That vitamin A exerts a direct local action on bone has been demonstrated by attaching small particles of crystalline vitamin A acetate to fragments of the parietal bone implanted in the brain of experimental animals. In a short time, bone resorption accompanied by osteoclastic activity occurs locally at the site of the vitamin particles. It has also been demonstrated in tissue cultures of isolated and explanted skeletal rudiments that a high vitamin A concentration leads to rarefaction and resorption of bone.[2]

[1] E. Mellanby, 'Nutrition in relation to bone growth and the nervous system', Croonian Lecture, *Proc. Roy. Soc.* Series B, **132**, 1944; 'Vitamin A and bone growth: the reversibility of vitamin A-deficiency changes', *Journ. Physiol.* **105**, 1947.

[2] N. A. Barnicot, 'The local action of vitamin A on bone', *Journ. Anat.* **84**, 1950; H. B. Fell and E. Mellanby, 'Effects of hyper-vitaminosis A on foetal mouse bones cultivated *in vitro*', *Brit. Med. Journ.* **2**, 1950.

14. THE HORMONAL CONTROL OF BONE GROWTH

Apart from dietetic factors, the normal growth and maintenance of the skeleton depend on influences exerted by the secretions of certain of the endocrine glands of the body (see p. 276). It has already been made clear that, among its other functions, skeletal tissue provides a reserve of calcium which can be drawn upon as needed for metabolic requirements elsewhere. The parathyroid glands are known to play an extremely important part in the calcium metabolism of the body by the elaboration of a specific hormone which is able in some manner to mobilize the calcium in the tissues and lead to its concentration in the blood. Hyperplastic tumours of the para-thyroid gland are for this reason associated with a widespread decalcifica-tion of bone, causing a pathological rarefaction (*osteitis fibrosa*). It was shown by Burrows[1] that the injection of parathyroid extract in rats produces the same condition. As the result of such an experiment, the cancellous bone in the limb skeleton undergoes extensive resorption in twenty-four hours, and is largely replaced by fibrous tissue. The osteoblasts show de-generative changes, becoming converted into spindle-shaped cells some-what resembling mesenchyme, and the osteoclasts (at least in the later stages of bone absorption) assume great activity. It still remains uncertain by what mechanism the parathyroid glands produce this result. Possibly they influence directly the cellular activity of the osteoclasts, for it has been observed that where a fragment of parathyroid gland is grafted in contact with a skull bone, the latter undergoes local absorption apparently by osteo-clastic activity. In any case, however, these experiments demonstrate that the parathyroid hormone (like vitamin A) exerts its efforts directly on bony tissue, and not secondarily, as some had supposed, by influencing the renal excretion of phosphates and thus lowering the phosphate level in the blood.[2]

In the limb bones, storage of calcium is mainly the function of the cancellous tissue in the metaphyses, and as a normal physiological process the element is liberated from time to time by the resorption of small quantities of the tissue through the mediation of the parathyroid hormone. This hormone, therefore, by its specific action on bone is essential for maintaining and regulating the calcium ion concentration in the blood plasma.

The influence of thyroid-gland secretion on skeletal growth is demon-strated by the effects of thyroidectomy in immature animals. Following such an operation, there is defective growth in length of the shafts of the long bones. This is the result of disturbances localized in the region of the epiphysis and the diaphysio-epiphysial junction, which lead to a slowing up of the velocity of growth. At the same time the duration of growth is not extended—hence the limbs remain proportionately short. In the skull, the growth of the maxilla may be considerably inhibited, while the mandible

---

[1] R. B. Burrows, 'Variations produced in bones of growing rats by parathyroid extracts', *Amer. Journ. Anat.* **62**, 1938; M. Heller, F. C. McLean, and W. Bloom, 'Cellular trans-formations in mammalian bones induced by parathyroid extract', ibid. **87**, 1950.

[2] N. A. Barnicot, 'The local action of the parathyroid and other tissues on bone in intracerebral grafts', *Journ. Anat.* **82**, 1948.

is relatively unaffected.[1] Suture closure is retarded or does not occur after thyroidectomy. There is no evidence that the influence of thyroidectomy on the growth of the skeleton is due to a specific hormonal effect—it is probably the result of the general decrease in metabolic rate which follows this operation, or possibly of a disturbance in the synergistic action which is believed to occur between the thyroid and the anterior lobe of the pituitary gland. The influence of the thyroid gland on general metabolism perhaps accounts also for the fact that hyperthyroidism leads to an increased excretion of calcium and phosphates, with a corresponding rarefaction of bony tissue.

The skeletal tissues are involved in the general over-growth of the body which occurs in excessive pituitary activity—a condition known as acromegaly. It has been questioned, however, whether the pituitary gland secretes any specific hormone directly related to ossification as such, though there is a tendency in this disease for certain bones to be more affected than others, e.g. the mandible and the extremities of the limbs. The bony changes may be simply part of the general effect of the growth-promoting or somatotropic hormone which is elaborated by the anterior lobe of the pituitary gland. In experimental destruction of the anterior lobe, the growth activity at the epiphyses is very markedly inhibited. The epiphysial cartilage undergoes degenerative changes, the cells die, the vascularity is greatly decreased, and the cartilage becomes sealed off from the diaphysis by a lamina of bone. Conversely, in experimental hyperpituitarism, the cartilage is stimulated to excessive growth and its cells proliferate more rapidly.[2] If the somatotropic hormone (which can be prepared in pure form) is administered to animals deprived of the anterior lobe of the pituitary, the effects of this deprivation are rapidly reversed, and growth at the epiphysial cartilage is once more resumed.

In so far as the internal secretions of the sex glands influence growth processes in general, these will also exert some control over the growth of bone.[3] This action, again, is not specific except in the case of skeletal formations which are definitely secondary sexual characters. For instance, the growth of the bony antlers in male deer is conditioned by the secretion of male hormone, and the latter also has a direct effect on the development of the os penis in those mammalian species in which this bone is present. In such cases, it is to be supposed that the osteogenic tissues are specifically sensitive to the influence of the hormone. A remarkable series of bone changes has been observed in birds in relation to the egg-laying cycle. During the maturation of the ovum, there is developed an extensive secondary system of bone in the marrow spaces of the long bones, called 'medullary bone'. This allows for the storage of the calcium needed later for the egg shell. When the latter is being formed, there is a phase of rapid

[1] T. W. Todd, R. E. Wharton, and A. W. Todd, 'The effect of thyroid deficiency upon bodily growth and skeletal maturation in the sheep', *Amer. Journ. Anat.* **63**, 1938.

[2] H. Becks *et al.*, 'Response to pituitary growth hormone of the tibia', *Anat. Rec.* **94**, 1946.

[3] For a general account of the influence of sex hormones on the skeleton, see N. V. Gardner and C. A. Pfeiffer, 'Influence of oestrogens and androgens on the skeletal system', *Physiol. Rev.* **23**, 1943.

destruction of the medullary bone, in the course of which the calcium is liberated. These changes, which are correlated with a cyclic activity of osteoblasts and osteoclasts,[1] can be induced in sexually immature birds by the administration of female hormone (oestrogen). It is interesting to note that similar changes can be induced in the male bird by an appropriate mixture of sex hormones.

Removal of the testicles during the growing period in man may have the somewhat unexpected result of leading to an overgrowth in the length of the limb bones. This is due to the fact that the obliteration of the epiphysial cartilage at the ends of the long bones, which normally occurs at full maturity, is retarded, and the growth period of the bones is thus prolonged. This accounts for the disproportionately long limbs which are often characteristic of eunuchs.[2]

15. THE VASCULAR AND NERVE SUPPLY OF BONE

Bone is well supplied with blood vessels which ramify freely in the periosteum, and, in the case of the shafts of long bones, by vessels in the medullary cavity. Fine branches from the periosteal network communicate with the small blood vessels which permeate the system of Haversian canals in the compact bone of the shaft, and a bony surface denuded of its vascular periosteal covering may undergo necrosis. Impairment of the circulation by interruption of some of the vessels of supply as the result of injury or fracture may also lead to necrosis, and in the healing of fractures bony union may be delayed. For this reason it is of great practical importance that the mode of vascularization of bones should be well understood by the orthopaedic surgeon.

The extremities of the shafts of the limb bones, i.e. the metaphyses, are particularly rich in blood vessels. They are derived from the periarticular arterial plexus which is found at the capsular attachment of all joints, and enter the bone by numerous and often fairly large foramina. These metaphysial vessels are said to be virtually 'end arteries', with little anastomosis between the main branches. In the immature bone, also, there is no anastomosis with the vessels in the epiphysis. In other words, the epiphysial plate of cartilage forms a complete barrier between the two vascular territories.

The medullary cavity of the long bones is mainly supplied by one or two conspicuous vessels which enter the middle of the shaft by what are commonly called 'nutrient foramina'. In the medulla, the vessels divide and course towards either extremity in a series of leash-like branches. Their terminal branches communicate freely with the blood vessels in the Haversian canals, and take an important part in the vascularization of the bony

---

[1] W. Bloom, M. A. Bloom, and F. C. McLean, 'Medullary bone changes in the reproductive cycle of female pigeons', *Anat. Rec.* **81**, 1941.

[2] That this explanation is probably not entirely adequate is suggested by the observation that in many lower mammals fusion of the epiphysis with the diaphysis seems never to occur normally—even when the animal is fully mature.

tissue. The nutrient foramen and canal of the shaft of a long bone are usually directed obliquely away from the growing end of the bone, i.e. in the upper limb towards the elbow, and in the lower limb away from the knee. This obliquity is a necessary consequence of the mode of growth in length of the bone. To begin with, the nutrient artery,[1] which is a branch of a main artery in the neighbourhood, runs directly into the shaft, entering it at right angles. As the shaft extends in length away from the growing end the artery is deflected by its attachment to the periosteum (which expands by interstitial growth), so that its entrance becomes more and more oblique in a direction towards the opposite extremity of the bone. Most of the long bones have but one 'nutrient foramen'. Some, such as the femur and the clavicle, commonly have two.

Fine medullated and non-medullated nerve fibres enter the bone with the blood vessels, and can be followed ultimately into the Haversian canals. There is some doubt how far these nerves are related to the blood vessels rather than to the bone itself. The periosteum and bony tissue are known by observation to be extremely sensitive to painful stimuli, which certainly indicates a sensory innervation. F. de Castro has described nerve fibres in developing bone entering the zone of ossification with the irrupting blood vessels.[2] It appears that many of the nerve fibres undergo degeneration as the bone matures, and indeed this can hardly fail to be the case with the continual reconstruction of the bony architecture which occurs during development and growth. However, Hurrell has demonstrated (with the use of a special silver technique) that adult bony tissue still contains quite a number of nerve fibres.[3] They can be traced into and along Haversian canals, and while some end blindly in the bone matrix, others terminate in close relation to bone cells.

It was at one time widely held that bones are supplied by 'trophic' nerves through which spinal centres in some way directly control their nutrition. This was believed to be the explanation of the marked bony atrophy which occurs in certain diseases (such as tabes dorsalis and leprosy) in which the peripheral nerves are affected by a degenerative process. It is probable, however, that other factors are involved in these conditions. Howell[4] has shown experimentally that a completely denervated bone does indeed show evidence of atrophy, such as a diminution in the diameter of the shaft and a deficient development of muscular ridges and eminences. Similar changes, however, have been recorded by Tower[5] in animals in which the motor and sympathetic innervation of a limb was left intact, while the musculature was rendered inactive and atonic by cutting the appropriate posterior

---

[1] The common phrase 'nutrient artery of a bone' is likely to lead to a misconception. This vessel is largely concerned with supplying the bone marrow as well as contributing to the supply of bony tissue.

[2] F. de Castro, 'Technique pour la coloration du système nerveux quand il est pourvu de ses étuis osseux', *Trab. Lab. Invest. biol. Univ. Madrid*, **23**, 1925.

[3] D. J. Hurrell, 'The nerve supply of bone', *Journ. Anat.* **72**, 1937.

[4] J. A. Howell, 'An experimental study of the effect of stress and strain on bone development', *Anat. Rec.* **13**, 1917.

[5] S. S. Tower, 'Trophic control of non-nervous tissues by the nervous system', *Journ. Comp. Neur.* **67**, 1937. See also K. B. Corbin and J. C. Hinsey, 'Influence of the nervous system on bone and joints', *Anat. Rec.* **75**, 1939.

roots of the spinal nerves and isolating the corresponding segments of the spinal cord. It appears, therefore, that the atrophic bony changes in these experiments are simply the result of muscular paralysis. This inference is supported by the observation that the atrophy of the bones of a limb resulting from its complete immobilization is not accelerated by simultaneous section of the main motor nerves.

VI

MUSCLE

THE property of contractility, which is one of the fundamental properties of protoplasmic substance, is developed to a highly specialized
degree by the cells of the muscular tissues. In the simplest types of
multicellular organisms these tissues are not fully differentiated, and in
many coelenterates, such as the Hydra, movements are effected by processes
of the superficial epithelial cells (which are therefore termed 'musculo-
epithelial cells'). In higher forms, separate and specific muscle cells are
differentiated to provide a more efficient mechanism for producing movements under nervous control.

In vertebrates there are three well-defined types of muscle, visceral,
somatic, and cardiac. Visceral or splanchnic muscle is structurally the simplest type of contractile tissue. It responds slowly to a stimulus and is
capable of sustained contraction. Somatic or skeletal muscle is much more
highly differentiated and contracts with great rapidity, but becomes more
easily fatigued. Somatic musculature is innervated by the cerebrospinal
system of nerves and, being (at least, in man) under the dominance of the
highest functional levels of the central nervous system in the cerebral cortex, is commonly referred to as voluntary musculature. Visceral muscle, on
the other hand, is involuntary since it is innervated by the autonomic nervous system and is not under the direct control of the will. It preserves the
ability to contract automatically, spontaneously, and often rhythmically,
even without the stimulus of a nervous impulse. It is, therefore, less dependent on nervous control than somatic muscle, the latter being completely
dependent for its activity in the living body on the initiative of the central
nervous system.

The designations 'voluntary' and 'involuntary' are not altogether satisfactory. The muscles of the pharynx and upper part of the oesophagus, for
example, resemble somatic muscles in their structure, but they are not voluntarily controlled in the strict sense. The ciliary muscle of the eye, whereby
the curvature of the lens is changed for focusing at various distances, is composed of visceral musculature, but it is sometimes spoken of as a voluntary
muscle since the focus of the eyes can be changed by an effort of will.

Cardiac muscle, in many of its structural features, is intermediate between somatic and visceral muscle. It is specialized to serve the automatic
rhythmic contractility of the heart as a whole.

Developmentally, almost all muscles are derived from mesodermal tissue
in the embryo. Exceptions to this rule are found in the muscles of the iris
in the eye and the smooth muscle of the sweat glands, which are apparently
derived from ectoderm.

Musculature is primarily designed for mobility. Somatic muscles serve
to adapt the organism to its external environment. Visceral muscles adjust

the internal environment by providing motive power in the processes of digestion, circulation, secretion, excretion, &c. The very ability to produce movement, however, implies the necessity of being able to restrain it. It is important to realize, therefore, that muscles are also essential for the limitation of movement. Some of the skeletal muscles are primarily concerned with maintaining the stability of joints by controlling and limiting normal movements and preventing undesirable movements. Similarly, the sphincters of visceral musculature in the alimentary canal serve the purpose of temporarily stopping the flow of its contents.

Somatic muscles in warm-blooded animals play a most important role in heat production. Even when they are not producing active movements they may be maintained in a state of partial contraction, and this implies a continuous metabolic activity (which, of course, is greatly increased in muscular exercise) associated with a process of oxidation and the liberation of heat. When the temperature of the body falls below a normal level, an attempt is made to generate more heat by the rapid muscular contractions which we know as 'shivering'.

1. THE STRUCTURE OF SOMATIC MUSCLE

Somatic muscle is composed of individual and separate cylindrical fibres bound together in a matrix of connective tissue. They vary in width from 10μ to 100μ, and in length from 1 mm. up to (in general) about 5 cm. In some of the longer muscles in the human body (e.g. sartorius) the fibres are said to reach a length of 15 cm. or more. Indeed, Lockhardt and Brandt[1] isolated one fibre 34 cm. long, and even that was broken at its ends. Unlike cardiac muscle fibres, those of skeletal muscle do not ordinarily branch. Branching has, however, been observed in the muscles of the tongue close to tendinous attachments, particularly in animals with very mobile tongues.

The surface of each muscle fibre is covered by an extremely tenuous membrane, the *sarcolemma*, which forms a resistant sheath enclosing the protoplasmic contents. The fact that the contents can sometimes be seen to be retracted from the sarcolemma in injured muscle fibres suggests that the latter is extraneous in origin—probably derived from the cellular tissue matrix in which the fibres lie. When a muscle fibre contracts, it becomes shorter and broader, its total volume remaining practically unchanged.

Each fibre is multinucleated, and may thus be regarded as a syncytium. The mature fibre, however, forms a structural unit with no other indication of a composite cellular origin. The nuclei are situated at the surface immediately under the sarcolemmal sheath and are surrounded by a narrow zone of granular protoplasm, and there may be as many as several hundred in one single fibre. Primitively, and in the embryo, the nuclei occupy a position in the middle of the fibre; this is often the case in the muscles of lower vertebrates. It may be supposed that they tend here to interrupt the continuity of the contractile mechanism of the muscle fibre, and that, with

[1] R. D. Lockhart and W. Brandt, 'Length of striated muscle fibres', *Journ. Anat.* **72**, 1937–8.

increasing functional efficiency, they become pushed to one side. Even in mammals some fibres still retain their nuclei in a more central position.

The characteristic feature of a somatic muscle fibre is its transversely striated appearance, and it is for this reason that somatic muscle is synonymously termed striated or striped muscle. In fresh or fixed and stained tissue, these transverse striae are very evident as a series of alternating light and dark bands extending through the whole thickness of the fibre (Fig. 49). Their precise nature was for long a matter of dispute. If a muscle

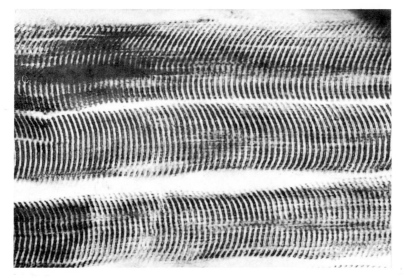

FIG. 49. Microphotograph of striated muscle fibres from the tongue of an ant-eater. The light and dark bands of striation are well shown. × 740.

fibre is teased after treatment with a corrosive (such as chromic acid), it can be split up into bundles of longitudinally disposed fibrillae, or *myofibrils* (*sarcostyles*) (Fig. 50). Even in ordinary sections stained with haematoxylin and eosin these fibrils may be visible as separate elements. Each of the fibrils, which are 1 μ or even less in thickness, shows alternating dark and light sections, and it appears that the transversely striated appearance of the whole muscle fibre is the result of the fact that the dark and light sections of all the component fibrils are aligned in approximately the same transverse plane (Fig. 51). The myofibrils are held together in a clear matrix which is called *sarcoplasm*, and they are responsible for the faint longitudinal striations which may be seen in the muscle fibre. Observations on living muscle suggest that in normal fibres the cross-striated substance of the myofibrils and also the sarcoplasm are in the form of a gel, but that, as the result of irritation or injury, the sarcoplasm may change to a sol state.[1]

During embryological development, the fibrils are first formed within the cytoplasm of small myoblastic cells, and when the latter are extended

[1] C. C. Speidel, 'Studies of living muscles', *Amer. Journ. Anat.* **65**, 1939.

into elongated fibres they become disposed at the surface of the latter; this peripheral position may still be maintained in the primitive musculature of invertebrates. In mammals, they normally extend throughout the thickness of the fibre, sometimes collected into bundles which in cross-section appear as polygonal areas (*Cohnheim's areas*) separated by clear zones of sarcoplasm.

It has been generally assumed that the myofibrils are the essential contractile elements of the muscle fibres. They have been observed to change

FIG. 50. Microphotograph of striated muscle fibres from the tongue of an ant-eater. The myofibrillae which make up each individual muscle fibre are shown. × 740.

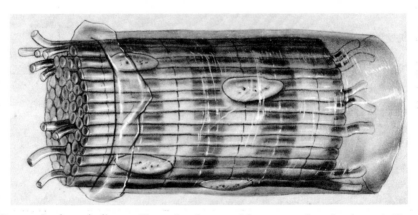

FIG. 51. A schematic diagram illustrating the essential structure of a striated muscle fibre. Each fibre is composed of a bundle of myofibrillae enclosed in a membranous sheath, the sarcolemma. The nuclei of the fibre are situated peripherally, immediately beneath the sarcolemma.

their appearance during contraction and to contract when isolated from the sarcoplasm, and they become progressively differentiated in their anatomical features in correlation with increasing functional efficiency. The differentiation of alternating dark and light segments reaches its acme in the wing muscles of insects in some of which it is possible for contractions to occur at the almost incredible rate of 1,000 per second. Insect muscle provided the material for the classic study of Sharpey Schäfer on the essential anatomical structure of striated muscle.[1]

In view of the fundamental part which myofibrils are believed to play in the mechanism of muscle contraction, it is necessary to consider their structure in some detail. Each fibril, as already noted, is made up of alternating light and dark portions. The former are only slightly birefringent, and are called *isotropic bands* (or *I bands*). The dark portions are strongly birefringent and are therefore termed *anisotropic bands* (*A bands*). Cutting across the isotropic bands is a thin dark line which was believed at one time by histologists to mark the position of a membrane (*Krause's*

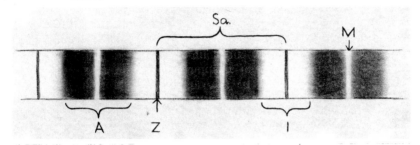

FIG. 52. Diagram showing the structure of a single myofibril at high magnification. *A*. Anisotropic band; *I*. Isotropic band; *Sa*. Sarcomere.

membrane) separating the length of the fibril into a series of partially isolated segments or *sarcomeres*. It is now termed the *Z disk* (from the German word Zwischenscheibe meaning 'intervening disk', that is, the disk intervening between adjacent sarcomeres). In the middle of the anisotropic band, a thin clear zone is sometimes seen—the *M disk* (from Mittelscheibe, meaning 'intermediate disk'; this is also frequently termed the *H zone*, after Hensen who originally described it in 1869) (Fig. 52).

Both the Z and M disks have been interpreted as due to an interference effect at the junction of zones of different refractile properties. On the other hand, it was supposed by the older histologists (on the basis of fixed and stained preparations) that the Z disk represents a membrane running in continuity from one fibril to another across the intervening sarcoplasm, so that it ultimately extends throughout the whole thickness of the entire fibre, dividing the latter into a series of flat plates. An old observation made by Kühne many years ago has been used as an argument against such a conception. He happened to find a minute living parasite—a nematode worm—inside a fresh muscle fibre, and he was able to see that it could move quite freely up and down the fibre without permanently interfering in any way with the appearance of the cross-striations. This, it was supposed, could hardly be possible if there are real structural membranes cutting across the muscle fibre. In later years Carey[2] claimed from his investigations on insect

[1] E. A. Sharpey Schäfer, 'On the minute structure of the muscle columns or sarcostyles which form the wing muscles of insects', *Proc. Roy. Soc. London*, **49**, 1891.
[2] E. J. Carey, 'Experiments on the pneumo-muscular system of aerial insects', *Amer. Journ. Anat.* **61**, 1937.

muscle that the contents of a muscle fibre are far more homogeneous and fluid and less organized than the descriptions of most histologists suggest to be the case. In insects, each fibre is permeated by a system of fine branching air canaliculi—extensions of the tracheal tubules by which tissue respiration is maintained in these forms, and in living muscle fibres this network of intracellular tubules can be observed to shift up and down over an extent of several cross-striations with each contraction. This seemed to prove quite definitely that the cross-striations are in no sense rigid membranous partitions. Nevertheless, the available evidence favours the view that there is an adhesive linkage of some sort whereby the Z disks of adjacent myofibrils are normally held in alignment, so that they are in this sense continuous through the whole width of the muscle fibre. Indeed, these interconnexions are believed to be concerned in the transmission into the interior of the fibre of the stimulus for contraction.[1]

Theories regarding the mechanism underlying the contraction of a muscle fibre have in the past led to several controversies on the question whether the transverse striations betoken a real structural and chemical differentiation, or whether they are no more than optical effects dependent on transitory changes in the physical properties of a comparatively homogeneous medium.[2] It had been found, for example, that if a muscle fibre is pressed against a layer of soft collodion, the transverse striations are reproduced on the imprint, and it was argued from this observation that they are entirely due to the refraction of regular varicosities and not to a real structural differentiation within the fibre. However, this explanation was shown to be inadequate for, if a fibre is stretched so that all surface varicosities disappear, the alternating light and dark bands are still found to be present. It has further been shown that, following treatment with some histological fixatives, the anisotropic bands swell and that this serves to explain the imprint of the cross-striations that have been observed on collodion.

That the striations do, indeed, have a structural basis of some sort is now well assured. They are demonstrable at an early stage of the histogenesis of striated muscle (appearing in myofibrils which develop as discrete and peripherally situated intracellular strands), the isotropic and anisotropic bands have different staining properties,[3] microchemical tests show that the bands differ in chemical composition, the number of striations (in spite of some statements to the contrary) remains constant under varying conditions of contraction, temperature, &c., and the striations present a comparable appearance whether seen in fixed and stained preparations, in living muscle fibres viewed by ordinary illumination or by polarized light, or in preparations studied with the electron microscope.

The existence of myofibrils as real structural entities in living muscle fibres was at one time called into question, for the suggestion had been

---

[1] A. F. Huxley and R. E. Taylor, 'Activation of a single sarcomere', *Journ. Physiol.* **130**, 1955.

[2] For reviews on the structure and composition of muscle fibres and its relation to contractile functions, see A. Sandow, 'Muscle', *Ann. Rev. of Physiol.* **11**, 1949; 'Physiology of voluntary muscle', *Brit. Med. Bull.* **12**, 1956.

[3] E. W. Dempsey, G. B. Wislocki, and M. Singer, 'Chemical cytology of striated muscle', *Anat. Rec.* **96**, 1946.

made by some anatomists that they are fixation artefacts dependent on the peculiar manner in which contractile cytoplasm coagulates. Here again, however, all the evidence now available clearly demonstrates their existence as real morphological units of the muscle fibre. For example, myofibrillae with cross-striations can be seen in muscle cells in living cultures, having the same appearance as in fixed and stained tissue,[1] and they have also been observed under certain conditions in the translucent tissues of living invertebrates.[2] Further, by the use of micro-electrodes it has been possible to produce contractions of single isolated groups of myofibrils in a muscle fibre, thus demonstrating their structural independence within the fibre.[3] Lastly, myofibrils as individual units have been clearly defined by electron microscopy.

The appearance of a single myofibril under the electron microscope is seen in Fig. 53. This photograph shows that the myofibril is composed fundamentally of longitudinal filaments (each varying from 50 Å to 100 Å in width) arranged in closely woven strands. The filaments are comprised of the essential protein of muscle, *actomyosin*, which is a combination of myosin with the fibrous protein actin. At the M and Z bands, the filaments appear to be firmly adherent across the width of the myofibril and each filament has a beaded outline, the 'beads' representing the macromolecular components of the muscle proteins linked together in chain formations. There is now little doubt that these protein chains are the ultimate contractile elements of striated muscle, but the mechanism underlying their contractile property has been much debated. Suggestions had been made that the different optical properties of the A and I bands might be the result of a difference in the folding or spiral coiling of the chains, and that contraction is explicable in terms of a rapidly reversible folding in the A bands. This theory, however, was discounted as the result of studies with the electron microscope, for by this technique it appears that the chains run a straight parallel course in both the A and I bands, and that this even disposition is undisturbed during contraction. The contrast in the birefringence of the bands must evidently have some other explanation, and it now appears, from the work of H. E. Huxley and J. Hanson[4] that this is to be found in the serial arrangement within the myofibril of the essential protein elements of the contractile substance. These are the proteins *myosin* and *actin* which together comprise about half the total protein in muscle. They are disposed in the filaments displayed by the electron microscope, and it has been demonstrated that these filaments are of two types, a coarser (containing myosin) and a finer (containing actin). Moreover, the two types are so arranged that in the resting muscle the myosin filaments are restricted to the A band of each sarcomere, while the actin filaments extend on either side from the Z disks to the margin of the M disk, thus partly overlapping and interleaving with the myosin filaments (Fig. 54). When a myofibril contracts, the A band remains at a constant

[1] M. J. Hogue, 'Studies of heart muscle in tissue cultures', *Anat. Rec.* **67**, 1936–7.
[2] C. C. Speidel, 'Studies of living muscles', *Amer. Journ. Anat.* **65**, 1939.
[3] R. Barer, 'The structure of the striated muscle fibre', *Biol. Rev.* **23**, 1947.
[4] J. Hanson and H. E. Huxley, 'The structural basis of contraction in striated muscle', *Symposia Soc. Exp. Biol.* **9**, 1955.

length and the I band progressively shortens until it is no longer visible. This appears to result from the fact that the arrays of actin filaments slide into the arrays of myosin filaments in interdigitating fashion, as shown in the diagram in Fig. 54. It is suggested, further, that the sliding movement of the two sets of filaments involves a successive linking and re-linking process (by analogy, a sort of 'zipper-like' action) between the myosin and actin molecules which are strung along the interleaved filaments. This hypothesis of the structural mechanism underlying the shortening of striated muscle which occurs on contraction is based on observations derived from various sources, not the least being the actual appearance seen under

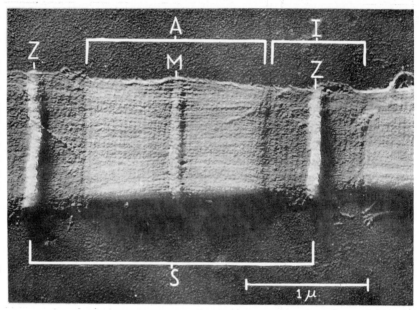

FIG. 53. The appearance of a myofibril as seen with an electron microscope (shadowed with platinum). ×32,200. (From Draper and Hodge, *Quart. Journ. Exp. Biol. Med. Sc.* **27**, 1949.)

the electron microscope. Nevertheless, that the existence of cross-striations does not provide an *essential* basis for the elementary property of contractility of skeletal muscle has been shown by the observation that, in tissue-culture experiments, myoblasts (i.e. embryonic muscle cells) become contractile before they acquire their cross-striations,[1] and it is also the case that muscle fibres which have undergone dedifferentiation in cultures may remain contractile even when they have lost their striations. On the other hand, there is evidence that their appearance during development is related to the degree of the contractile power of the muscle cells, and particularly to the rapidity of contraction.

[1] W. H. Lewis, 'The cultivation of embryonic heart muscle', *Contrib. to Embryology*, Carnegie Trust, **18**, 1926; G. M. Goss, 'First contractions of the heart without cytological differentiation', *Anat. Rec.* **76**, 1940.

Red and white muscle. For many years it has been recognized that striated muscle fibres can be divided into two kinds, red and white. There are, however, transitional forms between these two main types. In red muscle fibres the myofibrils seem to be less numerous, the cross-striations are less regular, and there is usually a greater amount of interstitial sarcoplasm. The latter is often more granular and thus somewhat obscures the cross-striations. Moreover, the nuclei are not strictly confined to a hypolemmal position but tend to be scattered in the substance of the fibres. In general, therefore, they are rather more primitive in structure than white fibres. Their red colour is due to a pigment called muscle haemoglobin or myoglobin (similar to, but not identical with, the haemoglobin of blood).

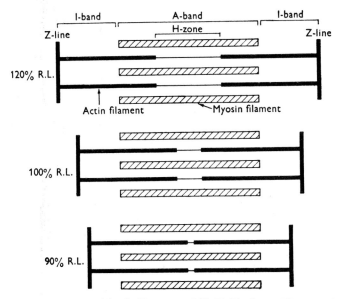

FIG. 54. Diagram constructed by J. Hanson and H. E. Huxley to illustrate their interpretation of the structural basis of the contraction of a myofibril. The arrangement of the filaments of actin and myosin are shown in a fibril at resting length (*R.L.*), stretched to 120 per cent. of the resting length, and contracted to 90 per cent.

Red fibres tend to predominate in muscles which are required to contract over long periods of time, and here they contract more slowly than white fibres and sustain their contraction for a longer duration. This may be an explanation of the greater proportion of sarcoplasm in each fibre (which is believed to be a nutrient medium for the myofibrils) and of the richer vascularity of muscles in which red fibres predominate. The muscle haemoglobin extracts oxygen from the blood haemoglobin and passes it on to the contractile elements of the muscle fibre. In other words, it provides for the storage of oxygen which can be drawn upon in sustained contraction.

In most mammals, including man, both red and white fibres are always intermingled in any single muscle, though the latter have been stated to predominate in muscles required for rapid movement. In some animals, individual muscles may be found to be composed almost entirely of one or other type of fibre. In the rabbit, for example, the semi-membranosus muscle is made up of white fibres, and the soleus of red fibres. Physiological studies have shown that the mechanism of contraction of the red and white fibres is fundamentally the same, the type of response to stimulation being different in degree only. However, quite apart from any abrupt differences in colour and histological appearance, the fibres in mammalian striped muscle are commonly disposed in groups which differ in their speed of contraction, the groups sometimes forming the different 'heads' of one muscle. In general the more slowly contracting fibres predominate in the deeper components of the muscle and these components also tend to be redder in colour.[1]

The organization of muscle fibres in muscles. In short muscles individual fibres may extend without interruption from one attachment to the other, but in muscles with long fasciculi the fibres predominantly have intrafascicular terminations—ending in tapering extremities. Very rarely in mammals a direct anastomosis may be found between adjacent muscle fibres, though this is quite commonly seen in the striated muscle of invertebrates. Each fibre has (outside the sarcolemma) a delicate sheath of fine connective tissue which is termed *endomysium*, and the individual fibres are collected into bundles or fasciculi ensheathed by a somewhat denser layer of connective tissue containing collagenous and elastic fibres, which receives the name of *perimysium*. Lastly, the entire muscle is covered by a sheath of connective tissue called the *epimysium*, which is in some places conspicuously thickened to form intermuscular septa separating adjacent muscles.[2]

The texture of a muscle is determined by the size of its component fasciculi, and in a muscle capable of delicately adjusted movements of great precision (e.g. the ocular muscles) the texture is correspondingly fine. Where only movements of a gross character and general strength are required, the fasciculi are coarse. It is probable that the connective tissue of the endo- and perimysium is more than merely an inert matrix. By the interconnexion of the sheaths of the individual muscle fibres and fasciculi the extent of their contraction is to some extent controlled and regulated. It has also been shown that the initial tension which is developed by extending a relaxed muscle is entirely due to its connective-tissue component.[3]

At the attachment of a muscle, the transition from muscular tissue into tendon is relatively abrupt. The ends of the muscle fibres here are often bluntly rounded, and there is a direct continuation of the connective tissue of the endo- and perimysium into the fibrous-tissue basis of the tendon. Embryological studies[4] have strengthened the conception that there is

[1] G. Gordon and C. G. Phillips, 'Slow and rapid components in a flexor muscle', *Quart. Journ. Exp. Physiol.* **38**, 1953.

[2] The toughness of meat is mainly determined by the amount of collagenous tissue in the endomysium and perimysium. Its proportion of the total protein in lean meat varies from 3 to 30 per cent.

[3] M. G. Banus and A. M. Zetlin, 'The relation of isometric tension to length in skeletal muscle', *Journ. Cell. Comp. Physiol.* **12**, 1938.

[4] M. E. Long, 'The development of the muscle-tendon attachment in the rat', *Amer. Journ. Anat.* **81**, 1947.

primarily no continuity between myofibrils and tendon fibrils at the musculo-
tendinous junction (as had been claimed by some anatomists to be the case).

2. THE FASCICULAR ARCHITECTURE OF MUSCLES

In simple fusiform or strap muscles, the fasciculi are disposed longitudinally, running from one attachment to the other. This axial arrangement
allows the maximal degree of movement. Direct observation and measurement have shown that under normal physiological conditions a muscle
is able to shorten to about 65 per cent of its fully relaxed length.

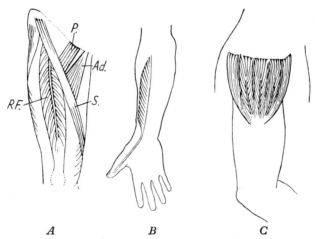

$$A \qquad\qquad B \qquad\qquad C$$

Fig. 55. Diagram showing the arrangement of muscle fasciculi in various muscles. *A*. The
front of the thigh, showing muscles with parallel fasciculi, pectineus (*P.*) and sartorius
(*S.*); a triangular muscle, adductor longus (*Ad.*); and a bipenniform muscle, rectus femoris
(*RF.*). *B*. The front of the forearm, showing a unipenniform muscle, flexor pollicis longus.
C. The lateral aspect of the shoulder, showing the deltoid muscle, which is a multipenni-
form muscle.

With fasciculi arranged in linear series, therefore, a muscle which is 6
inches long when fully relaxed can theoretically shorten by about 2 inches,
with a corresponding degree of displacement of the structures to which it
is attached. In other words, the range of movement which a muscle can
effect is proportional to the length of the muscle fasciculi. It is for this
reason that muscles which act on very freely movable joints have comparatively distant attachments, and therefore long fasciculi. For example, the
trapezius and latissimus dorsi muscles which move the shoulder region
extend in their origins up to the skull and down to the pelvis respectively.
It can be shown, indeed, that the length of the fasciculi which compose
muscles concerned in producing active movements by isotonic contraction
is broadly related to the range of movement which they are required to
effect. The precision of this correlation in individuals of different build
raises some interesting problems in regard to the factors which determine
the growth in length of muscle fibres during development. In this connexion, it is interesting to note that if the extent by which the tibialis
anterior muscle has to shorten, in order to produce the full range of

movement at the ankle joint, is altered experimentally in young rabbits, the growth rate of the muscle is correspondingly altered so that, at full growth, the final length of the muscle belly is adjusted to the new conditions.[1]

In many cases the component fasciculi are disposed obliquely with relation to the functional axis of the muscle, and are attached along one or both sides of a long tendon of insertion (Fig. 55). Muscles of this type are called unipenniform or bipenniform. Multipenniform muscles also occur (e.g. the deltoid) in which fasciculi are related in a bipenniform manner to a parallel series of tendinous plates embedded in the muscle. In these arrangements, the individual muscle fasciculi running from their origin into the tendon of insertion are relatively much shorter, and the range of movement which can be produced is correspondingly smaller.[2] On the other hand, they represent a device by which a larger number of muscle fibres can be packed in a given bulk of muscle. Since the *force* of contraction depends directly on the number of fibres involved (assuming that the average thickness of individual muscle fibres is constant), such a muscle gains in strength at the expense of movement. A penniform arrangement is therefore found in muscles which are required to exert a powerful pull through a small space.

3. THE SHAPE OF MUSCLES

Apart from the arrangement of muscle fibres and fasciculi, the shape of the complete muscle varies considerably. We may note first the simpler types with an attachment at either end. The attachments are commonly termed the *origin* and the *insertion*. The former is the relatively fixed end of the muscle, and the latter the relatively movable end. Thus the pectoralis major muscle is described as taking origin from the chest and being inserted into the arm, and its function is normally to move the arm in relation to the trunk. It will be realized, however, that under exceptional circumstances the usually fixed and movable ends of any muscle may be reversed. For example, if the body is suspended by the arms, the pectoral muscle may be used to raise the trunk towards the arms. The muscle is then said to act from its insertion. In many muscles it is hardly possible to state which are normally the fixed and movable attachments, and in these cases the application of the terms origin and insertion is mainly a matter of convention.

A fusiform muscle may be attached by a tendon at one or both ends of the fleshy belly. Tendons serve a purpose in concentrating the pull of the muscle on a small area, and they are also of considerable importance for the convenient packing of muscles, allowing them to act from a distance.

---

[1] G. N. C. Crawford, 'An experimental study of muscle growth in the rabbit', *Journ Bone and Joint Surg.* **36**, 1954.

[2] The amount of displacement of a tendon in a bipenniform muscle will depend on the length of the muscle fasciculi and the angle of their attachment to the tendon. If a is the angle and x the extent by which the fasciculi can be shortened in contraction, then the displacement $= x \cos a$. Haughton (*Principles of Animal Mechanics*, London, 1873) has estimated that the displacement effected by a bipenniform muscle 6 inches long with fibres of 1·84 inches attached to a central tendon at an angle of 55° could be produced by a simple fusiform muscle of a length less than 3 inches.

In the case of the hand, numerous muscles are required to produce the complicated and delicate movements of which the fingers and thumb are capable, but it would clearly not be feasible to accommodate all these muscles in the hand and wrist region, for by their bulk they would seriously interfere with the purposes for which the hand is required. They are therefore mostly packed away at the upper end of the forearm and connected up with the digits by long slender tendons. Lastly, tendons allow the line of the pull of a muscle to be changed, for they can alter their direction by turning over bony prominences or under fibrous retinacula. By doing so their leverage power is reduced, but the velocity ratio of the muscle is increased by increasing the angular movement of the joint in relation to the amount of shortening of the muscle belly.

A muscle may take origin by two or more 'heads', in which case it is termed *bicipital, tricipital, quadricipital,* or *multicipital.* This elaboration is sometimes the result of the fusion of distinct muscle elements to form a common tendon of insertion. Some muscles have two fleshy bellies with an intermediate tendon—these are termed *digastric.* They are usually formed by the fusion of two separate muscles which may have a different nerve supply. In these cases the intervening tendon is often regarded as the remains of a myoseptum which in the early embryo is interposed between adjacent muscle segments (see p. 26). The rectus abdominis muscle is intersected by a series of intermediate tendons in the form of transverse bands or *inscriptions,* which have been taken to represent a persistence of the embryonic metameric arrangement. However, this is almost certainly incorrect, for experimental embryological studies have now thrown considerable doubt on the common assumption that this muscle is derived from the segmentally arranged myotomes (see p. 143).

Many muscles are in the form of flat layers, and may be attached by means of sheets of strong fascia or *aponeuroses.* The latter are to be regarded as flattened tendons, having precisely the same structure and the same relation to the muscular tissue as ordinary cord-like tendons. Flat muscles may be triangular, quadrate, rhomboidal, and so forth, each shape modifying in some way the direction and concentration of the force of contraction. Around orifices muscle fibres may be disposed circularly to form *sphincters* whose function is to constrict or close the opening.

Most somatic muscles are attached directly to the bony or cartilaginous skeleton, and effect movements at adjacent joints by a leverage action of one or other type. Attachments of other kinds are, however, quite frequent. Some muscles arise from or are inserted into ligaments, fascia, tendons of other muscles (the lumbrical muscles of the hand and foot), mucous membranes (the muscles of the tongue), synovial membranes, viscera, &c. In the subcutaneous layers of the integument of most mammals, particularly in the more primitive types, extensive sheets of cutaneous musculature cover the head, trunk, and the proximal part of the limbs. This musculature is termed the *panniculus carnosus.* Its fibres are inserted directly into the skin, and by its action the latter is capable of considerable mobility. In man, the panniculus carnosus has largely disappeared, but portions of it are still represented by the facial musculature which, in the platysma sheet, extends down into the neck and sometimes even for a considerable distance over the upper part of the chest. The cutaneous attachments of the facial muscles provide the basis for movements of facial expression.

4. MUSCLE ACTION

The action of a muscle depends ultimately on the contraction of the individual fibres of which it is composed. It is well known that individual fibres conform to the 'all-or-none' principle—that is to say, when they contract at all in response to a nervous impulse they do so to their fullest extent. This conclusion has been reached by inference from the study of the relation of discontinuities in the contraction of whole muscles to the number of motor nerve fibres which are stimulated,[1] and by direct microscopic examination of isolated fibres.[2] The graduated nature of the response of a whole muscle is based upon the fact that in a partial contraction a proportion of its component fibres are fully contracted, the others remaining inert. As the force of contraction of the whole muscle is increased, more and more individual fibres come into action. A muscle has for this reason been compared to a vessel which goes full-steam on half boilers, and never half-steam on all boilers. The insulation of individual fibres in skeletal muscles forms an essential basis for the delicate gradations of movement of which these muscles are capable.

It was at one time supposed that in the maintenance of posture over a considerable period of time, involving a continuous contraction of certain muscles (as, for instance, in animals that sleep in a standing position), a rotation of fibre activity occurs, so that when one group of fibres becomes fatigued, others take on the work. However, no direct demonstration of this supposed mechanism has been obtained.

In considering the action of muscles as a whole, several points require emphasis. As already noted, a muscle may act from its insertion as well as from its origin, and a consideration of its possible functions should take this into account. Secondly, a muscle may be even more important for restraining than for producing movement. For example, the short muscles attached round the head of the humerus play little part in effecting movements at the shoulder joint—the position of their attachment allows them too little leverage power. On the other hand, they are extremely important for maintaining the stability of the shoulder joint. This joint allows considerable mobility in all directions, and its ligamentous capsule is correspondingly lax. The short muscles, in such a case, function as extensible 'ligaments' which can adapt their tension to varying positions of the joint and at the same time maintain a close apposition of the articular surfaces. Herein muscles have an advantage over capsular ligaments, which can only retain joint surfaces in contact in those positions in which they are fully

[1] K. Lucas, 'The all-or-none contraction of amphibian skeletal muscle', *Journ. Phys.* **38**, 1909.
[2] F. H. Pratt, 'The all-or-none principle in graded response of skeletal muscle', *Amer. Journ. Phys.* **44**, 1917.

stretched. When a muscle functions as a restrainer of movement, it increases its tension by isometric contraction, that is to say, its fibres contract with practically no shortening of the muscle as a whole. In a similar manner, muscles may be used for raising the pressure in cavities; for example, the muscles of the abdominal wall by their contraction raise the intra-abdominal pressure in the act of coughing or defaecation, and the buccinator muscle of the cheek is used to raise the pressure necessary for blowing or whistling.

Lastly, it should be noted that a muscle may contract in a part of its extent only, and the movement which it produces will vary correspondingly. The trapezius muscle extends in its origin from the skull down to the last thoracic vertebra, and it is inserted into the shoulder. Its upper fibres, contracting alone, will elevate the shoulder; its lower fibres will have the opposite effect of depressing it.[1]

In regard to their actions, muscles are commonly grouped into several categories. As long ago as 1777, John Hunter described the actions of muscles as being immediate and secondary, the former being concerned with the direct production of the movement required, and the latter with supporting actions accessory to the main action. A more detailed classification of muscle action was elaborated by Beevor in 1904, and this still finds general acceptance.[2] The direct movement which a muscle produces by pulling on its insertion is its primary action, and, working in this capacity, the muscle is termed a *prime mover*. Thus the brachialis muscle is a prime mover in flexion of the forearm, the quadriceps femoris in extension of the knee, and the pronator teres in rotating the radius on the ulna. In many cases the action of a prime mover can be inferred by a consideration of its attachments, its relation to the joint at which it produces movement, and the type of movement of which that joint is capable. There are, however, some exceptions. The action of the common extensor muscle of the fingers, for example, or of the psoas muscle in the region of the hip joint, is not readily evident from a study of the cadaver, and in these cases experimental or clinical observation may be required to elucidate their true function.

Muscular contraction, involving single muscles or groups of muscles, may be studied in the living body by electrical stimulation applied over the position of nerve trunks, or over motor points (i.e. the site of entry of motor nerves to individual muscles), though this does not necessarily indicate the normal action of a muscle in composite movements under natural conditions. Another method of studying the action of muscles in the living body (and patterns of muscular activity in composite movements) has been developed by the use of electromyography. With this technique it is possible to record the electrical action potentials which are produced whenever a normal muscle contracts. Such records may be obtained, in the case of a superficial muscle, by placing surface electrodes on the skin over the

---

[1] It is interesting to note that in some Primates this double action of the trapezius is expressed morphologically by its division into two separate parts, of which the lower is called the *depressor scapulae*.

[2] C. E. Beevor, *Croonian Lectures on Muscular Movements*, London, 1904.

muscle. The activity of a deeper muscle can be studied by means of electrodes inserted into its substance.[1] The application of this method involves a number of technical difficulties, not the least of which is the interpretation of the electromyographic records in terms of the actual distribution of activity within the muscle as a whole. It is also the case that, while some of the earlier workers had reported complete absence of electrical activity in resting muscles, later studies with a more sensitive recording apparatus have demonstrated that there is, in fact, a continuous, diffuse, low-grade activity.[2] These discrepant results have led to some discussion regarding the real basis of what is termed 'muscle tone'—whether the latter is the result of a continuous slight activation of the muscle fibres by the constant arrival of nervous impulses, or whether it simply reflects the passive internal elastic tension of the muscular tissue in general. It now appears that the former interpretation is probably correct.

It is remarkable that many individual movements at joints can be effected by one or other of several muscles, so that the paralysis of one muscle alone may be compensated by the action of those which remain intact. In such a case, muscles are sometimes called into play to produce a movement with which they are not normally concerned, but which is possible for them by reason of their attachments. Examples of this type of compensation are well known in cases of nerve injuries in man, and they illustrate very clearly the adaptability of voluntary muscular action. Reference may also be made to the observation, by the careful analysis of cinematographic records, that the normal gait of a rabbit is not disturbed after section of one flexor muscle alone of the knee joint—indeed noticeable changes only occur after three separate flexor muscles have been put out of action. These experiments also showed that the intact muscles contracted more strongly to take on the work of those which had been cut.[3]

It is important to note that in the living body no muscle ever normally acts alone, for the simplest movement requires the co-operation of a number of muscles. Those which act in an ancillary capacity to a prime mover are called *synergists* and *antagonists*. When a muscle contracts to produce a movement in one direction at a joint which is capable of movement in several directions, synergic muscles come into play in order to prevent the movements which are not required. For example, the flexor carpi ulnaris acting by itself will flex and adduct the hand at the wrist joint. If pure adduction is required, the extensor carpi ulnaris acts synergically with the flexor. The former alone is an extensor and an adductor. Working together, the flexor action of the one will be neutralized by the extensor action of the other, and only adduction occurs.

Synergists play an important part in fixing joints where no movement is required, and these are sometimes termed fixation muscles. In clenching the fingers vigorously, the biceps and triceps muscles will be found to

[1] W. F. Floyd and P. H. S. Silver, 'Electromyographic study of patterns of activity of the anterior abdominal wall muscles in man', *Journ. Anat.* **84,** 1950.

[2] H. Göpfert, 'Die Darstellung von Faseraktionen der ruhenden Muskulatur am Menschen', *Pflüg. Arch. Physiol.* **256,** 1952.

[3] D. Stewart, 'Variations from normal gait after muscle section in rabbits', *Journ. Anat.* **72,** 1937-8.

contract. They do so in order to fix and keep rigid the elbow joint, from the neighbourhood of which the digital flexors take origin, thus allowing the latter muscles to act from a firm base.

A muscle whose action is the opposite of that of the prime mover in any movement is called the antagonist. Thus, in extending the knee joint by contraction of the quadriceps extensor, the hamstring muscles act as antagonists, and in pronating the forearm the supinators are the antagonists. In any movement, particularly if it is slow and deliberate, the antagonist is extremely important functionally. If, for example, the quadriceps extensor contracts, the hamstrings do not remain in a passive and relaxed condition. If such were the case, the movement at the knee joint could not be kept under precise control; it would tend to be jerky and abrupt. On the contrary, the antagonist comes into action at the same time as the prime mover in order to steady the joint. As the prime mover actively contracts, its antagonist *actively relaxes* in a graduated manner to a corresponding degree, 'paying out the slack' so as to render the movement smooth and even. Another example of this action is well illustrated by the movement of bending forward from the waist in the standing position. Here the most important muscles are those of the back (sacrospinalis) which, as prime movers, are extensors of the spine. As the trunk is allowed to fall forward under the influence of gravity, the sacrospinalis muscles let it go gradually by progressive relaxation. The action of an antagonist is based upon what has been called by Sherrington[1] the 'reciprocal innervation of muscles'. Skeletal muscles are innervated by afferent as well as efferent nerves. When a muscle contracts afferent impulses lead to a central inhibition in the spinal cord of the motor neurons which innervate the antagonist. This inhibitory activity results in a diminution in the tone of the antagonist which proceeds *pari passu* with the contraction of the prime mover or protagonist.

From this brief consideration of muscle action it is evident that any muscular movement requires the simultaneous action and accurate coordination of a whole series of separate muscles. The co-ordinating mechanism is to be found in the central nervous system. In lower vertebrates the neural patterns which provide the basis for motor reactions are fixed and permanent, and therefore allow of no modification to adapt the organism to unusual circumstances. This may be illustrated by reference to some remarkable grafting experiments in amphibians.[2] Weiss found that if the fore-limbs of a salamander (*Amblystoma*) are transposed from one side to the other so that the position of the limbs in an antero-posterior plane is reversed, the muscles still contract in the same sequence as in a normal limb. In other words, when the animal moves with the intention of going forward, the normal hind-limbs execute the appropriate movement, but the grafted and reversed fore-limbs work in the opposite direction, so that no progression is possible. These animals never 'learn' to modify the inherent neural pattern which controls the movements of the limbs. The

[1] C. S. Sherrington, *The Integrative Action of the Nervous System*, London, 1906.
[2] P. Weiss, 'Further experimental investigations on the phenomenon of homologous response in transplanted amphibian limbs', *Journ. Comp. Neur.* 67, 1937.

same rigidity of pattern has been found in the rat for, if in this animal the flexor and extensor muscles of the foot are transposed, no re-educational adjustment by modifying the action of the transposed muscles seems to be possible.[1]

In higher mammals, the fixed neural patterns of the spinal cord may be controlled and modified to some extent by the greater development of higher functional levels of the nervous system to which they become sub-ordinated, particularly the cerebral cortex. These higher levels, it seems, are able to exert an excitatory or inhibitory action on the neural elements of the lower centres and thus provide the basis for individual adaptation and education of motor reactions. The extent to which this can be achieved in man is well shown in those cases where muscle transplantation is employed in order to overcome a paralysis. For example, if the extensor muscles of the wrist are paralysed, a tendon of a flexor muscle may be separated from its normal insertion, carried round the forearm and attached to the back of the hand to take the place of the paralysed muscle. Following an operation of this kind it is possible for the patient to re-educate himself —or, in other words, for the neural patterns of the nervous system to become readjusted—so that in future the transplanted flexor muscle will contract when extension is required. In general, this kind of motor re-education is not only possible, but it is often fairly successful, particularly if the transposed muscle normally acts in a synergic relation to the muscle for which it is required to compensate. The capacity for re-education also seems greater for the arm than the leg, since the movements of the latter are usually more automatic and less consciously controlled. The modifi-ability of the pattern of muscle action in man is sometimes expressed in the statement that, whereas in the spinal cord individual muscles and muscle groups are reflected in the neural pattern, in the cerebral cortex only move-ments are represented and not the individual muscles which effect them.

The progressive domination of motor reactions by the highest functional levels of the central nervous system in evolutionary development is of con-siderable significance for the physiotherapist. This domination has become more fully expressed in man than in any other mammal, and upon this fact depends his greater educability in regard to muscular control. But he has to pay a price for this advantage which he holds over the rest of the animal world. The harmonious co-ordination of his voluntary muscles is no longer a matter of automatic and unconscious control by lower centres. It is therefore apt to be disturbed quite easily by disharmonies which primarily have their origin in the conscious levels of the mind. The formation of bad muscular habits, which are expressed in faulty posture and in an uneconomical use of individual muscles, is a common cause of motor disabilities in modern man. Their treatment requires a detailed knowledge of muscle action in order to analyse the nature of the maladjustment and to deter-mine the type of remedial exercise which is necessary for the re-education of muscle control. It also requires a considerable psychological insight.

[1] R. W. Sperry, 'The functional results of muscle transplantation in the hind-limb of the rat', Journ. Comp. Neur. 73, 1940.

5. VARIATIONS OF MUSCLES

Muscles are notoriously variable, and it is quite common to find muscular anomalies in the course of routine dissection of the human body. The variations may be conveniently grouped in three categories—progressive, retrogressive, and atavistic.[1]

Progressive variations are represented by the tendency of some muscles to become increasingly complex. As an example of this we may refer to the deep flexor muscle of the fingers in the forearm—flexor digitorum profundus. In the course of evolution, the long flexor muscle of the thumb has already separated off as an independent element—flexor pollicis longus. In modern man there is a tendency for the part of the common digital flexor which is concerned with movement of the index finger to follow the same course. This portion may occasionally be found to be practically separated by a longitudinal cleavage from the remainder of the muscle, in association with the increasing functional individuality of the index finger. Possibly this is to be regarded as a stage in the evolutionary development of a new muscular element—a flexor indicis profundis.

Retrogressive variations are related to degenerative changes in a muscle, associated presumably with a loss of function. The palmaris longus of the forearm and the plantaris in the calf are examples of this type of variation. In primitive mammals these muscles form superficial flexors for the fingers and toes. In man the distal part of the palmaris longus has degenerated into the palmar fascia, while in the proximal part the fleshy belly is often shrunk to an insignificant size, with a long slender tendon connecting it with the palmar fascia. Quite frequently the muscle is absent altogether. In the leg the tendon of the plantaris has become separated into two parts by the prominence of the heel. The distal portion remains as the plantar fascia which is now attached posteriorly to the calcaneum, while the plantaris muscle itself is inserted directly into this bone. Retrogressive changes are also to be noted in the short muscles of the little toe, which in many cases are replaced to a considerable extent by fibrous tissue. Together with degenerative changes in the bones and the integument of the little toe which are sometimes evident, this has been taken to suggest that the digit is in the process of atrophy.[2] Degeneration in a muscle is usually manifested by the replacement of muscular tissue with fibrous tissue, but it is uncertain whether this is a direct transformation (metaplasia) of the muscle fibres.

Where, in the course of evolution, the range of movement which a muscle is required to produce is diminished, the muscle becomes partly replaced by tendon so that the length of the muscle fibres becomes adjusted to the new mechanical conditions. An extreme example of this is shown by the coccygeus muscle (in the pelvic floor). In lower mammals, this is inserted into the basal vertebrae of the tail. In man, some of these

[1] For a classical treatise on muscular variation, see Le Double, *Traité des variations du système musculaire de l'homme*, Paris, 1897.

[2] Loss of digits is a common evolutionary phenomenon in the adaptation and specialization of extremities for support and progression with a concomitant loss of prehensile functions.

vertebrae have become incorporated with the rigid sacrum, and the portion of the coccygeus muscle which thus acquires an attachment to the sacrum becomes entirely replaced by fibrous tissue and forms the sacro-spinous ligament.

Atavistic variations of muscles are not uncommon in the human body. Muscular elements which have been completely lost during evolutionary development occasionally make an appearance quite abruptly. For instance, a portion of the panniculus carnosus (see p. 131) may be found as a sheet of muscle stretching across the axillary fossa, forming the so-called 'axillary arch'. Another interesting example is the upper part of the coraco-brachialis muscle in the arm. Primitively this muscle consists of three parts, but only the middle portion persists in man; the lower component is represented by the fibrous tissue of the medial intermuscular septum of the arm, while the upper part has disappeared. In lower mammals the latter runs from the coracoid process of the scapula to the neck of the humerus, and because of its action is called the *rotator humeri*. This muscular element occasionally reappears in man, and its presence may be a source of difficulty requiring surgical treatment, for it sometimes interferes with free movement at the shoulder joint.

The manner in which a vanished muscle can put in an abrupt appearance as a complete entity suggests that muscles have a morphological individuality akin to that of skeletal elements. It is also remarkable that muscles which have apparently lost their function can persist in quite a well-defined form. This phenomenon is of the greatest significance to the comparative anatomist in his search for evidence bearing on the phylogenetic history of any particular species. The human ear is immobile in the normal person, and yet it continues to be supplied with well-organized muscles. It is difficult, from a consideration of their attachments, to assign any other function to these muscles except that of moving the ears. Hence it can only be inferred that, in his evolutionary history, man has been derived from an ancestral stock in which the ears were freely mobile. This argument also applies to some of the short muscles of the big toe, which in lower Primates are clearly adapted for prehensile movements of this digit. These movements are no longer possible in the human foot—yet the appropriate muscles remain well developed.

The conservatism of the muscular system in mammals is emphasized by the relative constancy with which a common pattern is reproduced in animals of widely different habitus. For this reason, it is usually not difficult to determine the homologies of individual muscles even in quite distantly related forms. Indeed, it is doubtful if there is any muscle element in the human body which is not also found in lower primates; even the peroneus tertius muscle of the foot—which was at one time regarded as distinctive of man and related to his erect mode of progression—has occasionally been found in the gorilla.

In adaptation to varying requirements in different types of mammals, muscles often shift their attachments to some extent. In man the pectoralis minor is inserted into the coracoid process of the scapula, whereas in most lower mammals it is attached to the head of the humerus. The flexor accessorius is confined to the sole of the foot in man, but comparative anatomical studies indicate that it originally took origin from the fibula. Evidence of such evolutionary migrations is sometimes provided in the human body by anomalous attachments.

6. TENDONS AND TENDON SHEATHS

It has already been mentioned that tendons serve the purpose of concentrating the pull of a muscle on a small area, allow muscles to act from a distance, and in some cases change the direction of the pull and so increase the velocity ratio of the muscle at the expense of its leverage power. Given the attachments of a muscle, the length of a tendon is determined by the length of the fleshy fasciculi of muscle fibres, which again is related to the range of movement required. Tendinous intersections in the substance of a fleshy muscle provide for a penniform arrangement of muscle fibres (*vide supra*, p. 130), and in some cases may represent the persistence of fibrous septa between the segments of a compound muscle.

In structure, a tendon is almost entirely composed of white fibrous tissue. Its collagenous fibres are arranged in closely packed parallel bundles and form a flexible and (under physiological conditions) practically inextensible cord through which the pull of a muscle is transmitted to its insertion. Between the bundles of fibres are single rows of fibroblasts which are flattened out in adaptation to the angular spaces in which they are packed. In a cross-section of a tendon, therefore, they appear star-shaped, with pointed processes projecting between the adjacent bundles. The great tensile strength of tendons is emphasized by the fact that a healthy tendon is rarely ruptured in its course by excessive tension, although the latter may lead to a tearing of the tendon from its muscle belly or from its bony insertion. It has been found from direct experiment that the average tensile strength of tendons in human cadavers varies from 8,700 to 18,000 pounds per square inch,[1] and that there are considerable differences in the strength of different tendons, even where they have approximately the same cross-sectional area.

Tendinous patches are frequently developed in muscles at places where the latter are subjected to pressure and friction. This occurs, for example, where the trapezius muscle overrides the medial end of the scapular spine. Again, the outer head of the triceps muscle is tendinous where it is compressed by the overlying deltoid, but as soon as it escapes from under cover of this muscle in passing down the arm it becomes fleshy. Similar tendinous areas can be produced experimentally in muscles which are normally fleshy. In a previous section (p. 109) it has been noted that a sesamoid bone may develop in a tendon where the latter is subjected to pressure in passing over a bony convexity. A slight degree of pressure will manifest itself in a local thickening only (sometimes termed a fibrous sesamoid); or this may become further organized into a cartilaginous or osseous sesamoid by a process of chondrification or ossification.

[1] A. E. Cronkite, 'The tensile strength of human tendons', *Anat. Rec.* **64**, 1936.

It may be presumed that the initial stimulus to the development of a tendon is a tensile stress imposed by a muscle on undifferentiated mesenchyme. This leads to a proliferation and an axial orientation of fibroblastic cells, followed by the deposition of parallel columns of collagenous fibres. Such, at least, seems to be the process involved in the regeneration of a tendon. Thus it has been shown that if a segment of tendon half an inch

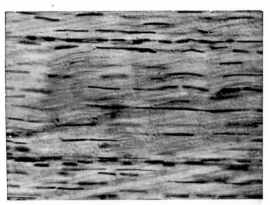

FIG. 56. Section of a normal tendon of a cat. (From D. Stewart.)

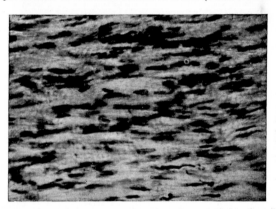

FIG. 57. Section of the regenerated tendon of a cat. (From D. Stewart.)

long is excised from the flexor muscles in a cat's leg (with preservation of the tendon sheath), the animal's gait returns to normal in six weeks with an apparent regeneration of the tendon.[1] The regenerated tissue in these particular experiments was unlike normal tendon to the extent that its fibres were arranged in rather wavy columns instead of being quite straight, they did not always run in parallel formation, and the constituent fibroblasts were less compressed. However, there is other evidence to show that, if such an experimental animal is allowed to survive for several months, the regenerated tissue becomes practically indistinguishable from true tendon.

[1] D. Stewart, 'An experimental study of the return of function after tendon section', *Brit. Journ. Surg.* **24**, 1936.

It is interesting to note that if, in an experiment of this kind, the muscle is also severed so that no tension is put on the cut tendon, the latter is not regenerated.[1]

Experimental studies on young rabbits have shown that tendons increase their length by interstitial growth, but the maximal growth occurs at the muscle–tendon junction and there is a gradient of decreasing rate of growth away from this region. If, in the case of the tibialis anterior muscle, the normal tension on the tendon is diminished by freeing it from the restraint of the extensor retinaculum beneath which it passes (so that the distance it needs to traverse to its attachment is shortened), the rate of growth of the tendon is correspondingly retarded. As a result of this experiment it is necessary for the tibialis anterior to contract over a greater distance in order to effect the same range of movement at the ankle joint. In adaptation to this requirement, as already noted, the muscle belly actually increases in length to an appropriate degree by accelerated growth.

The glistening appearance of a tendon is due to the fact that it is immediately surrounded by connective tissue of very loose texture, the meshes of which contain a glairy fluid similar to the synovial fluid of joints. In its movements a tendon must necessarily 'work out' a channel for itself in the connective tissue in which it lies, and in many cases these channels have well-defined walls which constitute definite *tendon sheaths*. In structure a tendon sheath is closely similar to a bursal sac or a synovial cavity of a joint. Flattened connective-tissue cells form a mesothelial lining to the sheath and also cover the tendon itself (like the parietal and visceral layers of a serous membrane). Practically, tendon sheaths serve to minimize friction. Clinically, they are of considerable importance because of the ease with which septic effusions can spread along them from a focus of infection.

In their embryological development, tendon sheaths appear very early—before it seems possible for mechanical factors due to the pull of muscles to take part in their differentiation.[2] When the tendon first begins to 'crystallize out' in an indifferent matrix, mesenchymal cells are seen to be arranged in a concentric manner round it. In the digital tendons of a 25-mm. human embryo, these cells begin to separate and, with the retraction of their interlacing cytoplasmic processes, this leads to the formation of a definite cavity lined by a flattened mesothelium. Not infrequently, two or more tendons come to occupy a common sheath which—as in the palm of the hand—may be quite extensive.

Tendons, being composed of comparatively inert and passive tissue, are poorly supplied with blood vessels. These enter at both ends, from the muscle and also from the periosteum at the site of attachment to bone. In the case of the long flexor tendons of the fingers, an additional vascular supply reaches them in the middle of their course by means of delicate reflections (or 'mesenteries') of the mesothelial lining of the tendon sheaths, which are termed *vincula vasculosa*.

---

[1] O. Levy, 'Über den Einfluß von Zug auf die Bildung faserigen Bindegewebe', *Arch. Entwicklungsmech.* **18**, 1904.

[2] R. T. Shields, 'On the development of tendon sheaths', *Contrib. to Embryology*, Carnegie Inst. **15**, 1923.

The sensory nerve supply to tendons is relatively abundant, the nerve fibres terminating either in free ramifications (which may be somewhat elaborate, forming what are called *Golgi tendon organs*), or in capsulated endings such as Pacinian corpuscles (see p. 321). The former are situated actually in the substance of the tendon, close to the musculo-tendinous junction, while the latter are found in the tendon sheath. These nerve endings provide an important mechanism whereby an increase of tension in the tendon initiates proprioceptive impulses which are conveyed to the central nervous system. In addition, tendons and their sheaths are supplied with a plexus of exceedingly fine nerve fibres of the type which appears to be particularly associated with the conduction of painful impulses. This explains the sensitiveness of tendons to painful stimuli such as pressure.

7. THE EMBRYOLOGICAL DEVELOPMENT OF STRIATED MUSCLE[1]

In the very early embryo, the paraxial mesoderm becomes segmented to form a series of rectangular blocks called *somites* arranged in a row along either side of the neural tube (see Fig. 12). These segments appear first at the anterior end in the region of the hind-brain in a three-week human embryo, and at five weeks the whole number (between 35 and 40) is complete. Each segment is innervated by its appropriate segmental nerve, and each is separated from its neighbour by a layer of connective tissue or *myoseptum*. The medial portion of the segment, called the *sclerotome*, is composed of loose tissue which takes part in the formation of the vertebral column. On the lateral surface is a layer of cells, the *dermatome*, which is believed to give rise to the connective tissues of the skin. The intermediate part of each segment is the *myotome* proper, from which certain elements of the skeletal musculature of the body are derived. When first formed, the cells of the myotome are spindle-shaped, their long axis being directed longitudinally. The myotomes very quickly lose their primary simple and regular arrangement with their transformation into the definitive musculature. This involves changes of several kinds which have been listed by Arey[2] as follows: (1) a change in direction of muscle fibres from their original cranio-caudal orientation in the myotome; (2) a migration of myotomes from their original paraxial position, either in part or as a whole; (3) a fusion of portions of successive myotomes; (4) a longitudinal splitting of myotomes into subdivisions; (5) a tangential splitting into layers, and (6) a degeneration of myotomes or parts of myotomes to form fascial structures. In the differentiation of a muscle, one or several of these processes may be involved.

For many years it was supposed that, with the exception of that derived from the branchial arch system, all the skeletal musculature is developed from the segmented mesoderm of the myotomes. It had been assumed, for example, that portions of the lower thoracic myotomes migrate forwards in the abdominal wall and become cleft tangentially to form the

[1] See J. D. Boyd, 'Development of striated muscle' in *The Structure and Function of Muscle*, ed. by G. H. Bourne, Academic Press, 1960.
[2] L. B. Arey, *Developmental Anatomy*, London, 1946.

muscular layers of the external oblique, internal oblique, and transverse muscles, that fusion of portions of several myotomes leads to the formation of such muscles as in the adult are supplied by a nerve whose fibres are derived from several spinal nerve roots, and that wherever a myotome or a portion of a myotome may migrate in its embryological history it takes its segmental nerve supply with it. It was supposed, therefore, that the developmental origin of any muscle in the adult could be inferred from a consideration of its nerve supply, and this criterion of nerve supply was taken to imply that all the limb muscles are derivatives of myotomes which have migrated from a paraxial position into the growing limb buds. That such a migration does occur to some extent in lower vertebrates (e.g. fishes) seems very probable, but critical studies in recent years have made it apparent that in the higher vertebrates it is very limited in its extent. Straus and Rawles[1] examined the problem experimentally by marking myotomes in chick embryos with fine particulate carbon and following their displacements during subsequent development, and also by intracoelomic grafts of the somatopleure of the body wall. They concluded that the myotomes are involved in the formation of the intrinsic back muscles, and the dorsal parts of the intercostal and abdominal musculature, but apparently very little more than this.

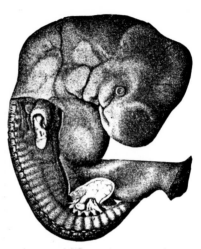

FIG. 58. Reconstruction of a 9-mm. human embryo, showing the segmental arrangement of the myotomes. ×6·5 (From Arey after Bardeen and Lewis.)

So far as studies of the human embryo are concerned, it has for some time been doubted whether the limb musculature is of myotomic origin (except, perhaps, for the most proximal elements of the limb girdles), and the view is now commonly accepted that it becomes differentiated *in situ* from the mesenchymal cells of the limb buds. Similarly, it has been suggested that the tongue musculature is not really the result of a forward migration of the occipital myotomes as was commonly supposed, but rather a local differentiation of the mesodermal tissue in the floor of the mouth.

The difficulty of following all the stages in the development of the muscles in the human embryo is very great, owing to the astonishing rapidity with which this process occurs. In a five-week human embryo, the myotomes are neatly arranged in segments alongside the neural tube. Two weeks later, the definitive arrangement of the musculature has already become established to a very remarkable degree, not only in the trunk and proximal part of the limbs but even in the hands and feet (Figs. 58 and 59).

[1] W. L. Straus and M. E. Rawles, 'An experimental study of the origin of the trunk musculature and ribs in the chick', *Amer. Journ. Anat.* **92**, 1953.

It may be argued that, although in man and other mammals the whole process of migration of paraxial myotomes into the limbs has not been actually demonstrated, analogy with lower vertebrates as well as details of nerve supply suggests that such a migration does in fact occur (at least in part). But unfortunately for the apparent simplicity of this attractive hypothesis it finds no support from experimental studies.

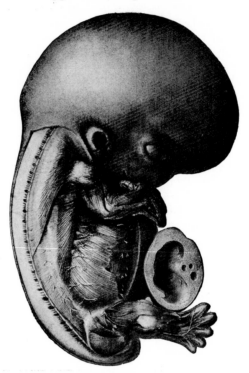

FIG. 59. Reconstruction of the superficial muscles of a 20-mm. human embryo. × 4·5. (From Arey after Bardeen and Lewis.)

In the head and neck regions there is an extensive system of striated muscle which is developed, not from myotomes, but from the splanchnic mesoderm of the visceral pharyngeal arches. This is the 'branchial' musculature, innervated by a special system of nerves associated with the pharyngeal arches. From the mandibular arch are developed the muscles of mastication, from the hyoid arch the facial muscles, and from the more caudal arches the pharyngeal and laryngeal muscles. There is nothing in the minute structure of these muscles in the adult which distinguishes them from muscles of myotomic origin.

It has been mentioned that the cells of the myotome are at first spindle-shaped. In this stage they (and also those mesenchymal elements which are destined to form skeletal musculature) are called *myoblasts*. These cells become gradually elongated and multinucleated. During the second month of foetal development, the myoblasts begin to acquire their transverse striations. Sections of fixed and stained tissue give the appearance of granules being deposited in the cytoplasm, becoming arranged in rows, and fusing to form myofibrillae. The myofibrillae are at first situated peripherally in the developing muscle fibres, the nuclei occupying a central position. At the tenth week, the fibres increase rather rapidly in size and, with their peripheral fibrillae and a single row of nuclei in the centre, appear tubular in structure. They are therefore called at this stage *myotubes* (Fig. 60). As the myofibrillae increase in number—which they are said to do partly by longitudinal division—they come to occupy the whole thickness of the fibre, pushing the nuclei out into a hypolemmal position.

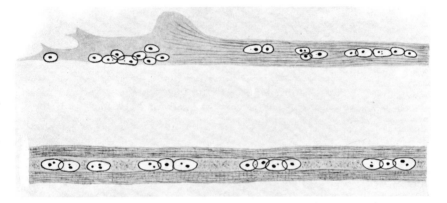

FIG. 60. Drawings of regenerating fibres of striated muscle. Above is seen the growing extremity of a muscle-fibre sprout, with the characteristic accumulation of nuclei at its tip. Below is a young fibre in the 'myotube' stage of development; the nuclei here are disposed along the central core, and the myofibrillae (in which striations are beginning to appear) are situated peripherally.

Hewer[1] has shown that in the human embryo the muscles of the tongue are the earliest to become differentiated histologically, and that at all phases of development the musculature of the hand and arm is in advance of that of the foot and leg.

It is probable that the full number of individual muscle fibres is acquired quite early during embryonic development, although it is difficult to get final and conclusive evidence on this point. According to MacCallum[2] (who counted the number of fibres in the sartorius muscle in embryos of different ages) muscle fibres show no multiplication in human embryos over 17 cm. in length (i.e. about 4 or 5 months old). Subsequent to this date, the increase in size of a muscle depends on the growth of individual fibres.

[1] E. E. Hewer, 'The development of nerve endings in the human foetus', *Journ. Anat.* **69**, 1934–5.
[2] J. B. MacCallum, 'On the histogenesis of striated muscle fibers and the growth of the human sartorius muscle', *Johns Hopkins Bull.* 1898.

8. GROWTH, REGENERATION, AND REPAIR OF STRIATED MUSCLE

Striated muscle fibres are highly specialized cells and under normal conditions probably do not undergo proliferation by cell division. When a muscle becomes hypertrophied as the result of exercise, the component fibres enlarge in size but do not increase in number. It has long been recognized that individual muscle fibres, if only partially destroyed, can undergo regeneration. In this way, muscular tissue which has been extensively damaged by toxic conditions or local bruising can recover its normal healthy state and tone. In lower vertebrates, the regenerative capacity of striated muscle may be very much greater. This is particularly the case in the process of limb regeneration.[1] If, for example, the limb of a urodele amphibian is amputated, the muscles left behind in the stump undergo extensive degeneration, the muscle fibres becoming mostly dedifferentiated into mesenchyme cells of embryonic type. During regeneration, new muscle is partly formed from sarcoplasmic buds which grow out of still undegenerated muscle fibres, and partly by the differentiation of mesenchyme cells. It is difficult, however, to determine whether the latter are direct descendants of dedifferentiated muscle cells, since they are not distinguishable morphologically from the mesenchyme which is formed by the dedifferentiation of other tissues in the amputation stump.

The phases in the regeneration of individual muscle fibres in the living animal have been studied in a very ingenious way by Speidel.[2] In this investigation, individual fibres in the translucent tails of tadpoles were kept under continuous observation for a period of several weeks. They were subjected to various types of injury, and their recovery watched from day to day. Badly injured fibres may undergo complete degeneration. Fibres which are less severely injured often dedifferentiate to the extent that they lose everything except a few nuclei and some sarcoplasm. From their remnants, myoblasts arise which form mother cells for the production of new muscle fibres in a few days. After treatment with heat, there may be complete loss of the cross-striated organization and new fibres subsequently develop within the old sarcolemma sheaths.

It is often assumed that the regenerative capacity of mammalian striated muscle is negligible. It has been observed, however, that regeneration can occur to the extent of repairing a very limited destruction of muscular tissue. For example, in certain diseases such as typhoid and pneumonia, the abdominal muscles may be affected by local patches of necrosis which in a few days are replaced by completely reconstituted muscle fibres. Experimentally induced necrosis in rabbits leading to the partial damage of a limb muscle has also shown that the latter can undergo regeneration very rapidly.[3] This process of regeneration involves first the removal of

---

[1] C. S. Thornton, 'The histogenesis of muscle in the regenerating forelimb of larval *Amblystoma punctatum*', *Journ. Morph.* **62**, 1938.

[2] C. C. Speidel, 'Growth, injury and repair of striated muscle', *Amer. Journ. Anat.* **62**, 1938.

[3] W. E. Le Gros Clark, 'An experimental study of the regeneration of mammalian striped muscle', *Journ. Anat.* **80**, 1946.

the necrotic muscle tissue by the action of macrophages, and secondly its replacement by the development of new fibres. The latter appear usually to be derived as outgrowths which extend in continuity from the stumps of old muscle fibres at the margin of the necrotic area (Fig. 60), though it remains possible that portions of these sprouting strands of nucleated sarcoplasm may become detached to form independent myoblastic elements. An important factor in the successful regeneration of necrosed muscular tissue is the preservation of the endomysial tubes of connective tissue which ensheathed the original fibres, for these tubes provide the pathways which, so to speak, guide the newly growing fibres. Incipient regeneration has often been observed in human muscles following local destruction due to injury but, because of their size, fibrosis in distant parts of the damaged tissue proceeds sufficiently rapidly to obliterate the endomysial framework before the regenerating fibres have time to arrive there. This is the reason why regeneration in cases of severe muscle destruction in man is usually not significant enough to lead to any marked functional recovery. That the endomysial tubes only play a passive role in facilitating the outgrowth of new muscle fibres is shown by the interesting observation that regenerating muscle fibres may sometimes be found to have gone astray and to be growing down a degenerated nerve, making use of the endoneurial tubes which have been vacated by the degenerated nerve fibres.[1]

Striated muscle fibres have been successfully cultivated in tissue cultures. It has been observed, for instance, that an abundant outgrowth of skeletal muscle occurs if myoblastic tissue from chick embryos is explanted in Locke's solution.[2] Adult human and rat skeletal muscle has also been cultivated in vitro, with the formation of myoblastic elements leading to regeneration and growth.[3] The fact that myoblasts can become differentiated into cross-striated muscle cells under these conditions shows that the process is independent of nervous or other extraneous influences in the developing embryo. From the cut ends of striated muscle cells in tissue cultures, protoplasmic buds grow out as in regenerating muscle fibres. These buds acquire transverse striations but before they do so they may become detached and undergo rhythmical contractions. The latter are very variable in rate, being sometimes as rapid as 120 a minute, and occasionally they are also seen in well-developed muscle fibres with cross-striations. These observations are of considerable significance in so far as they demonstrate that the essential property of contractility of skeletal muscle can develop quite independently of any nervous connexions.

9. THE NERVE SUPPLY OF STRIATED MUSCLE

Each skeletal muscle has a motor nerve supply which enters it at one or more relatively constant points. Each nerve of supply commonly divides

[1] F. K. Sanders, 'Invasion of nerve homografts by regenerating muscle fibres', *Journ. Anat.* **84**, 1950.

[2] W. H. Lewis and M. R. Lewis, 'The behaviour of cross-striated muscle in tissue cultures', *Amer. Journ. Anat.* **22**, 1917.

[3] I. A. Pogogeff and M. R. Murray, 'Form and behaviour of adult mammalian skeletal muscle *in vitro*', *Anat. Rec.* **95**, 1946.

into a series of twigs close to its termination and pierces the muscle usually on its deep surface, or at its margin. It has been stated that the point of entrance tends to correspond with the geometric centre of the muscle, a disposition which evidently favours the even distribution of nerve fibres to all parts of the muscle; however, this principle only applies in a very general way. In compound muscles whose fibres preserve their segmental sequence, e.g. the muscles of the abdominal wall, twigs from the corresponding segmental nerves enter in series. Most muscles, particularly

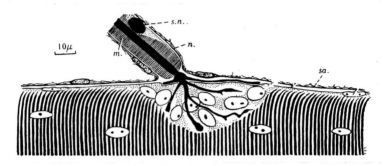

FIG. 61. Diagrammatic representation of a motor end plate. *m.* myelin sheath; *n.* neurilemma; *s.n.* nucleus of Schwann cell; *sa.* sarcolemma. (From E. Gutmann and J. Z. Young, *Journ. Anat.* **78**, 1944.)

those of the proximal part of the limbs, appear to be plurisegmental in origin, and their motor nerves are similarly composed of fibres from several spinal segments.

Within each muscle the branches of the nerve of supply form a rich plexus from which terminal medullated fibres emerge to innervate individual muscle fibres. The mode of termination of each nerve fibre is shown diagrammatically in Fig. 61. On reaching its appropriate muscle fibre, the nerve fibre loses its myelin sheath, while its delicate neurilemmal sheath becomes continuous with the sarcolemma. The naked axon then pierces the sarcolemmal sheath and at once breaks up into terminal arborizations which ramify in a local accumulation of granular sarcoplasm containing several nuclei. This structure is called the *motor end plate*. The nuclei of the end plate are derived from those of the muscle fibre, and, if the motor nerve is cut and allowed to undergo complete degeneration, they remain at first unaffected. In the past, there has been considerable difference of opinion regarding the precise relation of the nerve endings in the motor end plate to the contractile substance of the muscle fibre. The neurofibrillae of the nerve terminals were described as spreading out in the end plate to form an extremely fine periterminal network which ultimately runs into direct continuity with the sarcoplasm. This interpretation of the histological picture, however, has gained no credence, and the view is now firmly established that the nerve terminals end freely in contact with the sarcoplasm but separated from it by a plasma membrane. Apart from the fact that a theoretical consideration of the mechanism involved in the transmission of a nerve impulse to a muscle fibre makes it necessary to postulate such a membrane, the existence of the latter has been objectively demonstrated by electron microscopy.[1]

The pattern of distribution of motor nerve fibres to a muscle is a matter of some importance in the study of muscle action. If transverse sections are taken through a motor nerve at different levels along its course, the motor nerve fibres are found to increase progressively in number. This is due to the branching of the original nerve fibres (i.e. the axonal processes of single motor nerve cells in the spinal cord) on their way to the muscle. Within the muscle substance further subdivision takes place, so that each original fibre innervates a number of muscle fibres. The axonal process of a single motor nerve cell together with the muscle fibres which it supplies represents a functional neuromuscular unit and is termed a *motor unit*. In powerful muscles of coarse action the motor unit is large. For example, in a limb muscle the number of muscle fibres innervated by one axon is usually about 150 but may reach as many as a thousand, and the force of contraction which can be engendered by the excitation of a single motor nerve cell is correspondingly great. In finely grained muscles in which contraction requires to be delicately adjusted for movements of high precision (such as the muscles of the eye), the unit is much smaller, consisting perhaps of not more than ten fibres. Studies of whole preparations of muscles in which the nerves have been stained with methylene blue have made it clear that (at least in the muscles studied) the motor unit is not an anatomical entity in the sense that the muscle fibres of the unit form a compact group. On the contrary, the terminal branches of a single nerve axon may be distributed to widely scattered muscle fibres, so that there is an extensive interlocking of the component elements of several motor units. It will be realized, therefore, that the pattern of innervation of a muscle may be very complex indeed.

From time to time evidence of a physiological or histological nature has been brought forward which seems to indicate that the long muscle fibres of fusiform or strap muscles are commonly supplied with more than one motor end plate, and that in some cases these end plates may even be innervated from different segmental nerves. The evidence, however, is open to criticism, and it can probably be assumed that if multiple end plates do occur they are quite exceptional.

Besides efferent nerve fibres, muscles are also supplied with afferent fibres. While the terminations of the former occupy a hypolemmal position *inside* the muscle fibres (that is to say, they lie beneath the sarcolemma), the sensory endings are practically always on the surface or epilemmal. They may be relatively simple branching terminals with minute varicosities, embracing the muscle fibre around its circumference. More highly organized sensory endings occur in the form of *muscle spindles*. These structures are made up of several muscle fibres of small calibre, enclosed in a well-defined connective-tissue sheath from which they are separated by a zone of loose tissue sometimes termed a 'lymph space'. They may reach

[1] J. D. Robertson, 'The ultrastructure of a reptilian myoneural junction', *Journ. Biophys. Biochem. Cytol.* **2**, 1956. For a discussion on motor end plates, see E. Gutmann and J. Z. Young, 'The reinnervation of muscle', *Journ. Anat.* **78**, 1944.

L

a length of 4 mm.[1] The sensory nerve fibres supplying muscle spindles are myelinated and relatively thick, and they terminate by wrapping round the constituent muscle fibres in a spiral manner (Fig. 62).

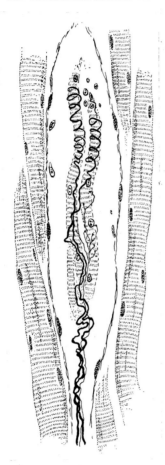

FIG. 62. Drawing of a muscle spindle from the anterior tibial muscle of a mouse. Note the diminutive muscle fibres within the spindle (intrafusal fibres) and the nerve fibres which wind round them in characteristic spirals. The whole spindle is encased in a sheath of connective is tissue, and beneath the latter a 'lymph space' × 350 (approx.).

Muscle spindles are concerned with the initiation of the proprioceptive impulses which are required for the control and regulation of muscular activity. They are particularly abundant in those muscles which are used in the maintenance of posture and are usually found close to the tendons. Their main function is to provide the receptor elements of a 'feed-back' mechanism which ensures that the contraction of a muscle is accurately related to the effort exerted. In the nerve trunk which supplies such a muscle, as many as 50 per cent of the fibres may be afferent. It has been erroneously supposed that muscle spindles are only to be found in those muscles concerned in the maintenance of posture—the so-called 'anti-gravity' muscles. However, they have also been observed in other functional types of muscle, e.g. the diaphragm and the extrinsic muscles of the larynx. They have also been described in the extrinsic ocular muscles of man and ungulates, though in certain other mammals such as the monkey and cat they appear to be absent here (at least in their typical form). 'Spindles' are disposed longitudinally in the muscle, and respond to the stimulus of tension. They are therefore *stretch receptors*. Since they are arranged in parallel formation with surrounding muscle fibres, they are not affected by a contraction of the muscle as a whole; they are only stimulated by passive stretching during relaxation. Herein they contrast with the stretch receptors in tendons, which are stimulated by the tension transmitted to the tendon during muscular contraction.

Motor nerve endings begin to mature during embryological development at the end of the seventh month, that is, at the time the muscle fibres begin to present a finished histological appearance. Before this time, however, motor nerve fibres are already in functional connexion with muscle fibres. Indeed, motor endings

[1] For a study of the anatomy of the muscle spindle, see D. Barker, 'The innervation of the muscle spindle', *Quart. Journ. Micr. Sci.* **89**, 1948.

of a primitive type are said to form as early as the seventh week of intra-uterine life, and spinal reflex arcs to be completed during the eighth week.[1] It is suggested that there is a close relationship between the maturation of the motor end plate and the commencement of normal functional activity.[2] At birth, for example, the nerve endings in the respiratory muscles (diaphragm and intercostals) are considerably better developed than those of the limb muscles. It is interesting to note, also, that sensory nerve endings in muscle fibres become differentiated before the motor end plates.

The possibility of a double motor innervation of striated muscle by sympathetic as well as somatic nerve fibres led to a considerable controversy some years ago, but it is now generally agreed that the histological evidence adduced in support of this conception was misinterpreted. Yet sympathetic activity undoubtedly has an influence on the contraction of striated muscle. Thus, stimulation of the sympathetic delays fatigue in a muscle which is forced to contract by repeated stimulation of its somatic motor supply (the Orbeli phenomenon) and accelerates recovery in a tetanized muscle. Moreover, this effect is independent of local changes in the circulation. It is not necessary to postulate a direct sympathetic innervation of striated muscle fibres to explain these phenomena, for they may be due indirectly to the liberation of adrenalin (or an adrenalin-like substance) at the endings of sympathetic fibres in the walls of the blood vessels of a muscle.[3]

Constancy of nerve supply of muscles. The constancy of the nerve supply of individual muscles is a question of considerable importance for the comparative anatomist. It has been a generally accepted opinion that this constancy can be relied upon for establishing the homologies of muscles in different animals by reference to their innervation. On this view, muscles which in different animals have the same relative position and attachments cannot be homologous if their nerve supply is quite different. Such arguments were based on the idea that the connexion between nerve fibres and their corresponding muscle cells is established at a very early date in embryonic life, and that wherever the muscle cells migrate in their subsequent development they will take their nerve supply with them. Support was given to this conception by the hypothesis that motor nerve cells are in direct protoplasmic continuity with their appropriate muscle cells in the very earliest developmental stages of the ovum, the axons of the motor neurons being formed by the 'drawing out' of these syncytial bridges as the cells become separated. This hypothesis, however, is opposed to the well-established neuron theory and is further discussed in the section dealing with nervous tissue (p. 366). Here it may be noted that there is evidence, derived from direct histological study and also from experimental observation, that motor nerve fibres grow into muscular tissue and establish their connexions secondarily. If this is the case, it is clearly possible that the

[1] W. F. Windle and J. E. Fitzgerald, 'Development of the spinal reflex mechanism in human embryos', *Journ. Comp. Neur.* **67**, 1937.

[2] E. E. Hewer, 'The development of nerve endings in the human foetus', *Journ. Anat.* **69**, 1935.

[3] E. Bülbring and J. H. Burn, 'Blood flow during muscle contraction, and the Orbeli phenomenon', *Journ. Phys.* **95**, 1939.

nerve supply may not necessarily be as constant as many anatomists have supposed. In fact, instances of a change of nerve supply in homologous muscles have been adduced from time to time.[1] Thus, in the foot the lateral plantar nerve may extend its territory in one species of mammal so as to innervate muscles which in other species are supplied by the medial plantar nerve. A similar type of variation may also occur in a single species. In the human body, for example, some of the small muscles of the thumb commonly innervated by the median nerve may receive their nerve supply from the ulnar nerve. The musculature of the fins in fishes may be innervated from one segmental nerve only, although the embryological evidence shows that it has been derived from several myotomic segments. In cases of this kind, it remains possible that the homologous muscles are still innervated by the homologous motor nerve cells in the brain stem or spinal cord, and that the peripheral nerve fibres follow an aberrant course in order to reach their muscles. This transference of nerve fibres may explain exceptional types of innervation. For example, it has been observed that in certain birds some of the fibres of the seventh cranial nerve emerge from the brain with the fifth nerve. Superficially this leads to an appearance of a change in nerve supply of certain muscles from the seventh to the fifth nerve, but the appearance is illusory for they are still innervated from the central nucleus of the seventh nerve. Haines has summarized a discussion of these problems by concluding that the nerve supply of muscles is sufficiently constant to permit use of this criterion in determining their homologies in a single class of vertebrates (e.g. in mammals). On the other hand, reliance on nerve supply may lead to error in attempting the comparison of muscles in widely separated groups such as mammals and reptiles.

Changes in peripheral nerve supply (that is to say, in the route followed by individual nerve fibres) are associated with the evolutionary migration of muscles. When the latter move from their original attachments, the fibres which supply them tend to run in the peripheral nerve trunks topographically related to them in their new position. Nevertheless, it is not uncommon to find motor nerves pursuing a lengthy and (as it seems) inconvenient course, which is difficult to explain except on the supposition that a migrating muscle has dragged its peripheral nerve supply along with it.

10. THE BLOOD SUPPLY OF STRIATED MUSCLE

Skeletal muscle is a richly vascular tissue, supplied by branches from neighbouring arteries which usually enter the muscle with the branches of the motor nerve. These muscular branches show some variation in their number and arrangement. In the substance of the muscle they form a plexus of which the finer ramifications run transversely to the line of the muscle fibres. Arteries and veins run together until the terminal arterioles and venules are reached; these come off from the parent vessels in alternate

---

[1] See R. W. Haines, 'A consideration of the constancy of muscular nerve supply', *Journ. Anat.* **70**, 1935–6; W. L. Straus, 'The concept of nerve-muscle specificity', *Biol. Rev.* **21**, 1946.

sequence. This arrangement allows the intervening capillaries to run in a relatively direct course from arteriole to venule and presumably serves the purpose of facilitating a very rapid removal of metabolites. The capillary vessels form a close-meshed plexus of rectangular pattern, running for the most part longitudinally between the individual muscle fibres, with transverse interconnexions crossing the fibres at short intervals. Capillary anastomoses are said to be particularly well developed in immediate relation to the motor end plates, suggesting that the neuromuscular junction is the site of considerable metabolic activity. Red muscle has a richer vascular supply than white muscle and the longitudinal capillaries run a tortuous course instead of being relatively straight. Moreover, the cross-connexions of the rectangular plesxu often show characteristic dilatations. It is suggested that these provide little reservoirs of blood which may continue to supply the muscle fibres with oxygen when the capillary circulation is compressed and impeded by sustained contraction.

Injected specimens show that the branches of intramuscular vessels are freely interconnected by arterial anastomoses, and it might be supposed therefore that a collateral circulation would be readily established following the interruption of one of the channels of blood supply. However, an experimental study of the vascularity of muscle has shown that the anastomoses are not very efficient, so that a part of a muscle can be functionally devascularized for several days by the ligature of one of its arteries of supply.[1] This observation is of some practical importance since it is probable that striated muscular tissue is unable to survive if it is deprived of all blood supply for more than a few hours.

11. VISCERAL MUSCLE

In comparison with somatic muscle, the fibres of visceral muscle are cells of simple shape and devoid of transverse striations. It is synonymously termed *unstriped* or *plain muscle*. On general functional grounds, it is difficult to define any absolute or final distinction between somatic and visceral musculature, but, broadly speaking, it may be said that the latter provides the motive power for all those mechanisms which are directly related to metabolic activities of assimilation, nutrition, and excretion. It plays an essential part in maintaining the normal physiological equilibrium of the body, while the adjustment of the organism to the external environment is secured by means of the somatic muscles. In the alimentary canal, it propels the ingested food, aids in the process of digestion by mixing it freely with the digestive juices, and carries on the waste products to be excreted. In the capsules of glands and in ducts it promotes the flow of secretory products. In the walls of blood vessels it serves to regulate the local distribution of blood according to the needs of the various tissues at the moment. In the ureter and bladder it is concerned with the excretion of urine. In the skin it plays a part in temperature regulation by controlling

[1] W. E. Le Gros Clark, 'Anatomical problems relative to the traumatic surgery of muscle', *Bull. War Med.* **6**, 1946.

the activity of the sweat glands. In the connective tissues of the body, particularly in the mesenteries of abdominal and pelvic viscera and in the orbit, it has a supporting function. In the duct of the testicle and in the oviduct and uterus it provides essential mechanisms for the continuance of the species.

Visceral muscle is therefore the muscle of the 'vegetative' functions of the body. Such functions, however, are also served in part by striated musculature. For example, this type of muscle controls the movements of the first few inches of the alimentary canal, in the abdominal wall it aids in maintaining the position of the abdominal organs and reinforces the activity of the intrinsic visceral musculature of the intestines, bladder, &c., it provides the motor mechanism of respiration, and it plays an important part in temperature control by the generation of heat.

FIG. 63. Plain muscle fibres. Note their fusiform shape, and the elongated central nuclei. For the appearance of plain musculature in section, see Fig. 71, which shows a cross-section of an artery of which the tunica media is composed almost entirely of plain muscle. × 400 approx.

Embryologically, there is a broad distinction between striped and unstriped musculature in that the former is mainly derived from the segmental paraxial mesoderm and the somatopleural mesoderm of the body wall, and the latter mainly from splanchnic mesoderm. This distinction is again not complete, for (apart from other exceptions) the striated muscle of the branchial arch system is also developed in splanchnic mesoderm. 'Branchial' musculature in its structure would be classed as somatic, while in its embryological origin and innervation it must be regarded as visceral. This mixture of contradictory features is reflected in its evolutionary history and in the functions which it still performs in man today. The jaw muscles of the mandibular arch, in so far as they are concerned with the reception of food and its preparation for swallowing, occupy a functional position between the somatic and visceral systems. The facial muscles of the hyoid arch were primarily muscles of respiration, and it has been suggested that the facial contortions which accompany acute respiratory distress in man still bear witness to this association. The muscles of the third and succeeding 'branchial' arches are largely concerned in the mechanism of swallowing whereby the food is passed into the main part of the alimentary canal.

In the human body somatic and visceral muscle may be intimately mingled to serve a common purpose. This is the case with the muscle that raises the upper eyelid (*levator palpebrae superioris*), and with the muscular strand (*gubernaculum testis*) which is associated with the descent of the testis into the scrotum.

The structure and disposition of visceral muscle. The fibres of visceral muscle are elongated, spindle-shaped cells with a central oval nucleus. They range in length from a few microns up to half a millimetre. On treatment with macerating fluids, fine longitudinal striae become evident in the cytoplasm, but these are much less conspicuous and definite than the myofibrils of a somatic muscle fibre, and they have not yet been certainly observed in fresh tissue. They are, however, clearly demonstrable by electron microscopy in the form of bundles separated by single rows of mitochondria. Occasionally a few irregular cross-markings are to be seen in the cytoplasm, which are interpreted as contraction bands fixed on the death of the cell. There is no true sarcolemmal sheath, and the individual cells of plain muscle are supported in a matrix of fine reticular tissue. There has been some uncertainty whether the cells are always really separate anatomical units, or whether they are—at least in some tissues—interconnected by fine protoplasmic bridges. Electron microscopical studies have certainly demonstrated the existence of such bridges in the uterine wall and in the intestinal tract of vertebrates, and a true syncytial connexion between individual cells is said to be more commonly found in invertebrates. The smaller muscle fibres, such as those found in the walls of the smaller blood vessels, may be rather irregular in shape—flattened and branched. The essential contractile substance of a plain muscle fibre is made up of the proteins actin and myosin, and studies with polarized light indicate that its ultra-structure is similar to the anisotropic disks of striated muscle (see p. 123).[1]

Visceral musculature is generally arranged in sheets or layers, and not in discrete fasciculi like somatic muscle.[2] In the connective tissues of the integument and elsewhere, the cells are scattered quite irregularly, either singly or in small groups. In the walls of hollow viscera and tubular structures, they form definite layers or coats. Where they serve merely to alter the lumen of a tube, the fibres are disposed almost entirely in a circular manner, and the lumen can then be constricted by their contraction and dilated by their relaxation. If, in addition, the muscular coat is required actively to propel the contents of the tube, it is separated into at least two distinct layers in which the fibres are disposed circularly and longitudinally respectively. This is the case in the whole length of the alimentary canal, where an inner circular coat is covered superficially by an outer longitudinal coat. This simple arrangement is disturbed and elaborated in saccular viscera where the movements are more complicated.[3]

Circular and longitudinal muscle coats are in a sense antagonistic, though they are both concerned in effecting the same general movement. In propelling the fluid contents of the intestine, for instance, the circular muscle undergoes a wave of contraction which forms a constriction ring travelling down the length of the canal, preceded by a dilatation. This is the basis of

[1] E. Fischer, 'The birefringence of striated and smooth mammalian muscle', *Journ. Cell. Comp. Phys.* **23**, 1944.

[2] An exception is found in the muscles which move the hairs—*arrectores pilorum*. These are arranged in definite bundles.

[3] The dependence of peristaltic movement on the co-operative action of the circular and longitudinal muscle is emphasized by the observation that in foetal guinea-pigs peristalsis of the intestine first occurs when both those muscular coats are developed. Earlier still, when the circular muscle alone is present, local stimulation of the intestine elicits only local constrictions. See W. F. Windle, *Physiology of the Foetus*, 1940.

the peristaltic movements of the intestine. A local contraction of the circular muscle, however, would tend to drive fluid contents in either direction were it not for the fact that the longitudinal muscle above the constriction ring simultaneously contracts so as to raise the pressure by shortening this part of the tube. At the site of a constriction ring, also, the longitudinal muscle fibres are anchored down momentarily, and, acting in their contraction from this basis, they pull up and locally dilate the wall of the lower part of the intestine over the contents which are being forced downward.

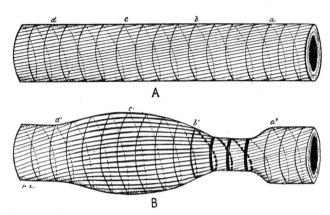

FIG. 64. Diagrams illustrating the relation of the spiral arrangement of smooth muscle fibres in the intestinal wall to peristaltic movement (according to Carey). *A.* The small intestine in a resting phase, showing the close and open spirals of the circular and longitudinal muscle coats. *B.* Diagram of a peristaltic wave. From *a'* to *b'* the contraction of the stronger inner coat causes a local constriction. From *b'* to *d'* the contraction of the outer coat leads to a shortening and dilatation of the gut. The contraction wave travels progressively down the intestine as the result of the effective reciprocal stretching of the muscle fibres which occurs distal to the wave. (From E. J. Carey.)

In the large intestine, the contents of the canal gradually become solid in consistency by the absorption of water through the mucosa. Here the longitudinal muscle coat is incomplete, being divided into three separate bands, the *taeniae coli*, between which the circular muscle coat is exposed to the surface. This arrangement allows for the continuous regurgitation of the fluid contents during peristalsis until water absorption is completed, while the solid contents, being actively gripped by the circular muscle, are forced on. In the terminal part of the large intestine—the rectum— where the contents have normally become quite solid, the longitudinal coat is again a continuous layer.

Anatomical studies have established the fact that both the longitudinal and circular muscle coats of the small intestine are really disposed in a left-handed helicoidal arrangement.[1] The circular fibres are wound in a close spiral, making a complete turn about every $\frac{1}{2}$ to 1 mm., while the longitudinal fibres form an open spiral making a complete turn every 50 cm. or so. It may be inferred, therefore, that peristaltic movements of

[1] E. J. Carey, 'Studies on the structure and function of the small intestine', *Anat. Rec.* **21**, 1921.

the intestine involve a left-handed screw action, and this has been verified experimentally. Carey points out that, of two contraction waves starting out from one point along the two muscle coats, the wave in the longitudinal coat will be transmitted farther in a given time than that in the more closely wound circular coat. Since a contraction of the longitudinal fibres leads to a shortening and dilatation of the gut, it follows that each wave of constriction due to the contraction of the circular fibres must necessarily be preceded by a wave of dilatation. It may be noted, moreover, that the dilating action of the longitudinal muscle will distend the spiral turns of the circular muscle, thus providing an effective stretching stimulus for the immediate subsequent contraction of the latter.

In certain parts of the alimentary canal and other tubular structures, the circular muscle coat becomes thickened locally to form *sphincters*. By their contraction and relaxation, these control the passage of material by closing or opening up junctional connexions. For example, the pyloric sphincter 'guards' the opening of the stomach into the small intestine, and the ileo-caecal sphincter controls the opening of the small into the large intestine. As we shall see, sphincteric mechanisms of a similar kind are also found in certain peripheral blood vessels, where they play a part in regulating the distribution of blood to the tissues (p. 203).

Like striated muscle fibres, unstriped fibres are highly specialized for the functions which they have to perform, and do not commonly undergo proliferation by cell division (though this has occasionally been observed). The hypertrophy during life of plain muscle (as, for example, in the uterus during pregnancy) is due almost entirely to the enlargement of individual fibres and not to an increase in their number. It has been estimated that, by this process, smooth muscle can increase its bulk eight times.[1]

12. THE NERVE SUPPLY OF VISCERAL MUSCLE

It has already been noted that somatic muscles depend in the living body for their contraction and the maintenance of their tone entirely on nervous connexions. A denervated somatic muscle is completely flaccid, and relaxation is therefore simply obtained by inhibition of the nervous impulses which stimulate contraction. Visceral muscle, however, retains powers of automatic and spontaneous contraction without the necessity of a neural stimulus. For this reason, it requires to be supplied by two types of nerves, one for producing contraction and one for producing relaxation.

The double motor innervation of the viscera is provided by the autonomic nervous system (p. 389), the contrasting roles being played on the one side by the thoracic autonomic nerves (i.e. the sympathetic) and on the other side by the cranial and pelvic autonomic nerves (i.e. the parasympathetic). The fibres which ultimately reach the muscle fibres are all post-ganglionic and non-myelinated. They end in extremely fine branches which break up into varicose terminations closely applied to the surface of the muscle cells. It is generally accepted that, unlike the motor nerves

[1] K. H. Lange, 'Über die Hypertrophie der glatten Muskulatur', *Morph. Jahrb.* **84**, 1939–40.

of striated muscle, they terminate in contact with the surface of the cell and do not penetrate into its substance. However, some histologists have described fine nerve fibrils actually terminating within the cytoplasm of a muscle fibre.[1]

Ganglion cells are usually abundant in the muscular coats of the viscera, forming peripheral plexuses. The latter provide an intrinsic neural mechanism through which the sequence and rhythm of muscular contraction can be initiated and co-ordinated independently of central connexions. For example, although isolated strips of plain muscle may undergo simple rhythmic contractions in the absence of any nervous tissue, the elaborate peristaltic movements of the small intestine cease if the activities of the peripheral ganglionic plexus in the muscular coat (myenteric plexus) are interrupted by the application of nicotine. On the other hand, they are not abolished by section of the nerve fibres running to the intestine from the sympathetic chain and collateral ganglia. That the myenteric plexus also controls the direction of contraction is indicated by the experiment of resecting a portion of intestine and replacing it in a reversed position. This leads to a blockage at the site of the operation, for the peristaltic movements in the reversed intestine now pass in a retrograde direction.

The pattern of the neural mechanism of the intestine determines that wherever a local stimulus is applied contraction of the circular muscle occurs immediately above and relaxation immediately below. It will be apparent, therefore, that once the passage of the contents has been started, it will tend automatically to continue down the length of the intestine.

Sensory nerve endings of varying degrees of complexity have been described in visceral muscle and there is some evidence that (at least in the musculature of the bladder wall) some of these endings serve the function of stretch receptors.[2]

13. CARDIAC MUSCLE

In many features, cardiac muscle fibres occupy a position morphologically between those of visceral and somatic muscle. They resemble the former in their derivation from splanchnic mesoderm, their innervation by the autonomic system, and in the central position of their nuclei, and they approach the latter in being cross-striated. Cases have been recorded of individuals who are able in some degree to change the rhythm of the heart beat by an effort of will, suggesting a functional approach to voluntary musculature, but the mechanism of nervous control is fundamentally quite different. Cardiac musculature is strictly limited to the heart (and the immediately adjacent part of the large vessels which enter and leave it), and its function is to maintain a regular rhythmic contraction which must continue uninterruptedly from the first few days of embryonic existence up to the end of life.

[1] C. J. Hill, 'A contribution to our knowledge of the enteric plexuses', *Phil. Trans. Roy. Soc.*, Ser. B, **215**, 1927.
[2] F. Kleyntjens and O. R. Langworthy, 'Sensory nerve endings in the smooth muscle of the urinary bladder', *Journ. Comp. Neur.* **67**, 1937.

Cardiac muscle fibres consist of elongated cells which frequently branch and in ordinary histological preparations are not defined by any clear cell boundaries. In other words, the whole extent of the heart muscle appears to form practically an uninterrupted syncytium. The fibres are joined end to end in continuous strands, and their side branches run freely from one fibre into another. As already mentioned, the nuclei are centrally placed inside the fibres. There has been some doubt whether the fibres have a sarcolemmal sheath; it is generally believed that this is so, but the sheath is much finer—and so less easily demonstrable—than in somatic muscle.

In the fibres are fibrils quite similar to the myofibrillae of skeletal muscle in their segmental arrangement of light and dark bands and in the changes

FIG. 65. Diagram illustrating the essential features of cardiac muscle. Note that the fibres have each a single, centrally situated nucleus, and that they appear to be in syncytial continuity by side branches. The appearance of the intercalated disks is also indicated.

which they undergo in contraction. They give rise to the characteristic cross-striation of the fibres as a whole, but this is rendered somewhat indistinct because they are embedded in a relatively large amount of granular sarcoplasm. On the other hand, the longitudinal striation is particularly conspicuous. The muscle fibres are separated into segments at short intervals by transverse bands which stain deeply with many stains. These are called *intercalated disks*. Their significance has long been a matter of argument, one view holding them to be equivalent to cell membranes separating one cell from another, and another view maintaining that they are analogous to the anisotropic disks of the myofibrils. In favour of the first conception was the argument that they stain deeply with silver salts, they occur at rather regular intervals, and they separate the fibres into short segments each of which usually contains one nucleus. Also, if cardiac tissue is macerated with strong alkalis, the muscle fibres break up completely so that these segments become separate and isolated. Against the view that they are cell membranes it was contended that the myofibrillae run in uninterrupted continuity through them from one segment to another, that they are not present in embryonic heart muscle, and that they increase numerically with age although there is no corresponding cell division. As with so many controversies of this kind, the matter now seems to have been settled by electron microscopy. Thus, magnifications of 80,000 diameters make it apparent that the intercalated disks extend uninterruptedly through the substance of the muscle fibre and that, after all, the myofibrillae do not bridge across the disks.[1] Cardiac muscular tissue is to be regarded, therefore, not as a true syncytium but as a tissue composed of individual muscle cells. Clearly, this conception is important in relation to the problem of the spread of the wave of contraction during the cardiac cycle. It should be noted that an intercalated disk does not always run as a straight transverse line through the fibre—commonly it extends across in a series of short steps. The evidence now available that the disks mark the position of actual partitions between adjacent segments of cardiac muscle fibres conforms with the evidence of direct observation that these segments may function independently. For example, in tissue-culture experiments it has been observed that two cardiac muscle cells, *apparently* in direct anatomical continuity, may contract rhythmically at different rates.[2]

The interconnexions by side-branches of cardiac muscular tissue is its most distinctive feature, and it presumably provides the anatomical basis for the spread of a wave of contraction over the whole heart. Thus, like the individual muscle fibres, the 'all-or-nothing' principle applies to the contraction of this organ as a whole. There are several reasons for supposing that the contraction wave is myogenic rather than neurogenic in origin. Perhaps the most convincing evidence is the anatomical observation that the heart starts beating rhythmically in the embryo before any nerve fibres have reached it and before any nervous tissue is incorporated in its walls.

Contraction in each cardiac cycle is initiated in a small mass of tissue (the *sinu-atrial node*) at the site of entry of the great veins into the heart. This is sometimes referred to as the 'pace-maker' of the heart, and it was first described in 1906 by Keith and Flack.

The sinu-atrial node consists of muscle cells which are somewhat similar to those of ordinary cardiac muscle except that they are more slender and fusiform, they branch very freely and anastomose in a plexiform manner, and their cross-striation is less distinct. They contain a high proportion of glycogen. The cells are mingled with nerve fibres and nerve cells, and are embedded in a connective-tissue basis. From this node the contraction wave spreads over the atria of the heart to the atrio-ventricular junction.

In lower vertebrates, the musculature of the atria is freely continuous with that of the ventricles around the wall of the atrio-ventricular canal. In higher vertebrates—including mammals—the atrio-ventricular junction is mainly composed of a ring of fibrous tissue, and only in one region is there direct muscular continuity. This is effected by a special bundle of 'conducting tissue'—the *atrio-ventricular* or '*A-V*' *bundle*. The latter begins in a circumscribed mass of tissue in the lower part of the interatrial septum called the *atrio-ventricular node*, similar in its structure to the sinu-atrial node. From here it extends down either side of the interventricular septum, ultimately to reach the papillary muscles of the ventricle. In some mammals, particularly carnivores and large ungulates, the bundle is composed of atypical cardiac muscle fibres which are conspicuous for their

---

[1] F. S. Sjöstrand and E. Andersson, 'Electron microscopy of the intercalated discs of cardiac muscle tissue', *Experientia*, **10**, 1954.
[2] W. H. Lewis, 'The cultivation of embryonic heart tissue', *Contrib. to Embryology*, Carnegie Inst. **18**, 1926.

large size. These fibres (called, after their discoverer, *Purkinje fibres*) are also characterized by the indistinctness of their transverse striation and the peripheral position of their myofibrillae. Similar fibres may be found in the atrial musculature, linking up the sinu-atrial with the atrio-ventricular node.

The whole of the conducting system is very vascular and richly innervated, and the terminal ramifications of its constituent fibres apparently pass without interruption into series with the ordinary cardiac muscle fibres. In many mammals, including man, the fibres of the conducting system are not very strongly contrasted with the rest of the cardiac muscle; while they are somewhat more branched and their striation is less distinct, they are not conspicuously larger in size.[1]

The A-V bundle evidently serves the purpose of transmitting the contraction wave to all parts of the ventricles almost simultaneously, and thus permits of a more even contraction in ventricular systole such as would not be possible if the atria were in broad muscular continuity with the ventricles. Some authorities have regarded the conducting tissue as the persistence of an embryonic and generalized type of cardiac muscle which has not undergone the differentiation characteristic of cardiac muscle fibres in general. However, the tissue can be distinguished as a separate and specialized system even at an early stage of embryonic development,[2] and comparative anatomical studies have shown that the system is only to be found in the warm-blooded vertebrates—mammals and birds. Thus from the evolutionary point of view it is to be regarded as a neomorphic development, perhaps facilitating the more rapid movements of the heart (in relation to its size) which are required in warm-blooded animals.[3] Since the A-V bundle provides for the normal rhythmical sequence of atrial and ventricular contraction, this sequence is grossly disturbed if it is damaged by disease.

It has been noted (p. 160) that cardiac muscle cells have been observed to contract rhythmically in tissue cultures in the absence of nervous tissue. This inherent property of rhythmic contractility, however, is not confined to cardiac muscle, for, as already mentioned, it has been observed in cultures of skeletal muscle fibres, and is also characteristic of plain muscle.

Cardiac muscle fibres do not appear to be capable of proliferation by active division; so far as is known, the increased bulk of the heart musculature which occurs in cardiac hypertrophy is due entirely to the enlargement of the individual fibres. Unlike skeletal muscle, also, cardiac muscle shows no evidence of a regenerative capacity following injury. In experimentally induced necrosis of the ventricular musculature in animals, for example, the fibres at the margin of the necrotic area remain quite inert, in strong contrast to the vigorous attempts at regeneration displayed by ordinary striated muscle fibres under similar conditions.[4]

[1] R. C. Truex and W. M. Copenhaver, 'Histology of the moderator band in man and other mammals', *Amer. Journ. Anat.* **80**, 1947.
[2] E. W. Walls, 'The development of the specialized conducting tissue of the human heart', *Journ. Anat.* **81**, 1947.
[3] F. Davies, 'The conducting system of the vertebrate heart', *Brit. Heart Journ.* **4**, 1942.
[4] R. G. Harrison, 'The regenerative capacity of cardiac muscle', *Journ. Anat.* **81**, 1947.

14. THE NERVE SUPPLY OF CARDIAC MUSCLE

While there remains no doubt of the myogenic basis of the rhythmic contraction of the heart, the *rate* of its contraction is under the control of the nervous system. Like visceral muscle generally, the heart has a double innervation from sympathetic and parasympathetic fibres. The parasympathetic fibres are supplied by the vagus.[1] Anatomically, the sympathetic fibres on reaching the heart are very fine and varicose, while the parasympathetic fibres are commonly thicker and more uniform. The former end in the musculature of all parts of the heart, while the latter appear to be mainly confined in their distribution to the atria and the atrio-ventricular bundle. Ganglion cells are found in the walls of the atria and in both sinu-atrial and atrio-ventricular nodes, but only occasionally in the ventricles (except in certain groups of mammals such as ruminants and cetaceans). They are related to the vagal fibres. The motor fibres end intra-protoplasmically after spinning a plexus round the cardiac muscle cells. Fine terminal fibres can sometimes be seen running from one fibre to another through the protoplasmic bridges which connect them together. In addition to motor fibres, there are also abundant sensory fibres which form fine plexuses in the non-muscular tissues of the heart. Experimental evidence indicates that they belong predominantly to the vagus, and their branched endings provide the sensory receptors which are involved in cardiac reflexes. Other afferent fibres from the heart (particularly those innervating the coronary arteries and parts of the ventricles) pass through the inferior cervical sympathetic ganglion and the dorsal roots of the upper thoracic nerves.[2]

---

[1] For an account of the nerve supply of the heart, see G. A. G. Mitchell, *Cardiovascular Innervation*, Livingstone, 1956.

[2] W. A. Nettleship, 'Experimental studies on the afferent innervation of the cat's heart', *Journ. Comp. Neur.* **64**, 1936.

VII

THE TISSUES OF JOINTS

THE manner in which one bone articulates with another in the human body varies very considerably. This variation is related to the presence or absence of movement at the joint, or to the type of movement which is required. Many joints are quite immovable; they represent growth lines at which the adjacent bones continue to expand by surface accretion, or they simply mark the lines of contact between the morphologically separate bony elements which enter into the composition of a single solid structure. Movable articulations range from joints at which little more than a slight play is allowed when they are subjected to stress, to ball-and-socket joints which allow of quite extensive mobility.

For convenience of description, some classification of joints is desirable. There are several ways of doing this but, on the whole, a broad functional division is the most practicable. Thus we may recognize immovable joints or *synarthroses*, freely movable joints or *diarthroses*, and, as an intermediate category, *amphiarthroses* or partially movable joints. All diarthroses are formed by cartilage-covered articular surfaces, separated by a joint cavity[1] which is partly lined by synovial membrane and contains a lubricating substance, the synovial fluid. In synarthroses and amphiarthroses, on the other hand, the articulating bones are united directly by fibrous tissue or cartilage, or both.

When the membranous precursors of skeletal elements are first laid down in the embryo, they are all joined directly with each other by mesenchymatous tissue. With further development of cartilage and bone, this tissue becomes compressed into a fairly compact plate—the *articular disk*—in which the component cells are flattened by the pressure of the opposed articular surfaces. Peripherally, the disk is continuous with a capsular layer which represents an extension of the perichondrium or periosteum from one articulating element to the other, and from which the fibrous capsule of the joint is subsequently differentiated.

In the case of diarthrodial joints, a cleft appears in the articular disk first in its peripheral zone and then extends and enlarges gradually to form the definitive joint cavity. The mesenchyme cells lining the cavity become differentiated to form a lining mesothelium—the *synovial membrane*. In some diarthrodial joints, remnants of the articular disk may persist and become organized into cartilaginous or compact fibrous tissue, which provides the basis for the development of intra-articular cartilages or menisci. When present, these structures are permanently interposed between the articular surfaces, separating them partially or completely. Most of the joint cavities in the human body have made their appearance by the tenth week of embryonic development.

[1] Under normal circumstances this cavity is little more than a potential cavity, since the opposed structures are merely separated by a thin film of synovial fluid.

Homologous joints in different mammals vary in the range and direction of their mobility in relation to different habits of life, and these functional variations are reflected in the anatomical configuration of the articulating elements. As an example, reference may be made to the mandibular joint. In some carnivorous mammals movement at this joint is limited to the vertical plane and the articular surfaces are shaped like a hinge. In herbivorous mammals, where side-to-side movements are required for chewing, the surfaces are flattened and more extensive. Yet it is always somewhat surprising to the student of human anatomy to find that in lower mammals the shapes of joint surfaces often resemble so closely those of man even where the functional requirements seem very different. In some cases profound modifications may occur in response to functional demands, and a joint which in one species of mammal allows movement may in another be completely fixed. For example, the inferior tibio-fibular joint is provided with a synovial cavity in those animals in which the demands for agility require a suppleness of movement at the ankle. In others, where the factor of stability is important, the lower ends of the tibia and fibula become united by cartilage or bone. The evolutionary conversion of diarthrodial joints into synostoses is also demonstrated by the cervical vertebrae of Cetacea. In these mammals all movement between the cervical vertebrae has been lost, and the individual bones have become fused together in a compact mass.

1. SYNARTHROSES

Besides the fact that they are not formed for the purpose of movement, these joints have certain other features in common. They represent a persistence of the embryonic stage of development when the articulating elements are united by a solid articular disk. The subsequent organization of this disk depends on the nature of the articulating elements. Where these are cartilage bones, they become joined by a plate of hyaline cartilage; in the case of membrane bones, the union is effected by fibrous tissue. Synarthroses are also similar in their lack of permanence. Many of them become obliterated as soon as the growth of the skeleton is completed, while others tend to disappear by fusion of the bony elements in early maturity or middle age.

Synchondroses. Immovable joints which are of an entirely temporary nature commonly become obliterated as soon as the growth of the articulating bony elements is complete. Such are *synchondroses*, in which the bony surfaces are united directly by a plate of hyaline cartilage (Fig. 66 B). An obvious example of this type of joint is the union between the diaphysis and epiphysis of a long bone by the epiphysial cartilage. The latter provides for the continued growth in length of the bone, as already described in Chapter 5. When maturity is reached, the cartilage is converted into bone by the extension of ossification from the diaphysial side, and the epiphysis becomes fused or synostosed with the diaphysis.

Synchondroses are also found in the base of the skull between the basi-occipital and basi-sphenoid bones, and between the ex-occipital and the petrous portion of the temporal bone. The cartilaginous plates here are to be regarded as persistent parts of the foetal chondrocranium, and they usually become obliterated by ossification by the age of 25 when the growth of the base of the skull is completed. The histological appearance of these cartilaginous plates is quite similar to that of an epiphysial cartilage. Another example of a synchondrosis is found in the joint between the first rib and the manubrium sterni. In this case, the cartilage commonly undergoes calcification in early adult life.

Sutures. In these articulations, denticulated bony edges are firmly interlocked and joined together by fibrous tissue (Fig. 66 A). Sutures are peculiar to the skull, and this is related to the fact that here the growing brain is enclosed in a rigid cranium, the vault of which is composed of several membrane bones. The latter (parietal, frontal, occipital, and temporal) come into contact at their margins as each bone grows peripherally from its centre of ossification. At birth the approximation of the individual bones is still incomplete, for at the bregma (the meeting-point on the top of the skull of the frontal and parietal bones) a large gap remains, closed only by fibrous tissue. This is the anterior fontanelle, which normally becomes obliterated by the end of the second year.[1]

The sutures in the mature skull vary somewhat in their complexity and also in the extent to which the articulating bony margins overlap. On the endocranial aspect of the skull, each suture runs a much straighter and more even course than on the outside.

The proposition (based on early experiments) that the position of sutural lines is morphologically predetermined before the process of ossification is complete now appears to be incorrect, at any rate in the earlier stages of ossification. It has been shown, for example, that if local damage is inflicted on the skull vault of rats *in utero* before the establishment of the sutures, or even of neonatal rats in which sutural contact has been effected, asymmetrical suture patterns develop as the result of a compensatory overgrowth of one bone into the normal territory of a neighbouring bone.[2] Corresponding sutures may show considerable differences in their complexity, and there may be scattered along them little separate ossicles (Wormian bones) which are formed by the detachment of ossifying spicules at the margins of the growing bones. In hydrocephalus, where the rapid distension of the brain interferes with the normal process of ossification of the membrane bones of the skull, these ossicles are particularly numerous and large.

Closure of sutures. The membrane bones of the skull expand in their growth to a considerable extent by deposition of new bone on their outer surface and absorption from their deep surface. Sutural lines do not, therefore, represent the only zones of growth of these bones. Nevertheless, there is a broad relation between suture closure and the growth of the skull as a whole. In conditions in which the brain ceases to grow, as in microcephalic idiocy, the sutures become obliterated early. Premature closure of

---

[1] In the case of nutritional deficiency (e.g. in rickets) the closure of the fontanelle may be considerably delayed.

[2] F. G. Girgis and J. J. Pritchard, 'Experimental alteration of cranial suture patterns', *Journ. Anat.* **90**, 1956.

certain sutures, also, may be associated with characteristic deformities of the skull. For example, if the sagittal (interparietal) suture becomes synostosed too early, increase in breadth of the cranium appears to be limited, and in order to provide for the expanding brain the skull grows abnormally in length, producing a variation in its shape known as *scaphocephaly*. Again, premature obliteration of the coronal (fronto-parietal) suture may be associated with an increase in height of the skull, or *acrocephaly*. However, the relation between these deformities and premature synostosis does not appear to be entirely consistent.

Normally, the cranial sutures start to become obliterated at about the age of 30. This process occurs first where the sutures are most simple, and the complex sutures are the last to disappear. It follows from this that synostosis first becomes evident on the endocranial aspect of the skull. The dates of the closure of individual sutures and the order in which it takes place have been worked out in considerable detail by Todd and Lyon.[1] From the statistical data compiled by these observers, it appears that the individual variability is too great to allow of more than an approximate estimation of the age by a consideration of the state of the sutures. There is some evidence that the closure on the endocranial aspect is more reliable from this point of view, and that suture closure tends to occur earlier in the male sex.

The factors which determine the time of synostosis are still quite obscure. There seems to be a relation to the growth of different parts of the brain, but the nature of this relation is not clear. The suggestion was at one time put forward that premature closure of the sutures might prevent the brain from attaining its normal size. It was even supposed that certain cases of microcephaly were due to this cause, and surgical operations were devised with the purpose of separating the bones again in order to allow a normal expansion of the brain. However, such operations met with doubtful success, and it is probable that premature synostosis in this condition is secondarily determined by an arrest in cerebral development. On the other hand, the deformities resulting from the premature synostosis of individual sutures (mentioned above) can be avoided or minimized by the surgical procedure of opening up the bones on either side of the suture.

Racial variations in suture closure have not been studied in great detail, with the exception of a comparison of Negroes and Whites by Todd and Lyon. This study showed no difference in the age or order of obliteration of normal sutures. An interesting racial difference, however, is shown in the abnormal persistence of a suture between the two halves of the frontal bone—the metopic suture. In man and the higher primates this usually becomes obliterated in early infancy. It rarely persists in primitive races (1 per cent in Australian natives and Negroes), but in the peoples of northern Europe it occurs in over 9 per cent of skulls. The suggestion has been made that this persistence is related to the increasing expansion of the frontal lobes of the brain, and is therefore to be regarded as an index of

[1] T. W. Todd and D. W. Lyon, 'Cranial suture closure', *Amer. Journ. Phys. Anthrop.* 7, 1924, and 8, 1928.

intellectual status. However, there appears to be no correlation between a persistent metopic suture and cranial capacity in individual cases, and it is well known that in many primitive mammals, including the lower Primates, the suture remains patent. Bolk[1] has advanced the theory that the condition is related to the extent of the attachment of the temporal muscles to the frontal bone. When this is large, it was supposed, the mechanical strain transmitted to the bone provides a stimulus which leads to early synostosis of the two halves. Available evidence from other sources, however, suggests that the closure of this and other sutures depends on intrinsic growth factors rather than on crude mechanical influences of stress and strain.

In lower mammals, as is well known, the order and degree of suture obliteration show a pattern characteristic for each group, and it has hitherto proved impossible to relate these patterns consistently to any one factor. It has been observed that, within the limits of the Primates, the date of closure of the cranial sutures in a comparative series becomes progressively delayed in relation to the growth of the body as a whole, leading from the Old World monkeys up to man.[2]

Besides synchondroses and sutures, two other types of synarthrodial joints have been defined. These are the *gomphosis*, in which the joint consists of a peg fitting into a socket (e.g. the articulation of the teeth with the jaws), and the *schindylesis*, in which a bony plate fits in a groove (e.g. the articulation between the vomer and the bony palate).

2. AMPHIARTHROSES

These joints differ radically from synarthroses in the fact that they have no relation to the growth of the articulating bones and are therefore completely permanent, and also in permitting some degree of movement. Two types are recognized, *syndesmoses* and *symphyses*, depending on the nature of the tissue which unites the articulating surfaces.

Syndesmoses. In these joints, the opposed bony surfaces are bound together simply by fibrous tissue (Fig. 66 c). The inferior tibio-fibular joint is usually quoted as an example of a syndesmosis, but, in so far as a small recess of the ankle joint projects up between the two bones, it may be said to contain a synovial cavity. In this joint, movement is limited to a slight amount of play which allows a trivial degree of separation between the two bones in dorsiflexion of the foot.

Symphyses. The essential anatomical features of a symphysial joint are (1) the articulating bony surfaces are covered by a layer of hyaline cartilage, and (2) these cartilaginous surfaces are united by fibrous tissue or fibro-cartilage (Fig. 66 D). The latter may form a relatively thick disk which, by virtue of its elasticity, allows a slight degree of play between the articulating bones. This is the case, for example, with the joints between the

[1] L. Bolk, 'On metopism', *Amer. Journ. Anat.* 22, 1917.
[2] W. M. Krogman, 'Ectocranial and endocranial suture closure in anthropoids and Old World apes', *Amer. Journ. Anat.* 46, 1930.

bodies of the vertebrae which are separated by intervertebral disks. The cumulative effect of slight movement between one vertebra and another is shown in the flexibility of the vertebral column as a whole. This is partly explained by the fact that in the centre of each disk is a mass of soft pulpy material (the *nucleus pulposus*) which, incidentally, is believed to be derived from the remnants of the notochord of the embryo. This central core not only augments the torsional elasticity of each disk as a whole, but,

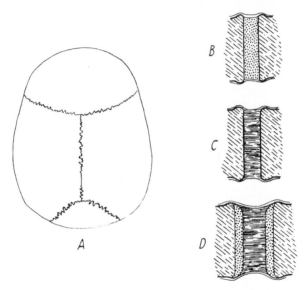

FIG. 66. Diagram illustrating different types of joints. *A*. Dorsal view of the skull showing sutural lines. *B*. Synchondrosis, in which the bony elements are united by hyaline cartilage (dotted area). *C*. Syndesmosis, in which the bony elements are united by fibrous tissue. *D*. Symphysis, in which the bony elements are covered by plates of hyaline cartilage, and the latter joined by fibrous tissue.

functioning in the manner of a water-cushion, it also adds to its resilience as a buffer, minimizing the effects of intermittent stresses transmitted along the length of the vertebral column.

The inter-pubic joint is another example of a symphysis. One interesting feature of this joint is that it usually contains a rudimentary cavity in the form of a vertical slit-like cleft in the middle of the fibro-cartilaginous plate. It has been observed that this cavity may in some cases become enlarged in women at the end of pregnancy, leading virtually to a transformation of the symphysis into an elementary diarthrodial joint. This change is evidently in adaptation to the requirement for a greater freedom of movement between the pubic bones to allow the passage of the foetus during parturition. In some lower mammals (e.g. the guinea-pig and gopher), the pubic bones become quite widely separated in the female in preparation for parturition, and are then connected only by a thin sheet of membranous tissue. In the guinea-pig the separation commences at the middle of pregnancy and, although the bones are approximated to some degree after parturition, they never become closely joined again.[1] This process is determined by hormonal influences. It has been shown, for example, that separation of the pubic bones can be produced in castrated *male* gophers by the injection of ovarian extract. Burrows has found that the same phenomenon can be induced in female mice and also in normal and castrated male mice by the action of oestrone, even though a separation of the pubic bones never occurs in the mouse under normal conditions.[2]

3. DIARTHROSES

The essential structure of a diarthrodial joint may be conveniently described by reference to a diagrammatic section through the elbow joint as shown in Fig. 67. From this figure, it will be seen that the actual articulating surfaces of these bones are covered by a thin layer of hyaline or articular cartilage. Elsewhere, the cavity of the joint is lined by a delicate synovial membrane which is somewhat similar in general macroscopic appearance to a serous membrane. The synovial membrane rests on a subsynovial layer of connective tissue, and outside this the whole joint is enclosed in a capsule of fibrous tissue.

In some joints a varying portion of the adjacent non-articular surface of the bone may be included within the joint cavity, in which case the synovial membrane extends over it from the margin of the articular cartilage for some distance before it is reflected on to the inside of the fibrous capsule, and the latter is attached to the bone at a corresponding distance from the actual articular surface. The synovial lining may also be complicated by the formation of folds which project into the joint cavity and increase the surface area of the membrane. The fibrous capsule usually develops local thickenings in the form of bands or ligaments which not only aid in maintaining the articular surfaces in contact, but also play a part in restraining movements in certain directions.

Synovial membrane. Early in foetal life, a diarthrodial cavity is completely lined by a continuous stratum of mesenchymatous tissue which covers even the articular cartilage. It is from this layer that the synovial membrane becomes differentiated. When, during the fifth month, active intra-uterine movements begin, it is soon rubbed off the articular surfaces by friction and pressure. At birth, the synovial membrane still encroaches to a slight extent on the margin of the cartilage, but with the development of a greater range of movement at the joints it almost entirely disappears here. Other regions of the cavity exposed to friction also become denuded of a mesothelial covering, and, indeed, it is hardly to be expected that such a delicate membrane could long withstand much wear and tear of this kind.

Sections through the synovial membrane show that it varies somewhat in its minute anatomy from one place to another. In most regions it consists

[1] T. W. Todd, 'The pubic symphysis of the guinea-pig in relation to pregnancy and parturition', *Amer. Journ. Anat.* **31**, 1923.

[2] H. Burrows, 'Separation of the pubic bones following the administration of oestrogens in male mice', *Journ. Phys.* **85**, 1935.

of a mesothelial lining of flattened cells covering a layer of loose connective tissue. In recesses where the membrane is completely sheltered from mechanical stresses, the cells are rounded or polygonal and may be disposed in two or three layers. Where the lining of the joint cavity is exposed to much pressure and friction, it is composed of dense, impervious fibrous

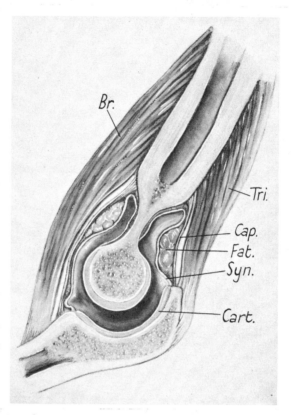

FIG. 67. Diagram of a section through the elbow to illustrate the main features of a diarthrodial joint. The joint cavity is lined by synovial membrane (*Syn.*) which extends to the margin of the articular cartilage (*Cart.*). Outside the synovial membrane, but deep to the fibrous capsule (*Cap.*), are subsynovial pads of fat (*Fat*). *Br.* Brachialis. *Tri.* Triceps.

tissue or fibro-cartilage. The surface of intra-articular fibro-cartilages is likewise devoid of a synovial mesothelium.

The nature of the cells of the synovial mesothelium is still in some doubt, but they are very similar in their appearance and reactions to fibroblasts. If a vital dye (trypan blue) is injected into the knee joints of rabbits, the histiocytes in the submesothelial tissue assimilate the dye in a characteristic manner, but the mesothelial cells (at least in earlier stages) only take it up in small quantities in the form of fine granules. Later on they react more strongly and assume a vacuolated appearance. Some of them also show phagocytic properties and may be found to contain extravasated red blood corpuscles as well as numerous granules of trypan blue. Others, under the influence of the irritation, become detached and set free as desquamated cells in the inflammatory effusion in the joint cavity. In general, under the conditions of these experiments the mesothelial cells maintain their distinctive characters. On the other hand, many of the histiocytes push their way through the mesothelium to reach the joint cavity, where they become heavily laden with the dye. The fact that, following destruction of a considerable area of synovial membrane, the latter can be extensively regenerated, suggests that the mesothelial lining is composed of but slightly modified connective-tissue cells.

It has been mentioned that, where the synovial mesothelium is deficient, the lining of the joint cavity is formed by dense fibrous tissue; the cavity is therefore nowhere in very free communication with the connective-tissue spaces outside. In other words, a diarthrodial joint is functionally a closed cavity even though its lining mesothelium is anatomically incomplete. It has been shown experimentally that true solutions, colloidal solutions, and fine suspensions are rapidly absorbed from a joint cavity, and it is probable that the absorption is mainly a physical process which does not involve any specific activity of the synovial cells. In the case of colloidal solutions and suspensions, the rate of absorption is proportional to the size of the particles. On reaching the subsynovial tissues, the smaller particles are taken up by blood capillaries and lymphatic vessels, whereas those above a certain critical size can only be removed by lymphatic vessels or not at all.[1]

The whole of the synovial membrane is concerned in absorption from the joint cavity, but this process occurs most rapidly where it overlies loose connective tissue and where it is reflected over subsynovial pads of fat. The latter are found in many diarthrodial joints, filling out folds and reduplications of synovial membrane (Fig. 67). They provide for an increase in the absorptive area of the synovial membrane, particularly since the mesothelium over them is often thrown into small complicated folds which in sections have the appearance of villi. In the knee joint they are especially conspicuous, and allow a very rapid absorption which readily accounts for the occasional appearance of general septicaemia in acute traumatic infections of this joint.[2] Experimental observations have repeatedly shown that absorption from synovial cavities is greatly accelerated by movement. This fact obviously has an important bearing on the treatment of chronic arthritic conditions accompanied by the effusion of fluid into joints.

Subsynovial pads of fat also serve the mechanical purpose of filling up the changing spaces which occur during the movement of a joint. For example, a pad of fat lies opposite the olecranon fossa on the back of the lower end of the humerus (Fig. 67). This fossa accommodates the olecranon process of the ulna in full extension of the elbow, but during flexion

---

[1] E. W. O. Adkins and D. V. Davies, 'Absorption from the joint cavity', *Quart. Journ. Exp. Physiol.* **30**, 1940.

[2] The complicated disposition of the synovial membrane in the knee joint is partly explained by the fact that it is developed from the coalescence of three separate joint cavities. Remnants of the partitions between these cavities persist as folds of synovial membrane.

the pad of fat is pressed into it by the overlying triceps. Similar subsynovial pads of fat are found in relation to the coronoid and radial fossae on the front of the lower end of the humerus.

Still another function ascribed to synovial folds and pads of fat is that of 'pad oilers' which serve to ensure a continuous film of the lubricating synovial fluid over the articulating surfaces.[1]

It should be noted that secondary communications may be established between synovial cavities of joints and bursal sacs in the immediate neighbourhood. For example, the cavity of the shoulder joint is frequently

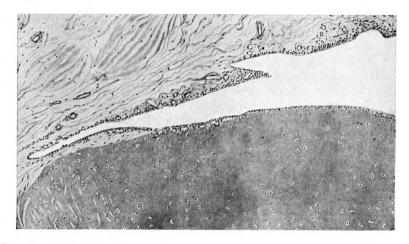

FIG. 68. Section through the margin of the articular cartilage of a human patella showing the synovial membrane of the knee joint. × 50. (From A. G. Timbrell Fisher.)

connected with a bursa underneath the subscapularis muscle through a gap in the fibrous capsule. Similarly, the knee joint almost always becomes continuous with a supra-patellar bursa which is developed independently beneath the quadriceps extensor tendon.

Articular cartilage and synovial fluid.[2] The cartilage covering the articular surface of the bones in a diarthrodial joint is usually of the simple hyaline variety (p. 67), and in this case is to be regarded as a persistent unossified layer of the cartilage from which the whole bone has been developed. When the articular surfaces are formed from membrane bones (e.g. in the mandibular joint), they are found to be covered with a dense layer of fibrous tissue or fibro-cartilage. Hyaline articular cartilage is completely avascular and devoid of a nerve supply. As already mentioned, it has no covering of perichondrium or synovial membrane, and it forms a smooth glistening surface which, lubricated with synovial fluid, allows movements to occur with a minimal amount of friction. The coefficient of friction of a dried joint has been found to be about fourteen times as great as that of a joint lubricated with synovial fluid, and it is also the case

[1] D. V. Davies, 'Observations on the volume, viscosity and nitrogen content of synovial fluid', *Journ. Anat.* **78**, 1944.

[2] For a review, see E. Gardner, 'Physiology of movable joints', *Physiol. Rev.* **30**, 1950.

that, if the viscosity of the fluid is reduced experimentally (by means of hyalase), the wear and tear of articular surfaces is much accelerated.[1]

The growth and nutrition of articular cartilage present certain problems of interest. The cartilage cells at the surface are considerably flattened, and show an appearance which is usually taken to indicate degeneration and gradual disintegration. The deeper zone of the cartilage, where it lies in contact with actual bone, is calcified. Proliferation of cartilage cells, therefore, presumably occurs in the intermediate zone in order to replace the continual wearing away of cartilage at the surface. This attrition is less extensive than might at first sight be supposed, for the articular surface is protected by a thin film of synovial fluid which serves as a very efficient lubricant.

From the point of view of its reparative power and of its reactions to irritants, articular cartilage may be divided into central and peripheral areas.[2] The latter are in immediate proximity to the vascular synovial membrane through which they can derive nutriment more readily than the central areas. They are therefore able to react much more vigorously when exposed to injury. It has been shown that an experimental lesion in the peripheral areas is followed by the active formation of new cartilage. This is partly the result of proliferation of the cartilage cells themselves, and partly due to the proliferation and metaplasia of the cells of the synovial membrane which is immediately adjacent. An incision in the central articular area, on the other hand, is followed by no formation of new cartilage of normal type—the lesion is simply repaired by the deposition of fibrous tissue or by only a partly chondrified matrix. In certain types of chronic arthritis, also, the central cartilage may succumb to toxic influences to the extent that it becomes entirely eroded from the surface of the bone, while the peripheral areas respond by active growth, producing a characteristic 'lipping' of the articular margin.

The nutrition of the central area of articular cartilage may theoretically be derived from two sources, the synovial fluid and the blood vessels in the subjacent bone. If the latter take any part at all in this process it must be to an almost insignificant extent, for, as already noted, there is a zone of calcified cartilage covering the bone and this could hardly permit the diffusion of nutrient material. It is generally agreed, in fact, that the cartilage derives its nourishment probably entirely from the synovial fluid.

Synovial fluid is a glairy, viscous fluid of which there is normally a very small amount in each joint forming a thin film covering the articular cartilage and synovial lining. The suggestion has been made that it is mainly a disintegration product of the cartilage, in which case it must be supposed that in the process of being worn away this tissue automatically supplies its own lubricant and its own nutrient pabulum. While it may be accepted, however, that the fluid will certainly contain disintegration products of articular cartilage, there is evidence to show that it is essentially formed by the synovial mesothelium. The viscosity of synovial fluid is

[1] C. H. Barnett, 'Wear and tear in joints', *Journ. Bone and Joint Surgery*, **38** B, 1956.
[2] A. G. T. Fisher, 'Physiological principles underlying the treatment of injuries and diseases of the articulations', *Lancet*, **1**, 1923.

largely determined by the mucopolysaccharide-protein complex which is its main constituent, but there is still some doubt about the origin of this substance. It is probably derived from the modified connective-tissue cells which constitute the synovial mesothelium but, apart from some dubious histological evidence, there is no reason to suppose that it is the product of a definite secretory activity. It has been shown that the concentration of the diffusible constituents of the fluid closely approaches that of blood plasma, and that diffusible substances introduced into the circulation rapidly pass from the blood into it.[1] It seems, indeed, that the synovial fluid is, at least in part, a dialysate of blood plasma. The composition and viscosity of synovial fluid vary from joint to joint in the same animal, possibly in relation to different mechanical requirements. The part played by the synovial fluid as a viscous lubricant has been discussed by Davies (op. cit.) who points out that, for fluid lubrication, the viscosity of the lubricant should vary directly as the load and inversely as the velocity of movement which is required at the joint. The fact that the volume of synovial fluid in joints is actually in excess of that theoretically required for perfect lubrication accords with the conclusion, previously mentioned, that it is also concerned with nutritional functions.

It appears that, from the biochemical point of view, synovial fluid is quite adequate for maintaining the nutrition of articular cartilage, and that this is certainly the case is proved by the observation that isolated fragments of cartilage, which have been detached by injury and lie free in a joint cavity, may not only survive but continue to grow (at least for a time).[2]

Varieties of diarthrosis. Diarthroses are classified mainly by reference to the type of active movement which they permit. Active movements of joints are those which can be performed voluntarily by direct muscular action. But it is of some importance to note that, in some diarthrodial joints, a slight degree of movement in directions which are not possible voluntarily can be effected passively by forcible manipulation. This permits a certain amount of play in these directions, and presumably serves a protective function if the joint is subjected to violent stresses. The following varieties of diarthrosis may be recognized:

(1) *Enarthrosis*, or ball-and-socket joint, in which a spheroidal articular surface fits into a corresponding concavity. In no case are the articular surfaces completely spherical in curvature; the radius of curvature is not identical in all diameters, so that they are to a slight extent ellipsoidal. An enarthrodial joint allows movements in all directions, that is to say, *flexion* and *extension, abduction* and *adduction,* and *rotation.* A composite movement involving all these elementary movements except rotation is termed *circumduction.*

(2) *Condylarthrosis* or condyloid joint, in which the articular surfaces are more conspicuously ellipsoid, and which allows flexion and extension, abduction and adduction, but no active rotation. Since these movements

---

[1] F. A. Cajori, C. Y. Crouter, and R. Pemberton, 'The physiology of synovial fluid', *Arch. Int. Med.* **37**, 1936.

[2] T. S. P. Strangeways, 'The nutrition of articular cartilage', *Brit. Med. Journ.* **1**, 1920.

take place about two axes only, this type is sometimes called a bi-axial joint. Examples are found in the wrist joint and the metacarpo-phalangeal joints of the fingers. Like ball-and-socket joints, a condyloid joint also allows circumduction.

(3) *Saddle-shaped joint* or joint of reciprocal reception. In this type of joint the articulating surfaces are each concavo-convex in opposite directions, and the movements which can take place between them are rather complicated. Like a condyloid joint, it allows flexion and extension, abduction and adduction, but the curvature of the articular surfaces also permits of a slight rotatory movement. Saddle-shaped joints combine considerable strength with fairly free mobility. One of the best examples of this variety is the carpo-metacarpal joint of the thumb.

(4) *Ginglymus* or hinge joint, which only allows movement about a transverse axis, i.e. flexion and extension. The articular surfaces are trochlear or pulley-like in shape. Examples of this type are the elbow joint and the interphalangeal joints.

(5) *Trochoid* or pivot joint, which only allows movement about a longitudinal axis, i.e. rotation. Two pivot joints are found in the human body—the superior radio-ulnar joint, and the joint between the first cervical vertebra and the odontoid process of the second. In each case, the articulation consists of a peg or disk fitting within an osseo-ligamentous ring.

(6) *Arthrosis* or plane joint, in which the only movement is a slight degree of gliding of one articular surface over the other. Movement is further limited by a tight fibrous capsule and frequently also by interosseous ligaments joining the opposed surfaces of the articulating bones. The articular surfaces are approximately flat, or show but a slight degree of curvature. Examples are found in the intercarpal and intertarsal joints.

Not all joints conform to one or other of these categories alone—in some, more than one mechanism may be involved. The mandibular joint is an instance of such a compound joint, for it contains two synovial cavities, at one of which a hinge movement occurs and at the other a gliding movement. It is therefore termed a compound ginglymo-arthrodial joint. The knee joint, similarly, is a complex mechanism involving a combination of movements characteristic of several of the elementary types.

4. LIGAMENTS

The fibrous capsule which encloses a diarthrodial joint is composed mainly of collagenous fibres which in general run directly from one bone to the other. More or less well-defined bands of fibres are usually differentiated as local thickenings of the capsule to form intrinsic ligaments which further strengthen the joint and play a part in restraining movements in certain directions. It may be supposed that these ligaments are developed in response to tensional forces which, as we have seen, determine the deposition of white fibrous tissue in accordance with mechanical requirements (see p. 47). It is, in fact, stated that the capsular ligaments of diarthrodial joints become fully differentiated only after birth, presumably in relation to the commencement of fully active movements, but there is little information on the details of this developmental process.

Besides the intrinsic capsular ligaments, the movements of some joints are functionally controlled by extrinsic ligaments which may be quite independent of the capsule, e.g. the coraco-clavicular ligaments in relation to the acromio-clavicular joint, and the coraco-acromial ligament in relation to the shoulder joint. Extrinsic contributions to the fibrous capsule may also be made by the tendons of adjacent muscles. For example, an expansion from the semi-membranosus muscle of the thigh helps to form the posterior oblique ligament of the knee joint, while the patellar ligament in front of this joint is really the tendon of the quadriceps extensor muscle in which the patella is developed as a sesamoid bone. Lastly, certain ligaments have been regarded as the persistent remains of tendons left behind after the phylogenetic degeneration or migration of muscles, but it seems doubtful whether such a morphological interpretation has ever survived a close analysis. The coraco-humeral ligament is sometimes quoted as an instance of this variety. It has been supposed that this ligament really represents a part of the tendon of the pectoralis minor muscle which in many lower mammals is inserted into the humerus but which, in man, only reaches the coracoid process. However, as an individual anomaly the muscle is sometimes found in the human body to be attached to the humerus, and in this case the coraco-humeral ligament is still quite well developed. Similarly, the ligament is also found in lower Primates in which the muscle has a humeral insertion.[1] The various attempts to homologize certain ligaments in man with muscles and tendons characteristic of lower vertebrates date back to the time when morphologists, stirred to enthusiasm by the concept of human evolution, diligently sought for additional evidence of man's kinship with other mammals. In many cases they unfortunately allowed this enthusiasm to take the place of critical analysis.

The disposition of local thickenings of the fibrous capsule to form individual ligaments is related to the type of movement at the joint. With few exceptions, ligaments are practically non-extensible under physiological conditions,[2] and they are commonly arranged so that they are taut when the joint is in a position of greatest stability, that is, when the articular surfaces are in maximal congruence with each other. In hinge joints which allow only movements of flexion and extension (e.g. the elbow joint or the ankle joint), the capsule is thin and lax in front and behind so as not to impede these movements, while on either side there are relatively strong and well-differentiated ligaments which help to prevent any lateral movement. The laxity of the capsule is directly related to the possible range of movement. In the shoulder joint where the range is very great, the capsule is loose enough to allow the head of the humerus (when it is partially abducted) to be drawn away from the articular surface of the scapula to

---

[1] F. G. Parsons, 'The joints of mammals as compared with those of man', *Journ. Anat.* **34**, 1899–1900.

[2] In isolated preparations in which a fresh ligament is exposed to a tensile force it can be shown to have some degree of extensibility (see J. W. Smith, 'The elastic properties of the anterior cruciate ligament of the rabbit', ibid. **88**, 1954). But, of course, the normal restraint of muscles and other tissues has been removed in such experiments.

a distance of an inch or so after the muscles have been severed, but with the fibrous capsule still intact. On the other hand, in the hip joint (which, like the shoulder-joint, is a ball-and-socket mechanism) the range of movement is much more limited in relation to requirements of greater strength, and the capsule is correspondingly thicker and tighter.

Besides true ligaments, textbooks of anatomy describe as 'ligaments' certain fibrous bands or synovial reflections which are not functionally concerned with limiting the movements of the joints with which they are related. In this category is the ligamentum teres of the hip joint, a cone-shaped reflection of synovial membrane extending from the central non-articular area of the acetabulum to the middle of the head of the femur. This so-called ligament represents a part of the synovial lining which has secondarily become included within the joint cavity. It is too weak in structure to play the part of a true ligament, but it may play a part in the lubrication of the joint by directing the flow of synovial fluid towards the main pressure-bearing area of the joint surfaces.[1] Another example of a mechanically functionless ligament is the 'spheno-mandibular ligament' of the jaw. This weak fascial strand is simply the remnant of the sheath enclosing the cartilage of the mandibular arch (Meckel's cartilage) which disappears during embryonic development.

5. INTRA-ARTICULAR MENISCI

Certain diarthrodial joints contain disks or menisci of fibro-cartilage which are interposed between the articular surfaces of the bones. In some cases these are complete so that the joint has two separate synovial cavities and there is no direct contact between the articulating bones—e.g. the sterno-clavicular joint and the mandibular joint. In other cases the menisci are deficient centrally, and annular or crescentic in shape, e.g. the semi-lunar cartilages of the knee joint. Developmentally, they are to be regarded as persistent organized portions of the embryonic articular disks which are found in all diarthrodial joints. Structurally, they are composed of very dense fibrous tissue, with a varying proportion of elastic fibres and occasional cartilage cells, but the latter may be altogether absent.

The real significance of intra-articular menisci is not fully apparent in spite of the many suggestions which have been put forward to 'explain' their presence in joints. No doubt they compensate for the incongruity of the articular surfaces between which they are interposed, but it is difficult in some cases to see why there should be this incongruity. In the knee joint, the semilunar cartilages seem to serve the purpose of resilient buffers minimizing the shock of impacts transmitted from the tibia to the femur, but, by splaying out, they also adapt their contour to the varying curvature of the different parts of the femoral condyles as the latter glide over the top of the tibia. It is interesting to note that the semilunar cartilages of the knee joint are capable of rapid regeneration after removal. This has been demonstrated experimentally in animals,[2] and has also been found to occur

[1] C. H. Barnett, D. V. Davies, and M. A. MacConaill, *Synovial Joints*, Longmans, 1961.
[2] R. Walmsley and J. Bruce, 'Replacement of semilunar cartilages after excision', *Journ. Anat.* **72**, 1938.

in man.[1] The process of regeneration is the result of a cellular reaction in the synovial membrane, leading to the production of dense fibrous tissue which forms the basis of a new meniscus.

Another function ascribed to intra-articular menisci is that they have some relation to the type of movement which takes place at the joint. In the mandibular joint, for example, two kinds of movement occur—a hinge movement and a gliding movement—and the joint cavity is divided by a disk into two parts, each of which is concerned with one of these movements. However, such a close functional relationship is negatived by the observation that in some carnivores in which the movements permitted at the joint are almost entirely limited to those of a hinge the meniscus is still constantly present. A mechanical significance of a different kind has been suggested by MacConaill, who points out that menisci bring about the formation of wedge-shaped films of synovial fluid in relation to the weight-transmitting parts of joints during movement, in accordance with what might be expected on the basis of physical theories of lubrication.[2] According to this idea they are to be found in joints where the articular surfaces have large radii of curvature and thrusts are likely to bring about a premature approximation of the surfaces.

The fact that, unlike articular cartilage, menisci are supplied with nerve fibres implies the possibility that they have a sensory function, allowing the muscular control of the joint to respond more promptly and with greater precision to rapid pressure changes within the joint cavity. In a complicated articulation such as the knee joint, the advantage of such a nervous mechanism can be readily appreciated.

It must be admitted that none of these 'explanations' can be satisfactorily applied to all instances of intra-articular menisci. The sterno-clavicular joint, for example, has a complete meniscus; yet it is hardly concerned with the transmission of weight, and its movements are of quite a simple type. On the other hand, the ankle joint has no intra-articular meniscus although, like the knee joint, it has to transmit the weight of the whole body and is equally susceptible to the shock of impacts.

The difficulty of finding a common functional factor for menisci in general leads to the consideration of purely morphological interpretations which have been advanced in regard to the source of these structures. The disk of the mandibular joint has been identified with a skeletal element of the reptilian jaw, or with the remains of the tendon of the lateral pterygoid muscle, while the disk in the sterno-clavicular joint has been homologized with a skeletal element in the pectoral girdle of lower vertebrates. The evidence in favour of such interpretations, however, is not always convincing.

This brief discussion on intra-articular menisci suggests that several factors, functional or morphological, are probably concerned with their development, and that from this point of view they have not all an equivalent significance.

[1] I. S. Smillie, 'The regeneration of semilunar cartilages in man', *Brit. Journ. Surg.* **31**, 1944.
[2] M. A. MacConaill, 'The function of intra-articular fibro-cartilages', *Journ. Anat.* **66**, 1931–2.

6. THE LIMITING OF MOVEMENT AND THE MAINTENANCE OF THE STABILITY OF JOINTS

Diarthrodial joints provide the opportunity of movement between one bone and another, but, particularly in the trunk and lower extremity, they must also allow of a temporary stabilization in certain positions for the transmission of weight. A number of factors are concerned with limiting the movements of a joint and maintaining its stability; these are primarily the shape of the articular surfaces and the action of ligaments and muscles. Other factors which are often mentioned as being concerned with holding joint surfaces in firm contact are atmospheric pressure, and the molecular cohesive force of the synovial fluid film which covers the articular surfaces. However, these are relatively insignificant factors.

Bony eminences in the immediate neighbourhood of articular surfaces, and irregularities of the articular surface itself, will obviously play a part in limiting movement. In the elbow joint, extension is brought to a stop by the olecranon process of the ulna fitting into the olecranon fossa on the back of the lower end of the humerus, while in the sacro-iliac joint the irregular nodulated articular surfaces of the sacrum and ilium interlock and only allow a very restricted mobility. In ball-and-socket joints it will be realized that the deeper the socket, the more limited is the range of movement. Compare, for example, the shoulder joint with its remarkably shallow glenoid cavity and the hip joint with its deep cup-shaped acetabulum. In the ankle joint, dorsiflexion is limited by the fact that the upper articular surface of the talus is broader in front than behind; consequently, as the foot is raised up at the joint, the talus becomes more and more tightly wedged between the tibial and fibular malleoli.

Besides these examples, in joints in which, at first sight, the articular surfaces would seem to permit almost complete freedom of movement in every direction, a close inspection will show that their curvature is such that complete congruence is only obtained in positions where the greatest stability is required. As will be realized, perfect coaptation of the articular surfaces will necessarily contribute to the rigidity of the joint. Walmsley has noted an interesting example of this mechanism in the case of the hip joint.[1] The head of the femur is really an ellipsoid in which the transverse diameter is slightly greater than the vertical diameter. During extension of the hip, the joint capsule becomes progressively twisted and shortened so that the head of the femur is gradually screwed more tightly into the socket of the acetabulum. In complete extension the articular surfaces become fully congruent and no further movement in this direction is possible. In other positions there is always (except at one zone of contact) a slight space between the head of the femur and the acetabulum, which is occupied by synovial fluid and part of the acetabular pad of fat.[2] During extension the latter is gradually extruded from the joint. In the knee joint,

[1] T. Walmsley, 'The articular mechanisms of diarthroses', *Journ. Bone and Joint Surgery*, **10**, 1928.
[2] It should perhaps be emphasized that during movement there is, strictly speaking, no *direct* contact between the cartilage-covered surfaces of a joint, for a synovial fluid film of microscopic dimensions is always interposed.

also, it has been shown that maximal congruence between the articular surfaces (including those of the semilunar cartilages) only occurs at full extension—the position of greatest stability.[1]

The part played by ligaments in restraining joint movements depends on the fact that they are relatively inextensible bands[2] disposed in such a way as to be pulled taut at the extreme limit of the movements to which they are functionally related. Except in these positions, they are therefore not capable of holding the articular surfaces in close contact. In weight-transmitting joints, the ligaments in general are all tightened up when these joints reach a temporary stabilization by the articular surfaces coming into maximal congruence. In the case of the hip and knee joints, this occurs in complete extension. In joints where no movement is possible about one or more axes, the ligaments related functionally to these axes are so arranged that they remain taut in all positions of movement about other axes. In hinge joints such as the elbow and ankle, for example, the collateral ligaments are taut during the whole range of flexion and extension, and in condyloid joints any tendency to medial or lateral rotation is partly prevented by obliquely disposed ligaments which likewise remain taut during other movements. It is possible for the restraining influence of ligaments to be modified in individuals by special exercise and habits of movement. Professional acrobats can increase the range of movement at certain joints to a remarkable degree by appropriate training, and there is also evidence that the range of movement in some joints (e.g. the carpo-metacarpal joint of the thumb) varies with occupational habits.

From the practical and clinical point of view, muscles no doubt provide the most important mechanism for maintaining the stability of joints. They obviously have the great advantage over ligaments in being able, by progressive contraction or relaxation, to keep articular surfaces in firm contact in all positions of the joint. In other words, they can function as contractile 'ligaments'. Their importance is demonstrated clearly enough in muscular paralysis, for in such a condition the related joints become flail and allow a much greater range of passive movement than is usual. Dislocations of joints are in the majority of cases the result of some indirect violence applied at a time when the articular muscles are relaxed and 'off their guard'. In such conditions it may require relatively little force to dislocate a joint, whereas joints actively supported by muscular contraction can be displaced only by considerable violence.

Certain muscles (sometimes termed 'articular muscles') are largely concerned with maintaining joint stability, that is to say, their purpose is not to effect movements but to limit them. This is the case with very freely movable joints where firm ligaments would necessarily impede mobility in some directions. In the shoulder joint, as already noted, the capsule is so loose as to be hardly capable of contributing to its stability. The joint is therefore closely surrounded by a group of short muscles which keep the

---

[1] M. A. MacConaill, 'The function of intra-articular fibro-cartilages', *Journ. Anat.* **66**, 1931–2.

[2] Exceptions, however, are found in certain ligaments which contain a preponderance of elastic fibres in their composition, e.g. the ligamenta flava of the vertebral column, and the capsular ligaments of the joints between the auditory ossicles.

head of the humerus firmly applied to the shallow glenoid cavity of the scapula in all positions, and so facilitate the proper action of the main effectors of movement, i.e. the long muscles such as the pectoralis major, latissimus dorsi, and deltoid.

The action of articular muscles depends on the reflex maintenance and regulation of a tonic contraction, the anatomical basis of which is supplied by afferent nerve fibres from the joint and its neighbourhood and from the muscles themselves. In flexion of a joint, afferent impulses initiated by the contraction of the flexor muscles lead to a central inhibition in the spinal cord of the motor neurons which innervate the muscles of the extensor side. The contractile tone of the latter is thus gradually diminished and they become progressively relaxed as the movement proceeds (see p. 135). Stretching of the capsule and of ligaments also stimulates sensory end organs in these structures, leading to an increased contraction of muscles appropriate for preventing excessive movement. It will be realized that any interference with this reflex mechanism will tend to distort the harmonious co-operation of the articular muscles and to result in defects in joint movement.

By way of summarizing the factors maintaining the stability of joints, reference may be made to those which contribute to the arch of the foot. In the first place, the arch is determined by the manner in which the tarsal and metatarsal bones fit together in their articulations. It is strengthened by a series of ligaments and finally supported by muscles. All these factors play their part in the maintenance of the arch, but there is reason to suppose that, at least during the period of growth, the muscular support is of the greatest practical importance. If the leg muscles should prove for some reason to be inadequate for sustaining the arch against the weight of the body in the growing individual, the excessive strain which is then imposed on the ligaments may soon cause these to give way. Eventually, with the consequent falling of the arch and displacement of its bony components, the latter may become actually distorted in shape. While, therefore, in the earliest stages of flat-foot the condition can be adequately treated by strengthening the muscles, in the latest stages an open surgical operation may be required in order to correct the deformity.

7. THE MORPHOGENESIS OF JOINTS

At the beginning of this section a very brief and simplified account of the embryological development of joints was given. Some further details of this process may now be considered. The articular disk of mesenchymal tissue which is primarily formed between the extremities of cartilaginous skeletal elements undergoes absorption in diarthrodial joints, apparently in relation to the pressure of the growing cartilages. This growth leads to a compression of the disk centrally, and to the appearance of a cavity in its circumferential portion. The cavity expands gradually and extends towards the centre of the joint. However, the matter does not end here, for the central part of the disk may itself undergo chondrification so that the articulating elements become joined together by direct cartilaginous union. This phenomenon is said to occur in late human embryos, and, according to Whillis, the union persists up to the time when foetal movements begin to occur, its breakdown being due to the movements themselves, or to a relative shrinkage of the cartilage.[1] Such a secondary cartilaginous union of articular surfaces has been denied by some authorities who ascribe the appearance of fusion to a histological artefact. However, Fell and Canti, in their studies of the differentiation of joints *in vitro*, also found that cartilaginous tissue extends across the site of the future joint cavity at certain stages of development,[2] and the fact that the formation of a complete definitive joint cavity has not been observed in an explanted limb bud suggests that movement may be a factor in determining its final appearance.

The existence of cartilaginous continuity between articulating elements was taken by Fell and Canti to indicate that joint formation is at least not due to the presence of non-chondrogenic tissue at the site of the articulation. They found, indeed, that if the region of the future joint is excised from an explanted limb bud, and the cut ends of the latter approximated, a new joint (though not normal in shape) is formed at the site of the excision. It appears, therefore, that while the actual formation of a joint may depend on extrinsic factors, or else on intrinsic factors which are not rigidly localized, the precise shape of the articular surface is certainly determined by definitely localized morphogenetic factors.[3] These tissue-culture experiments demonstrate that at any rate the *initial* stages in the development of a joint can occur in the absence of all movement. A similar inference follows from the experiments of Braus, in which fore-limb buds in amphibian larvae were grafted on to other parts of the body.[4] In this case the accessory limb may develop with no nerve supply and is therefore quite paralysed, yet a shoulder joint is formed in which the main morphological features appear to be normal. On the other hand, it has been noted in a previous chapter (p. 105) that the moulding of the articular surfaces to their final and normal shape can be modified experimentally after it has been acquired. For example, the shape of the head of the femur, which is normally spheroidal, can be made more cylindrical in animals by limiting movements of the hip joint to one plane.[5] It is also well known that the articular surfaces of chronically dislocated joints may become altered to a remarkable degree in response to new mechanical conditions.

Finally, it is a matter of considerable importance in orthopaedic work that new joints, or *pseudarthroses*, may occasionally be formed between the ends of a fractured bone which have failed to unite. The broken ends become expanded, covered with cartilage, and enclosed in a fibrous capsule, in all respects reproducing the main features of diarthrodial joints. The

---

[1] J. Whillis, 'Development of synovial joints', *Journ. Anat.* **74**, 1940.
[2] H. B. Fell and R. G. Canti, 'Experiments on the development *in vitro* of the avian knee joint', *Proc. Roy. Soc.* B, **116**, 1934.
[3] For a discussion on this problem, see P. D. F. Murray, *Bones*, Camb. Univ. Press, 1936, and also E. Gardner, op. cit.
[4] H. Braus, 'Gliedmassenpfropfung und Grundfragen der Skelettbildung', *Morph. Jahrb.* **39**, 1909.
[5] R. Fick, 'Über die Entstehung der Gelenkformen', *Abh. Preuß. Akad. Wiss.*, 1921.

capacity of skeletal tissues for forming new articulations under abnormal conditions is used by surgeons in the treatment of joints which have been immobilized by fibrous or bony anchylosis. This type of operation, known as arthroplasty, involves the mobilization of the bony elements, the reconstruction of their extremities so that they can suitably fit together, and their separation by a sheet of fascia doubled in to cover both ends in order to prevent direct union again. With the early establishment of passive and active movement after the operation, a permanent mobile articulation can be reproduced.

VIII

BLOOD VESSELS

THE need for some mechanism which allows the rapid transport to every part of the body of nutritive material and oxygen is met in all vertebrates by a circulatory system of blood vessels. Through these vessels blood flows in a continuous circulation, propelled by the pumping action of the heart. In small water-living invertebrates of a simple type such a system is not required, since in these forms it is possible for nutritive material to pass directly to the tissues within by surface absorption, or by absorption through the alimentary tract, while oxygen can be taken up directly from the surrounding watery medium.

When the size of an animal reaches a certain limit, and especially when a body cavity is developed so that the lining of the alimentary canal becomes separated from the body wall, a circulatory system of some type becomes a necessity. In its primitive formation in invertebrates, this system is not composed entirely of continuous closed vascular channels—it is rather an open lacunar system in which a fluid plasma percolates freely through tissue spaces, and is distributed from one part of the body to another in no definite direction by general muscular action. With increasing specialization, the circulation becomes more and more a closed system whose component parts are differentiated to form arteries or supplying channels, veins or draining channels, and capillaries through the thin walls of which the functional relationship between the circulating blood and the cells of the tissues is actually established.

The circulation in all vertebrates is maintained by the action of a rhythmically contractile heart. It is also aided in some lower types of animal by the spontaneous contraction of peripheral arterioles, and, indeed, this is the principal mechanism of the circulation of the blood in those animals which have blood vessels but no heart (e.g. in the primitive chordate *Amphioxus*). In many invertebrates the circulatory system remains partly open at the periphery, the blood being discharged into tissue spaces to be collected up again by definite venous channels. Even in mammals, this primitive arrangement is believed by some to persist in certain tissues, e.g. the spleen and bone marrow. This certainly seems to be the case in the spleen of the mammalian foetus where an open circulation has been clearly demonstrated.[1] Apart from such possible exceptions, however, the mammalian circulatory system consists of a closed system of tubes the endothelial lining of which is complete and uninterrupted from the heart, through the arteries into the capillaries, and back again through the veins.

[1] O. J. Lewis, 'The development of the circulation in the spleen of the foetal rabbit', *Journ. Anat.* **90**, 1956.

1. CAPILLARIES

In their essential structure, capillary blood vessels are fine channels whose walls are composed primarily of endothelium. In this sense, all blood vessels begin their development in the embryo as capillaries; even the heart, in its initial formation, is simply an endothelial tube. While in arteries and veins the walls become augmented by the addition of muscular and fibrous tissue, the capillaries retain their embryonic structure.

The endothelium of capillaries is formed of flattened pavement cells, adherent edge to edge along crenated margins which can be demonstrated by treating with silver nitrate. It was accepted for many years that the marginal adhesion of the cells is determined by an intercellular cement substance. However, although such a substance is not demonstrable by electron microscopy, the junctional cell membranes do display an 'electron-dense' material of some sort forming what is sometimes called an 'adhesion plate'.[1] The cells are elongated in the direction of the axis of the blood vessels, and each has an oval nucleus centrally situated (Fig. 69). In some parts of the body (as, for example, in the intestinal villi and in the liver) the outlines of the endothelial cells cannot always be clearly demonstrated by ordinary methods of staining. In the liver and the glomerular capillaries of the kidney (and perhaps also in endocrine glands), the endothelial cells of the finer blood vessels are perforated by pores that allow the blood plasma to gain more direct contact with the surrounding parenchymal cells.

Endothelium provides an extremely smooth lining to the whole of the cardiovascular system, minimizing friction in the flow of the blood and, in the smaller vessels, allowing the corpuscles to slip easily along its surface with no tendency to adhere. This character changes rapidly in response to any injurious stimulus.[2] If, for example, capillary endothelium is exposed to heat, to chemical irritants, or to mechanical injury, it becomes rather swollen and sticky, so that the corpuscles cling to its surface. It then softens, and leucocytes begin to push their way actively through it to get outside the vessel. This exodus of leucocytes appears to take place mainly at the venous end of the capillary system, that is, from minute venules rather than from capillaries proper, and it occurs through temporary gaps which the leucocytes make along the intercellular junctions of the epithelium. Lastly, red corpuscles may become extravasated through these weak spots in the endothelial wall. If the endothelium of a blood capillary does become ruptured as the result of gross injury, healing by proliferation of endothelial cells takes place again very rapidly. The changes in the consistency and apposition of endothelial cells described above form an important element in the vascular reactions to inflammatory conditions in the body.

Capillary endothelium is immediately covered by a perivascular layer of delicate connective tissue which usually separates it from direct contact with the cells of surrounding tissues. To this layer the term *perithelium* is sometimes given. The interstices of the perithelium contain 'tissue fluid',

[1] H. W. Florey, *General Pathology*, Lloyd-Luke, 3rd edn., 1962 (see chapter on Inflammation).
[2] E. R. Clark and E. L. Clark, 'Changes in blood vascular endothelium', *Amer. Journ. Anat.* **57**, 1935.

which is the essential medium through which a functional relationship is established between the blood circulating in the capillary system and the cells of the tissues nourished by it.

The diameter of capillary vessels varies quite considerably, depending partly on the state of their activity. This variation is very conspicuous in tissues where the metabolic level shows abrupt and wide changes. In

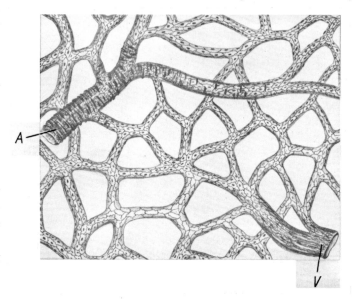

FIG. 69. Diagram representing the essential features of a capillary network. A terminal arteriole (*A*) is shown entering the capillary bed, and leaving it is a small venule (*V*). (For the appearance of capillary vessels in section, see Fig. 39.) × 250.

muscle, for example, the total capacity of the capillaries may be increased 750 times as the result of vigorous muscular contraction.[1] In a resting muscle, on the other hand, most of the capillary vessels are empty and temporarily collapsed. On the average, in man and other mammals the diameter of a capillary is slightly larger than that of the red corpuscles which flow along it—e.g. 8–$10\ \mu$. Some anatomists have described capillaries which have normally such a narrow calibre that they will not allow the passage of blood corpuscles, and only transmit the fluid plasma. It seems certain, however, that such statements are the result of observations on vessels which are temporarily distorted by external pressure or by their own contraction.[2]

The density of a capillary network is related to the metabolic level of the

[1] A. Krogh, *The Anatomy and Physiology of Capillaries*, 1929.

[2] In lower vertebrates in which the nucleated red corpuscles are relatively large, they can traverse the smaller capillaries only by becoming bent and distorted. This they can readily do by virtue of their plasticity. For discussions on the permeability of capillaries and the relation of this to capillary structure, see J. F. Danielli and A. Stock, 'The structure and permeability of blood capillaries', *Biol. Rev.* **19**, 1944; R. Chambers and B. W. Zweifach, 'Intercellular cement and capillary permeability', *Physiol. Rev.* **27**, 1947; H. W. Florey, op. cit.

tissue in which it is situated. Highly active tissues have a high degree of vascularity—i.e. the capillary bed is dense. In relatively inert tissues, the capillaries are sparse and scattered. This relation is well illustrated in the central nervous system, for here the capillary density tends to parallel the density of interconnexions between nerve cells and fibres (see p. 408). It is important to note that the 'vascularity' of any tissue can only be accurately determined by an estimation of capillary density based on an actual count of capillary vessels in a given volume of the tissue. Estimates based on the general calibre of vessels of supply in the immediate proximity may be very misleading.

The contractility of capillaries. We have described a capillary blood vessel as consisting of an endothelial lining covered by a perithelial layer of delicate connective tissue. On this definition, a small blood vessel containing in its wall muscle fibres or compact collagenous tissue must be distinguished as a minute artery (*arteriole*) or vein (*venule*). Failure to make such a terminological distinction has sometimes led to some confusion in recording the reactions of capillaries.

There is good evidence that in lower vertebrates true capillaries are contractile, i.e. they can alter their calibre by spontaneous contraction. This has been observed in the transparent tails of amphibian larvae and also in the area vasculosa of chick embryos, and there seems little doubt that such contraction can occur without the presence of muscle fibres in the vessel wall, and independently of extrinsic circulatory factors. Hence it is necessary to assume in this instance that the endothelial cells have themselves the property of contractility. In mammals, however, it is still a matter of controversy whether such is the case. Prolonged observations by Clark and Clark of blood vessels in the rabbit's ear led them to the conclusion that if there is any active contractility of true capillaries in this animal, it is so slight as to be negligible as a factor controlling the capillary circulation.[1] The technique which these workers used for their studies has already been briefly described (p. 6). It permits the examination of the smallest blood vessels under high magnification, so that true capillaries can be distinguished from small arterioles in which muscular tissue is present.

The results of studies with the transparent-chamber technique suggest that previous records attributing active contractility to capillaries in mammals have perhaps been based on observations on blood vessels which were really minute arterioles and not true capillaries. In this connexion, it should be noted that size is not by itself a distinctive criterion, since terminal arterioles may be even smaller in calibre than the capillaries which they feed. Further, in the transition from arterioles to capillaries the muscular coat of the former does not always terminate abruptly and, at their junction, scattered muscle fibres of atypical form may be present which are not to be detected under low magnification.

[1] Among the numerous papers by E. R. Clark and E. L. Clark and their collaborators dealing with the reactions of small blood vessels, the following may be consulted in this connexion: 'The ingrowth of new blood vessels into standardized chambers in the rabbit's ear', *Anat. Rec.* **50**, 1931; 'Observations on living preformed blood vessels in the rabbit's ear', *Amer. Journ. Anat.* **49**, 1932; 'Changes in blood vascular endothelium', ibid. **57**, 1935.

Even in the walls of true capillaries certain cells have been described which are supposed to control the calibre of these vessels by their contractile properties. These are the *Rouget cells*, first described by their discoverer in 1873. They are closely applied to the endothelial wall of capillaries and they have fine branching processes which wrap round the vessel in an intricate fashion. In appearance, Rouget cells (or *pericytes*, as they have been called) are quite similar to connective-tissue cells. Rouget concluded from his observations that they are really a kind of smooth muscle cell, for he believed that by the contraction of their processes they are responsible for local constrictions of the capillary vessels. Many years later, similar observations were made by Vimtrup,[1] who reported that, in Amphibia, local contraction of capillaries always started at the site of one of the Rouget cells. However, attempts to corroborate this inference by other investigators have on the whole led to negative results.

Clark and Clark,[2] studying capillary reactions in amphibian larvae (salamander and tadpole), found that contraction of these vessels may occur before any Rouget cells have developed in their walls, and even when they are present there may be no relation between their location and the site of contraction. Moreover, it was sometimes observed that, during active contraction in the immediate neighbourhood of a Rouget cell, the capillary endothelium actually draws away from the latter, leaving a perceptible space between them. This, of course, is contrary to what might be expected if the Rouget cell itself is the immediate agent producing the capillary contraction. A similar observation was made by Florey and Carleton[3] on the mesenteric vessels in mammals, and these authors also noted that Rouget cells were too few and scattered to be credited with the contractile functions which had been attributed to the capillaries. A further complication has been more recently introduced into this problem by the observation of Clark and Clark on the changes which occur in the transformation of a capillary into an arteriole during the development of new blood vessels. The Rouget cells proliferate rapidly and, arranging themselves transversely, actually become transformed into smooth muscle cells. Then, as soon as they are innervated by a vasomotor nerve, they become contractile.[4]

It is not easy to reach a final decision on this question of the contractility of true capillaries in mammals, that is to say, the contractile power of the endothelium as distinct from that of extra-endothelial elements. Experimental evidence has been offered that stimulation of the cervical sympathetic does lead to a localized contraction of capillaries in the rabbit's ear (after a latent period of 15–20 seconds).[5] On the other hand, in a general

---

[1] B. Vimtrup, 'Beiträge zur Anatomie der Capillaren', *Zeitschr. f. deskr. Anat.* **65**, 1922, and **68**, 1923.

[2] E. R. Clark and E. L. Clark, 'The relation of "Rouget" cells to capillary contractility', *Amer. Journ. Anat.* **35**, 1925.

[3] H. W. Florey and H. M. Carleton, 'Rouget cells and their function', *Proc. Roy. Soc.* B, **100**, 1926.

[4] E. R. Clark and E. L. Clark, 'Microscopic observations on the extra-endothelial cells of living mammals', *Amer. Journ. Anat.* **66**, 1940.

[5] A. G. Sanders, R. H. Ebert, and H. W. Florey, 'The mechanism of capillary contraction', *Quart. Journ. Exp. Phys.* **30**, 1940.

review of the problem, based on many years of personal experience, Clark and Clark insist that in mammals the control of the peripheral circulation resides in vessels which have muscular walls and an intact nerve supply.[1] They criticize contradictory accounts of other observers on the ground that no description is recorded of the circulatory conditions in adjacent arterioles and vessels in cases where an apparent contraction of a capillary was noted, and point out that passive changes in calibre may occur secondarily to alteration in the blood flow in the arterioles and venules with which the capillaries communicate, as the result of the elasticity of the capillary endothelium. In acute experiments, also, a further difficulty in the way of direct observation is that the capillary endothelium is easily damaged by manipulation, leading to a swelling of the endothelial cells and a narrowing of the lumen of the vessel which might be mistaken for active contraction of the capillary wall. In summing up the numerous careful observations which are now available in the literature, it is probably fair to say that, if the capillary endothelium of mammals is capable of active contraction, this property is relatively insignificant from the physiological point of view.

2. SINUSOIDS

In some parts of the body, true capillaries are replaced by sinusoids. The latter are characterized by the irregularity of their lumen (which shows uneven dilatations instead of an even calibre), by the absence of perithelial connective tissue (so that the endothelium lies in immediate contact with the cells of the surrounding tissues), by the presence of phagocytic cells of the macrophage type among the true endothelial cells, and (according to some anatomists) by actual deficiencies, or pores, in the endothelial wall. These features allow of a much more intimate relationship between the circulating blood and the tissue to which it is supplied. The dilatations of the sinusoids permit of a slowing up of the circulation, while the phagocytic cells, where they are present, sometimes play the part of scavengers by removing effete corpuscles from the blood and destroying them (see p. 56). Sinusoids in the human body are found (among other places) in the liver, the adrenal gland, and haemopoietic tissue such as red bone marrow. Apart from their structural peculiarities, they arise developmentally in a manner quite different from true capillaries. While the latter are developed as discrete endothelial tubes which subsequently join up with each other to form a plexus, sinusoids are differentiated from what is at first a large venous space filled with blood. In the case of the hepatic sinusoids, for example, this is formed by the termination of the vitelline veins. Into the space, reticulating columns of hepatic cells grow and soon break it up into anastomosing sinusoidal channels. A meshwork of sinusoidal vessels is thus formed by the partitioning of an originally single blood cavity.

[1] E. R. Clark and E. L. Clark, 'Calibre changes in minute blood vessels', *Amer. Journ. Anat.* **73**, 1943.

3. ARTERIES

Arteries are essentially conducting channels through which blood is conveyed from the heart to the capillary bed. Starting with the aorta and pulmonary artery (which in the human adult are a little over an inch in diameter), they become progressively smaller as they branch in their peripheral course. With their continuous ramifications, the sum of the diameters of the lumen of all the branches also increases progressively. In correlation with this it follows that the rate of flow of blood through the peripheral vessels steadily diminishes as they get farther away from the

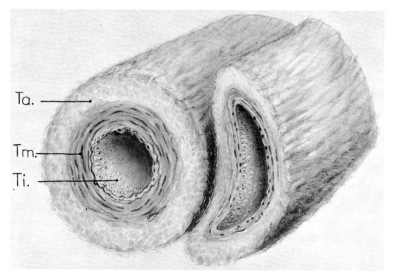

FIG. 70. A diagram constructed to show schematically the contrast between the cross-sectional appearance of an artery and its corresponding vein. The relative thickness of the arterial wall has been exaggerated in order to emphasize this contrast. *Ta.* Tunica adventitia. *Tm.* Tunica media. *Ti.* Tunica intima.

heart. Thus, the total cross-sectional area of capillary vessels in the body has been estimated to be approximately 800 times as large as the cross-sectional area of the aorta at its commencement. In relation to this, it may be noted that the velocity of the circulation diminishes from about half a metre a second in the aorta to half a millimetre in capillary vessels.

In structure, an artery consists of a tube lined by endothelium, with a wall of variable thickness which is composed of elastic and muscle fibres intermingled with white fibrous tissue. In the endothelial coat or *tunica intima* the endothelial cells are disposed longitudinally (as elsewhere in the vascular system generally). They rest on a thin sub-endothelial layer of elastic tissue which forms a fairly definite membrane (*lamina elastica interna*). The fibro-muscular coat can usually be divided into a relatively thick *tunica media* in which muscular or elastic tissue predominates, and a *tunica adventitia* or *tunica externa* composed mainly of collagenous fibres. The tunica adventitia grades without any sharp demarcation into the general connective tissue in which the blood vessel lies. Possibly it may play a minor part in preventing over-distension of an artery.

The structure of the arterial wall, however, varies to a considerable extent in relation to the size of the vessel, particularly with regard to the tunica media. In the large arteries elastic tissue predominates; in the smaller vessels muscular tissue forms the greater part of its thickness. This structural difference is related to functional requirements. The larger arteries are exposed to the force of the systolic contraction of the heart, as

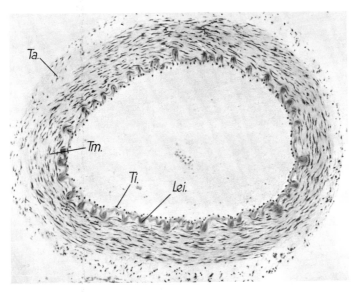

FIG. 71. Microphotograph of a section through a medium-sized artery (from the surface of the brain). Internally are seen the endothelial nuclei of the tunica intima (*Ti.*) and immediately outside this a well-developed lamina elastica interna (*Lei.*) The latter is thrown into folds by the partial contraction of the vessel. The tunica media (*Tm.*) is composed almost entirely of plain muscle. The tunica adventitia (*Ta.*) is relatively thin.
× 90.

the result of which they show a pulsation. By virtue of the abundant elastic tissue in their walls, they undergo expansion with each systole, and during diastole their recoil continues to drive the blood on peripherally. It follows that as the vessels get smaller the blood flow is gradually converted from an intermittent series of propulsions related to the rhythmic contraction of the heart to a steady and continuous stream. When the capillary vessels are finally reached, no pulsation is evident under normal conditions. In certain pathological conditions, e.g. aortic valvular disease, and possibly, also, with extreme peripheral vasodilatation, pulsation may be transmitted to the capillary vessels.

The amount of elastic tissue in the walls of an artery approximately corresponds to the pressure of blood within it. In the smaller arteries, the presence of muscular tissue in their walls enables them, under the influence of the autonomic nervous system, to contract or dilate and so regulate the distribution of blood to the areas which they supply. Unlike the larger arteries, therefore, they are more than merely passive conducting tubes.

It will be realized that the structure of the arterial wall changes quite gradually as the vessels become smaller. For purposes of description, however, it is customary to recognize three main categories, large arteries, medium arteries, and small arteries or arterioles.

In large arteries the tunica media is almost entirely composed of elastic tissue in the form of fenestrated membranes arranged concentrically. Scattered muscle fibres may, however, be found among the inner layers.

FIG. 72. An arteriole and venule from the subcutaneous tissue. The nuclei of the circular muscle fibres are seen disposed transversely on the wall of the arteriole, and deep to them the elongated nuclei of the endothelium disposed longitudinally. The endothelial nuclei of the venule are more rounded, and they are also more conspicuous since they are only covered by a thin layer of collagenous tissue (which is not stained in this section). On either side of the arteriole is seen a fine strand of vasomotor nerve fibres. × 220.

The sub-endothelial layer of elastic tissue is rather thick, and the tunica adventitia, which also contains elastic fibres, is not sharply distinguished from the tunica media.

In medium-sized arteries, muscular tissue forms a conspicuous element of the tunica media, predominating more and more as the vessels become smaller (Fig. 71). The plain muscle fibres are disposed circularly for the most part, but in vessels which run a tortuous course and are liable to be displaced with movements some of the fibres also run longitudinally. The circular disposition of muscle fibres has been shown in many instances to be really a close spiral; indeed, it seems probable that this is the common arrangement.[1] There is some evidence, also, that on the right side of the body the spiral usually runs in a clockwise direction, and the reverse on the left side.[2] In the larger of the medium-sized arteries, elastic fibres are

[1] K. C. Strong, 'A study of the structure of the media of the distributing arteries', *Anat. Rec.* **72**, 1938.

[2] B. S. Schultz, 'Über die schraubenförmige Struktur der Arterienwand', *Morph. Jahrb.* **83**, 1939.

still very evident in the tunica media. While these are also mostly arranged concentrically, some are disposed in a radial direction, and it is probable that these radial elastic fibres are partly responsible for producing dilatation of the vessel when the circular muscle fibres are relaxed. The adventitial coat of medium arteries is relatively thick and is composed of an inter-mixture of elastic and white fibrous tissue. The former may be sufficiently condensed to deserve recognition as a *lamina elastica externa*.

The smallest arteries—less than half a millimetre or so in diameter— are called *arterioles* (Fig. 72). In these vessels the tunica media is entirely composed of plain muscle. They lead on by repeated ramification to capillary vessels, and the smallest arterioles may be no larger in diameter than many capillaries. The change from a terminal arteriole to a true capillary vessel is marked by the disappearance of muscle fibres in the wall of the vessel— this disappearance sometimes being gradual and sometimes fairly abrupt.

Arterioles are functionally of the greatest importance in regulating the flow of blood through the capillary bed by the contraction and relaxa-tion of their muscular wall. This activity, which is controlled through the vasomotor system of nerves, allows a precise adjustment of local regions of the circulation in response to the changing requirements of the tissues. Apart from the regulation of blood flow by intermittent contraction and relaxation, it has been shown by the transparent-chamber technique that arterioles in the rabbit's ear normally undergo spontaneous rhythmic con-tractions at the rate of one to four a minute, the rhythm varying from one vessel to another and even in different parts of the same vessel. These rhythmic contractions evidently play a part in promoting the circulation through the capillary bed. They cease during anaesthesia and also during natural sleep, and it has been shown that they are initiated by impulses which reach the vessels from the sympathetic nervous system, for, in the growth of newly formed arterioles, spontaneous contractions are not evident until they become innervated by the extension of growing nerve fibres.[1]

In the erectile tissue of the penis, the terminal arteries show certain specialized features. The tunica intima is thickened in places to form ridges which partially occlude the vessels when they are empty. It is suggested that this allows a rapid dilatation when erection occurs. The arteries run a tortuous course (for which reason they are sometimes termed *helicine arteries*) and, instead of passing on into a capillary bed, they open into cavernous blood spaces. When the latter become distended they compress the veins near the surface of the penis and produce an engorgement of the whole organ; this is one of the factors concerned in the phenomenon of erection.

Main arterial paths. A general survey of the arterial system of the body makes it clear that, in general, arteries pursue the shortest and the most direct course in order to reach their objective, and that this course is partly ·determined by mechanical convenience. The main arteries in the

[1] E. R. Clark, E. L. Clark, and R. G. Williams, 'The new growth of nerves and the establishment of nerve-controlled contractions of newly formed arterioles', *Amer. Journ. Anat.* **55**, 1934.

limbs run along the flexor surfaces where they are less likely to be exposed to tension in movements of the adjacent joints. They avoid passing through actual muscular tissue which would compress them during active contraction—where they penetrate a muscle they pass rather through tendinous arches which protect them from this pressure. Arteries which are required to adapt themselves to varying positions, in association with movements of the structures which they supply, usually pursue a characteristically tortuous course so that they are to some degree extensible—e.g. the facial artery which runs along the mobile facial muscles, and the splenic artery which must adapt itself to the varying forms of the stomach behind which it passes to reach the spleen. In many cases, however, the course which an artery follows makes it evident that this is determined by morphological factors as well as by mechanical considerations.

The angle at which branches leave a main arterial stem certainly depends to a considerable extent on haemodynamic factors.[1] It is possible to calculate theoretically what the angle of branching should be in order to allow for the least possible loss of energy in the circulation, so that the fall of pressure between a point in the main vessel and a point in its branch is of minimal quantity. It follows from such calculations that branches of equal size should make equal angles with the same main stem, that the larger the branch the smaller the angle and the more the parent artery is deflected at the point of its origin, and that branches which are so small that they hardly affect the size of the parent stem should come off at a relatively large angle—usually between 70° and 90°. Actual measurements show that the angle of branching conforms in a general way to mathematical expectation, but by no means always so. The 'ideal' angle of bifurcation of the common iliac arteries, for example, should be 75°—but it is found to vary from 60° to 75°. Again, the intercostal branches of the thoracic aorta are all of a comparable size, and yet the angle at which they arise varies considerably. In this case differential growth of the aorta has led to a deflection of the upper intercostal arteries.

Arterial anastomoses. In many parts of the body the terminal distributing arteries in one area are connected quite freely with similar vessels in an adjacent area. These anastomotic connexions permit of an equalization of pressure over the vascular territories which are linked up in this way, and they also provide alternative channels for the blood supply of any particular area. In the limbs, arterial anastomoses are particularly free in the neighbourhood of joints, where movements are liable to interfere momentarily with free circulation along any one vascular channel. At the base of the brain the anastomosis between the main arteries of supply is so free as to form a complete arterial circle—the *circle of Willis*—which serves to equalize the distribution of blood to all parts of the brain.

It is of particular interest to note that a free anastomosis between arteries of large calibre does not necessarily mean that, under normal circumstances, the blood stream follows all the channels which appear to be open to it.

---

[1] D'Arcy W. Thompson, *Growth and Form*, Cambridge, 1942; C. D. Murray, 'The physiological principle of minimum work applied to the angle of branching of arteries', *Journ. Gen. Phys.* **9**, 1925.

Thus, the main components of the circle of Willis are normally each confined in their supply to a particular territory of the brain even though they
are in free communication with each other by wide arterial anastomoses. At
the base of the brain the two large vertebral arteries unite to form a single
median vessel, the basilar artery. Yet it has been observed that the blood
entering this main channel from one vertebral artery continues in a stream
of its own without intermingling with that from the opposite side, and it
fills only the ipsilateral branches of the basilar artery.[1] However, if the
normal balance of pressure is altered, the circle of Willis becomes a *functioning* series of anastomoses and the blood distribution to the brain then
changes. It is of importance to note that, as with other anastomotic interconnexions in the body, those of the circle of Willis show a considerable
degree of individual variation, so that their functional efficiency may differ
widely in different individuals.

From the clinical point of view arterial anastomoses are highly important,
since they provide a basis for the establishment of a collateral circulation
in the event of the main supply channel being interrupted by accident or
disease. They also require to be taken into consideration by the surgeon
when the ligature of an artery is contemplated. Where the anastomosis is
free, as, for example, in the mesenteric vessels supplying most of the
intestinal canal, a vessel may be ligatured with impunity, for in this case
the blood supply to the particular area involved can be derived from
adjacent vessels through anastomotic connexions. In some regions, however, the anastomosis of a distributing artery may be poor or variable, and
if such a vessel were ligatured the vitality of the tissue which it supplies
would be seriously endangered. Another practical point in connexion with
arterial anastomoses is that a small channel of this type may, as an anomaly,
be enlarged, while the normal main stem is reduced or absent. In this case
the main blood supply is found to be derived from an 'aberrant' vessel,
and the possible existence of such aberrant vessels sometimes requires to
be taken into account by the operating surgeon.

In certain organs and tissues the distributing arteries have normally no
direct anastomotic connexions with neighbouring vessels. Such arteries
are termed *end arteries*, and it will be apparent that, if such an artery is
interrupted, its territory of supply will be completely and permanently
cut off from the circulation. One of the best examples of an end artery is
the central artery of the retina. A blocking of this vessel by a thrombosis
or embolus leads to blindness in the eye involved, since there is no alternative route through which the retina can get an adequate blood supply.

Recent studies of peripheral blood vessels have shown that true end
arteries are not so common as had been supposed. In the substance of the
brain, for instance, the small arteries in the grey and white matter anastomose quite freely through the capillary bed, and also to a limited extent
even by direct connexions. The anastomoses, however, are not sufficient to
permit of the establishment of a collateral circulation if the main vessel
supplying a particular area of brain tissue is interrupted, so that, from the

[1] D. A. McDonald and J. M. Potter, 'Direct observation of stream lines in the basilar
artery', *Journ. Physiol.* **109**, 1949.

practical point of view, such arteries are functional 'end arteries', even if such a description is incorrect anatomically. In other words, if a vessel entering the brain is blocked, the area which it supplies becomes for practical purposes completely devascularized and eventually undergoes necrosis. This also applies to many other structures (such as the spleen and lung) in which the supplying vessels have been described as 'end arteries'. In muscles, the terminal branches of the arteries of supply actually appear, in injected specimens, to anastomose with comparative freedom; yet the obliteration of one branch may temporarily restrict the blood supply to its own territory to such an extent as to lead to a local necrosis.

Constancy of vascular supply. The relative constancy of arterial supply is a question which has occupied the attention of morphologists from time to time. As we have noted, the vascular supply of any organ or tissue is determined partly by morphogenetic factors and partly by mechanical convenience. In so far as the latter predominates in the production of any vascular pattern, the main channels of blood supply to a particular morphological element may be expected to vary from one species of animal to another in accordance with variations in its relative size and position. This, in fact, does occur to a certain extent, so that attempts to homologize structures in different animals by reference to their blood supply are open to fallacy. It is known that in the evolutionary development of the brain, for example, homologous vascular territories, supplied primarily from one source, may tap another source so that they come to be supplied in different species from different parent stems.[1] Nevertheless, comparative anatomy provides many illustrations of the constancy of vascular supply, and of the persistence of vascular channels even though the course which the latter come to follow may appear mechanically disadvantageous.

The suggestion is sometimes made that the distribution of the branches of a main artery is determined by functional considerations as the result of which structures taking part in a common activity tend to be vascularized by branches of the same arterial stem. In so far as this principle obtains, it is no doubt related to the fact that such functionally connected structures inevitably make simultaneous demands on the circulation, and it may therefore be a matter of convenience that they should be supplied from the same source. Examples which have been cited to illustrate such a functional distribution of arteries are the internal maxillary artery which supplies the jaws, teeth, and masticatory muscles and has therefore been called the 'artery of mastication', the ascending pharyngeal artery which supplies the palate and the pharynx and has been termed the 'artery of deglutition', and the posterior cerebral artery or 'artery of vision' which supplies the visual cortex, the lower visual centres in the thalamus and mid-brain, and part of the optic tracts. However, it is not possible to apply very closely this conception of arterial distribution, intriguing though it appears to be on first acquaintance. Such examples as seem to illustrate it

[1] Even if the origin of the main stem of vascular supply to any particular region varies in different species, the peripheral branches may still be homologous in the sense that they are derived from an homologous part of the primordial capillary network. This is indeed indicated by the fact that the vasomotor nerves which supply them tend to preserve a segmental distribution.

are probably to be explained mainly on the basis of the natural proximity of structures which are functionally associated with each other.

4. VEINS

Compared with their corresponding arteries, veins have a considerably larger lumen and much thinner walls (Fig. 70). The larger lumen is related to the fact that the venous system is required to transmit in a given time the same volume of blood as the arterial system, although the rate of flow in veins is much slower. The thinner walls, which are associated with the relatively low venous pressure, render veins easily compressible, and this, as we shall see, is a factor of some importance in promoting the venous circulation.

In structure, the wall of a vein is somewhat similar to that of an artery. Apart from the fact that it is much thinner, however, it shows a strong contrast in the poverty of muscular and elastic tissue (the place of which is to a great extent taken by white fibrous tissue), and in the frequent lack of any clear distinction between the tunica media and the tunica adventitia.

The tunica intima comprises the endothelial lining in which the individual cells tend to be less regularly disposed than in arteries. Some elastic fibres may be found in the sub-endothelial tissue, but they do not form a conspicuous elastic membrane. The composition of the tunica media and tunica adventitia varies very considerably from one vein to another. Circular muscle fibres are commonly found in small amounts in medium and large-sized veins, mingled freely with collagenous fibres. In some vessels, e.g. the pulmonary veins, the deep veins of the penis, and the uterine veins during pregnancy, this muscular tissue is fairly conspicuous. In others it is completely absent—these include veins in the brain and spinal cord, the intracranial venous sinuses of the dura mater, the veins of the retina, and the veins of bone marrow.

The tunica adventitia usually consists mainly of white fibrous tissue. In large veins it may be very thick relatively to the other coats, and it may contain elastic tissue and longitudinally arranged muscle fibres. These variations in the composition of the tunica adventitia undoubtedly represent functional adaptations, though their relation to the latter is not always clear. The collagenous fibres in the wall of the inferior vena cava and certain other veins are arranged as a regular criss-cross spiral network, a disposition which is apparently related to the need for functional changes in length. According to Franklin,[1] the presence of a well-developed longitudinal coat of elastic and muscle fibres in a vein such as the inferior vena cava not only allows of changes in length (associated, for example, with movements of the diaphragm), but also protects the vessel against external pressure. He notes, however, that in many animals (e.g. dog, cat, and rabbit) the inferior vena cava contains practically no muscular tissue in its wall. Elastic tissue will clearly allow of a rapid return to normal shape when a vein has undergone deformation by external forces. Where a longitudinal

[1] K. J. Franklin, *A Monograph on Veins*, Baltimore, 1937.

muscle coat is present, its fibres loop round the openings of tributaries of
the main stem in a characteristic way, and there is some evidence to show
that by their contraction and relaxation they can constrict or dilate the
openings and so regulate the volume of blood which is being returned by
the tributary vessels. Such a mechanism has been described in the hepatic
veins where they join the inferior vena cava. Another type of sluice-valve
mechanism occurs in the veins of the penis, where very thin-walled tributaries drain into veins with thick muscular walls. By their contraction the
latter can throttle back the venous return and so produce the turgidity
necessary for erection.[1]

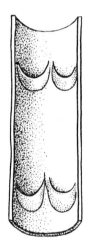

FIG. 73. Diagram re-
presenting a short
length of vein which
has been opened to
show the disposition
of the venous valves.

In aquatic mammals, which are accustomed to
diving for prolonged intervals, the venous return to
the heart is controlled by special muscular mechanisms. In the porpoise and seal, for example, the inferior vena cava is enclosed by a definite sphincter
of striated muscle derived from the diaphragm and
innervated by the phrenic nerve. Contraction of
this sphincter controls the entry of blood into the
right atrium of the heart during submersion, and
the venous blood is temporarily accommodated in
large suprahepatic venous sinuses situated between
the liver and the diaphragm. Further sphincteric
mechanisms are provided in these animals by local
thickenings of the circular plain muscle in the walls
of the hepatic veins.

The ultimate venous radicles (or venules) differ
from true capillaries by the fact that their wall is
supported by a thin but definite layer of condensed
connective tissue. It is of some importance to realize
that, in spite of this, the walls of venules possess a
degree of permeability which was commonly supposed
to be limited to the capillary endothelium.[2] Experimental observations have suggested that (in skeletal
muscles) there is a gradient of permeability in capillaries, increasing to a
maximum at the junction with venules, and in some tissues the permeability of the venules may greatly exceed that of the capillaries. In other
words, diffusible substances in the blood can pass directly from venules
into the surrounding tissues, and it is possible also for absorption to take
place in the opposite direction.

Valves. Most veins are characterized by the presence of valves
arranged at intervals along their course, and so disposed that they
direct the blood flow proximally. They are formed by a reduplication of
the lining endothelium, between the layers of which is an extremely thin
stratum of fibrous and elastic tissue. The endothelial cells on the surface

    [1] L. J. Deysach, 'The comparative morphology of the erectile tissue of the penis',
Amer. Journ. Anat. **64,** 1939.
    [2] P. Rous, H. P. Gilding, and F. Smith, 'The gradient of vascular permeability', *Journ.
Exp. Med.* **51,** 1930.

towards the lumen are disposed longitudinally (as elsewhere in blood vessels), but on the opposite surface their long axes are transverse. The cusps of the valves are crescentic in shape and usually arranged in pairs, though (particularly in the large veins of big mammals) there may be three or four in one group.

It is self-evident that valves serve the function of preventing a back flow of blood in a distal direction. This is apt to occur at the junction of tributaries with a main stem, especially since the pressure in a small vein is sometimes lower than in a larger vessel. Consequently, valves are most constantly found in main vessels immediately distal to the entry of a tributary, and also in the tributary itself close to its termination.[1] Many veins, however, are completely devoid of valves, e.g. cerebral veins, and veins of bone marrow. This seems to be related to the fact that such vessels are not exposed to intermittent and local compression by surrounding structures, for, as emphasized by Kampmeier and Birch,[2] valves are in general confined to veins subject to external pressure or liable to be influenced by muscular movements, e.g. in the limbs, or in mobile viscera.

Venous valves first appear in the human foetus at about $3\frac{1}{2}$ months; by $5\frac{1}{2}$ months the full quota of valves has probably been acquired. There is some evidence that their number may diminish after birth and with advancing age, but owing to the individual variability in the number and distribution of valves in different veins, many more observations are required in order to establish this with certainty.

Venous paths. Veins very commonly run with arteries. Most medium arteries (e.g. of the size of the brachial artery or smaller) are accompanied by two veins (*venae comites*), one on either side, which are closely bound down to the artery by fibrous tissue and connected with each other at frequent intervals by cross-connexions. Larger arteries are usually each accompanied by a single large vein, and, in this case, the vessels are closely bound together by a condensation of connective tissue which may sometimes be defined by dissection as a sort of tubular sheath. Undoubtedly such vascular sheaths are often described and figured in systematic textbooks of anatomy with a precision which is not to be found in nature—nevertheless they do provide a relatively resistant covering to the vessels which may be of some importance in the venous circulation (*vide infra*, p. 202).

The close topographical relationship between arterial and venous paths appears to be a secondary arrangement, for in the early stages of vascular development they do not usually follow a common route to and from their territory of supply. This is particularly well emphasized in the limbs, where, as we shall see, the main artery is primarily axial in position, while the blood is returned by superficial marginal veins. It may be presumed that, with further differentiation of the limbs, it becomes mechanically

---

[1] The position of valves can often be recognized in the course of dissection by a local bulging of the vein, which gives rise to a knotted or beaded appearance of the vessel as a whole.

[2] O. F. Kampmeier and C. L. F. Birch, 'The origin and development of venous valves', *Amer. Journ. Anat.* **38**, 1927.

disadvantageous for the venous drainage to take place entirely by the super-ficial vessels, and alternative deep routes are developed along the same fascial planes already used by the arteries. Although these new channels become relatively large, at least in the proximal part of the limbs, the more primitive superficial veins still persist as quite important alternative routes for the venous return. In some parts of the body there is no close associa-tion between arteries and veins. This is the case, for example, in the brain, where quite different routes are followed by the afferent and efferent blood supply of many areas. Here, the primary vascular arrangement is not disturbed by mechanical influences of muscular movement and local external pressure such as occur in the limbs.

The angle at which a venous tributary enters a main stem is to some extent related to haemodynamic factors. The relation, however, is less consistent than is the case with the branching of arteries because the venous blood pressure is so much lower; in some instances, indeed, tributaries enter their parent stems at what seems to be a disadvantageous angle. Thus the left testicular vein joins the left renal vein at approximately a right angle, and it has been suggested that the resistance which the venous return encounters at this abrupt junction may be partly responsible for the common occurrence of a left-sided varicocele.

In the brain, the superior cerebral veins enter the sagittal venous sinus in a direction opposite to that of the blood flow in the latter. It has been suggested that this is a mechanism which provides for a raising of the pressure in the peripheral venous tributaries, and so guards against the collapse of the thin-walled and easily compressible cerebral veins as the result of any temporary rise of intracranial pressure. There is some evi-dence, however, to suggest that this unusual venous pattern is the result of displacement of the tributaries during growth of the cerebral hemispheres.[1]

In general, venous paths in the adult are much more variable than arterial paths. This is undoubtedly related to the fact that haemodynamic factors play a less important part in their differentiation from the primor-dial diffuse vascular network of the early embryo (see p. 208). On the venous side of the circulation, the partial persistence of this network is shown by the frequent presence of cross-anastomoses between one vein and another.

Portal systems. In their course proximally, certain veins may break up again into a plexus of capillaries or sinusoids instead of continuing directly into main venous trunks. Such an arrangement is termed a *portal system*, and it provides a means whereby the products of absorption from a venous drainage area can be immediately transferred to the tissues of a neighbouring organ, without the necessity of passing through the heart and lungs and being distributed widely over the body by the arterial system. The portal circulation of the liver is the main example of this vascular mechanism.

From all parts of the alimentary tract, including organs such as the pancreas and spleen, blood is conveyed to the liver by the portal vein. Here the vein breaks up into sinusoidal vessels so that the blood is brougth

[1] J. E. A. O'Connell, 'Some observations on the cerebral veins', *Brain*, **57**, 1934.

into the most intimate relation with hepatic cells, allowing of the absorption of certain products of digestion. Once more, the blood is collected up by venous tributaries which transfer it along the hepatic veins to the inferior vena cava, and so to the heart.

In lower vertebrates (e.g. Amphibia) in which the kidney is derived embryologically from the mesonephros, there is a renal portal circulation. This has completely disappeared in mammals, except for a transitory appearance which it may show during embryonic development.[1]

Anatomical factors concerned in the peripheral venous circulation. The flow of blood along veins is primarily dependent on a pressure gradient, the pressure diminishing progressively from the periphery to the heart. This is the result of pressure changes related to the cardiac cycle, and also of the negative pressure in the thorax produced by respiratory action. Auxiliary mechanisms may be brought into play to assist further the venous circulation in peripheral vessels. Of these, the most important is muscular action.

Wherever veins run in immediate relation to muscles, they are liable to compression during muscular contraction. This is particularly the case where the tissues are enclosed in a resistant sheath of deep fascia which holds down contracting muscles under considerable tension (see p. 61). Since the veins are here liberally supplied with valves, their compression will serve to drive the blood in a proximal direction. It will be realized that this mechanism is of great importance in promoting the venous return from the lower extremity where, in the standing position, the circulation has to overcome the effects of gravity. Further, it will obviously be most effective in the case of intermittent muscular contraction. Prolonged or sustained contraction, indeed, may have the opposite effect by obstructing the venous return along the deep veins. In this case, the blood is diverted into the superficial veins by the numerous connexions which link up the superficial and deep vessels through the deep fascia, and the superficial veins may then be loaded beyond their normal capacity. It follows that pathological dilatation or varicosity of superficial veins may be associated with excessive muscular exercise. It is also the case that deficient muscular activity or poor muscle tone may predispose to varicosities in peripheral veins in the lower extremity since, under these conditions, muscle action does not play its normal role in promoting the venous circulation.

There is an evident relation between yawning and the venous circulation. Where the muscular system is entirely relaxed, as in sleep, the venous return no longer receives the aid of muscular activity. The act of yawning (accompanied by a stretching of the limbs) accelerates the return, partly by the deep inspiration which momentarily increases the suction action of the thorax while at the same time raising the intra-abdominal pressure, and partly by the vigorous contraction of muscles all over the body which helps to empty the peripheral veins. In this connexion it is interesting to note that animals which do not lie down to sleep, or do not completely relax

[1] In frogs, the presence of this portal system has been of some practical importance experimentally since it has provided a method of investigating the functions of the renal tubules.

their musculature during rest, do not stretch or yawn when they awake.

Another factor which must be taken into account in considering the blood flow in peripheral veins is the existence in many parts of the body of arteriovenous anastomoses which provide an alternative short-circuiting route linking up arterioles with venules (*vide infra*, p. 203). When these anastomotic connexions are opened by relaxation of their muscular walls they permit of a direct flow from the arterial to the venous system, with a corresponding rise of pressure in the peripheral venules. Under these conditions pulsation may even be transmitted to the veins.

The question now arises whether peripheral veins are capable of rhythmic contraction by virtue of their own musculature, and, if so, whether this may provide an additional impetus to the venous return of blood. It is well established that the plain muscle in the walls of veins reacts to mechanical stimuli and to stimulation of the sympathetic nervous system.[1] Observations have also been recorded of pulsatile veins, of which the most quoted examples are those of the bat's wing. In this case, however, the venous pulsation is evidently the secondary effect of numerous arteriovenous anastomoses which are present here in the peripheral circulation, and this is probably also the explanation of the 'inherent pulsation' which has been described in venules elsewhere. There is, in fact, no conclusive evidence that the veins can accelerate the blood flow within them by their own intrinsic activity.

The immediate proximity of many veins and arteries has suggested the possibility that arterial pulsation may directly influence the venous flow by a mechanical compression of the accompanying veins. That such a rhythmic compression does occur has been demonstrated with certainty, though it is difficult to assess its relative importance. In the case of *venae comites* which are firmly bound down to the artery by numerous cross-connexions, so that, in effect, they form a venous plexus closely ensheathing it, and in the case of main arterial and venous channels which are enclosed in a common vascular sheath of resistant and inelastic fascia, it may be inferred on anatomical grounds that the pulsation of the artery must have a not inconsiderable effect in driving the blood along the accompanying veins. This is certainly the case where, as occurs in certain animals, a peripheral artery breaks up into a plexiform leash of small channels to form a so-called *rete mirabile* intimately related to accompanying veins (see p. 218).

5. VASA VASORUM

In the case of all medium-sized and large arteries and veins, the walls of the vessels require their own vascular supply to maintain their nutrition. This is provided by small arteries which are derived either from branches of the main stem or from neighbouring arteries. They are termed *vasa vasorum*, and in microscopic sections they can be seen to break up into a capillary plexus in the deeper layers of the tunica adventitia. They probably

---

[1] For a discussion on the relation of veins to the nervous system, see K. J. Franklin, *A Monograph on Veins*, Baltimore, 1937.

do not usually penetrate more deeply than the middle of the tunica media, though in veins they are said sometimes to extend almost to the tunica intima. In both arteries and veins, however, it appears that the nutrition of the tunica intima and the deeper layers of the tunica media is maintained (at least partly) by diffusion from the lumen of the main vessel through the lining endothelium.

From time to time it has been suggested that certain degenerative changes which may affect the walls of arteries (e.g. those associated with senility and arteriosclerosis) are the result of interference with the normal blood supply from the vasa vasorum. However, while this is probably an important factor in pathological affections of the vascular system related to certain specific infections, there is no good evidence that it is applicable to arteriosclerotic changes in general. It has also been postulated that constriction of the vasa vasorum as the result of abnormal activity of the vaso-constrictor fibres of the sympathetic system can secondarily lead to degeneration of the muscle tissue in arterial walls, and its replacement by fibrous tissue. But, again, while the anatomical basis for such a mechanism certainly exists, it remains doubtful to what extent it actually contributes to these vascular changes.

It may be noted that lymphatic capillaries are present in the tunica adventitia and tunica media of all larger blood vessels. These drain into perivascular lymphatic plexuses.

6. ARTERIOVENOUS ANASTOMOSES

The widespread existence of minute anastomoses directly connecting arterioles with venules is now well recognized. The once common conception that the peripheral circulation can only be completed by the percolation everywhere of the blood through the capillary bed is thus not entirely correct, for in many tissues there are abundant arteriovenous anastomoses which can effect a short-circuiting of the blood from arterioles to venules under certain conditions. The importance of such a mechanism for regulating the supply of blood to these tissues can hardly be over-estimated.

In studies of the circulation in transparent chambers in rabbits' ears, the characteristics of arteriovenous anastomoses have been described in some detail.[1] They arise as side branches from the terminal arterioles and after a very short course run directly into a small venule. The muscular wall of these anastomotic cross-connexions is remarkably thick, and is particularly well supplied with vasomotor nerves. Observation on the living vessels has shown that they contract vigorously on stimulation of the sympathetic nerves which supply them, and in the normal animal they also show a rhythmic contractility similar to that of arterioles but usually with a higher frequency. The thick muscular coat of the anastomoses is to be regarded functionally as a sphincter. When it contracts, the blood is

[1] E. R. Clark and E. L. Clark, 'Living arteriovenous anastomoses as seen in transparent chambers introduced into the rabbit's ear', *Amer. Journ. Anat.* **54**, 1934. See also R. T. Grant, 'Observations on direct communications between arteries and veins in the rabbit's ear', *Heart*, **15**, 1930.

directed along the arteriole into the capillary bed. When it relaxes the blood is 'by-passed' directly into the venule, and the capillary circulation is temporarily diminished or completely shut down (Fig. 74).

Arteriovenous anastomoses may therefore be expected to be a common feature of the circulation in those tissues whose metabolic activity is of an

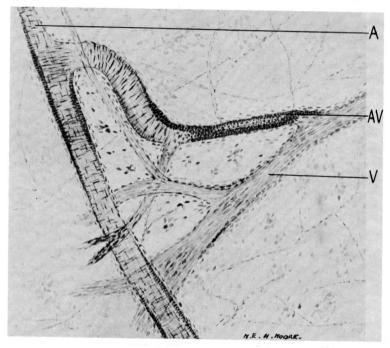

Fig. 74. Drawing of an arteriovenous anastomosis from the subcutaneous tissue of a rabbit's ear. The anastomosis (*AV*) is seen directly connecting an arteriole (*A*) with a venule (*V*). Note the relatively well-developed muscular wall of the anastomotic vessel.
× 300.

intermittent nature. This indeed seems to be the case. They are quite numerous, for example, in the mucous membrane of the gastro-intestinal tract.[1] During active digestion, the anastomotic channels are closed down so that the blood must circulate through the capillary vessels which take part in absorption, and in a fasting period they are relaxed since there is no need at this time for a complete flooding of the capillary bed. It may be noted, incidentally, that the opening of these anastomoses will raise the venous pressure in the portal system.

Abundant arteriovenous anastomoses are also found in the skin. Here their main function is that of temperature regulation by permitting a much more rapid circulation than is possible through the fine network of the capillary bed. There has been some doubt as to the precise significance

[1] T. E. Barlow, 'Vascular patterns in the alimentary canal', *Visceral Circulation*, Ciba Foundation Symposium, London, 1952.

of this effect for temperature regulation. Thus, observations that the anastomoses in the rabbit's ear open up when the animal as a whole is heated
suggested that they help to lower the body temperature by increasing the
blood flow through a large exposed area of skin. On the other hand, the
fact that local cooling also results in a dilatation of the anastomoses makes
it probable that their main function is to maintain the warmth of exposed
tissues, when the ambient temperature is low, by accelerating the blood
flow locally.[1] Arteriovenous anastomoses are especially numerous in
the more exposed parts such as the fingers and toes, nose, lips, and ears.[2]
In the skin of cold-blooded vertebrates, they are not present. As already
noted, they have been found to be particularly numerous in bats' wings,
over the membranous surface of which heat loss must be quite considerable.
It is interesting to note, also, that a plentiful supply of arteriovenous
anastomoses with an extremely rich innervation has been recorded in the
tongue of the dog, and it is suggested that this is related to the method
of eliminating heat from the body by panting, which is characteristic of
these animals.[3] Other regions in which arteriovenous anastomoses have
been described include the thyroid gland (where the function is probably
related to the fluctuating activity of different groups of thyroid follicles),
the nasal mucosa, the carotid and coccygeal bodies, and sympathetic ganglia.

In addition to controlling the local circulation, arteriovenous anastomoses
may assist the venous return of blood. Not only do they raise the pressure
in the peripheral venules by their relaxation, but by their active and
rhythmic contractions they provide a further impetus to the circulation.
Pathological conditions associated with venous stasis, therefore, such as
varicosities, may owe their origin not to any inherent primary defect in the
veins themselves, but to some functional interference with the nervous
mechanism controlling arteriovenous anastomoses.

It has been found that if, in rabbits' ears, the circulation is suddenly
increased by some local stimulus, new arteriovenous anastomoses may be
developed in the course of two or three days in order, apparently, to cope
with the augmented blood flow.[4] When the circulatory conditions have
once more become stabilized, the anastomoses revert in structure to ordinary capillary vessels, or they may retract and disappear entirely. It
appears, therefore, that these anastomoses provide a means whereby the
peripheral circulation can adapt itself quite rapidly to new and changing
conditions.

In the foregoing description, it has been mentioned that the muscular

[1] P. M. Daniel and M. M. L. Prichard, 'Arteriovenous anastomoses in the external ear',
Quart. Journ. Exp. Physiol. **41**, 1956.
[2] The plenitude of arteriovenous anastomoses in the external ear is related to the fact
that large ears do not necessarily indicate a high degree of auditory acuity. The inordinate
size of the ears in some mammalian species (e.g. the desert fox) is evidently associated
with their temperature-regulating functions. It is interesting to note, also, that young
rabbits are said to grow larger ears when exposed to high environmental temperatures,
as a result of which heat loss is presumably facilitated (C. A. Mills, 'Influence of environmental temperatures on warm-blooded animals', *Ann. N.Y. Acad. Sci.* **46**, 1945).
[3] M. E. Brown, 'The occurrence of arteriovenous anastomoses in the tongue of the
dog', *Anat. Rec.* **69**, 1937.
[4] E. R. Clark and E. L. Clark, 'The new formation of arteriovenous anastomoses in the
rabbit's ear', *Amer. Journ. Anat.* **55**, 1934.

wall of arteriovenous anastomoses is well developed. The muscle fibres are in many cases of a modified type, simulating an epithelium. Indeed, two different types of anastomosis have been described, one in which the tunica media contains ordinary smooth musculature, and one in which the muscle cells are epithelioid in character.[1] The latter type is very tortuous in its course, and occurs in closely packed groups. It has been suggested that the epithelioid cells expand and contract in response to chemical stimuli, thereby shutting and opening the anastomotic channel. However, the evidence for such a mechanism is not clear.

Direct arteriovenous communications may occur without the intervention of specialized connecting channels. Such a connexion is found in the placenta, where arterioles open directly into the venous blood spaces in which the villi of the foetal chorion are suspended. By this arrangement the villi are exposed directly to fully oxygenated blood.

7. THE DEVELOPMENT AND GROWTH OF BLOOD VESSELS

All blood vessels arise embryologically as a differentiation of mesenchymal tissue. In lower vertebrates they are first evident in the mesodermal covering of the yolk sac. Here local patches of proliferation give rise to isolated clumps of cells which are termed *angioblasts*. These send out sprouts which, uniting with each other, form a network of interlacing strands or *angioblastic cords*. While this process is going on, the angioblastic cells themselves undergo differentiation. The peripherally situated cells are flattened out as the result of pressure, partly caused by the growth of the central cells, and partly by the accumulation of fluid plasma among the latter.[2] In this way the angioblastic cords become converted into a reticulum of hollow tubes of which the peripheral cells form the lining endothelium (Fig. 75). The central cells remain for a while in clumps adherent to the endothelial lining, forming what are termed 'blood islands', but, as the plasma further increases in quantity, they become detached and set free within the embryonic vessels. They are the precursors of blood corpuscles. There is reason to believe that red corpuscles are developed from angioblastic cells within the lumen of the vessel, while white corpuscles are differentiated extravascularly from the same type of cell and find their way secondarily into the vessel. The white cells, it may be noted, make their appearance later than the red and are not really abundant in the human embryo until the latter is about one month old. Meanwhile the heart and aortic tubes are developed in the intra-embryonic mesoderm and, when these establish a connexion with the vascular network in the wall of the yolk sac, a closed system of vessels is finally completed and the circulation begins.

It was generally supposed by the older embryologists that, with the exception of the aortae, all the blood vessels within the body of the embryo

[1] v. Schumacher, 'Über die Bedeutung der arteriovenösen Anastomosen und der epithelioiden Muskelzellen', *Zeitschr. f. mikrosk.-anat. Forschung.* **43**, 1938.
[2] The plasma is said by some anatomists to be formed at first *intracellularly* by the angioblastic cells.

are formed by direct extensions into the latter from the vitelline plexus. Experimental evidence first threw serious doubt on this conception.[1] It was shown, for instance, that if the area vasculosa of the yolk sac of a developing chick is separated from the body of the embryo at a time when the latter is not yet vascularized, blood vessels appear subsequently in the embryo in the normal manner.

Detailed studies of human vasculogenesis have also demonstrated beyond doubt that vessels make an appearance simultaneously and independently in the wall of the yolk sac, the body stalk, and the chorion; there is some evidence, indeed, that they appear slightly earlier in the chorionic villi than in the body stalk and the yolk sac.[2] It is generally agreed that, as soon as the differentiated vessels have become linked up to form a closed system in the embryo and the circulation has begun, any further extension is brought about by a sprouting of pre-existing capillaries; in other words, after this stage of development vascular endothelium can no longer be formed by the differentiation of mesenchyme cells but only by the proliferation of already differentiated endothelial cells. There are, however, a few exceptions to this general statement.

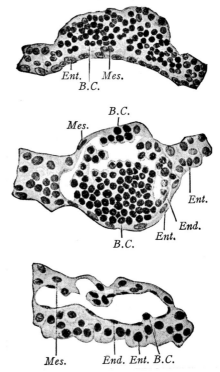

FIG. 75. Sections of the wall of the yolk sac of an early human embryo showing three stages in the development of a blood vessel. *B.C.* Blood cells. *End.* Endothelium. *Ent.* Entoderm. *Mes.* Mesoderm. (From Arey, after Evans.)

Hughes[3] has shown, from his studies on the area vasculosa of the chick embryo, that the peripheral extension of the area, even after the circulation is established, probably takes place to some extent by the conversion of mesenchyme cells into endothelium as well as by cell division within the vessel walls. He

[1] A. M. Miller and J. E. McWhorter, 'Experiments on the development of blood vessels in the area pellucida and embryonic body of the chick', *Anat. Rec.* **8**, 1914; F. P. Reagan, 'Vascularization phenomena in fragments of embryonic bodies completely isolated from yolk sac blastoderm', ibid. **9**, 1915.

[2] J. L. Bremer, 'On the earliest blood vessels in man', *Amer. Journ. Anat.* **16**, 1914. See also A. T. Hertig, 'Angiogenesis in the early human chorion and in the primary placenta of the Macaque monkey', *Contrib. to Embryology*, Carnegie Inst. **146**, 1935. This author found that in a pre-somite monkey embryo vascular primordia were already present in the chorionic villi, and they appear *subsequently* in the body stalk and yolk sac. He also adduces evidence that these primordia, as well as primary mesodermal tissue, are derived directly from trophoblastic epithelium.

[3] A. F. W. Hughes, 'Studies in the area vasculosa of the embryo chick', *Journ. Anat.* **72**, 1937.

found, for example, that while in most of the vessels in this region increase in diameter is brought about by proliferation of the endothelial cells, and is therefore accompanied by a corresponding increase in endothelial nuclei, in rapidly enlarging veins cell division in the endothelium (estimated on the basis of counts of mitotic figures) is not sufficient to account for more than one-fifth of the actual increase in the number of endothelial cells. It is presumed, therefore, that in this case the remaining four-fifths of the endothelial cells are derived from mesenchyme cells lying in contact with the vessels. There is no evidence, however, that new blood vessels can appear or increase their diameter in the *later* stages of embryonic development otherwise than by proliferation of the endothelium of pre-existing vessels. This is quite certainly the case in post-natal life.

The extension of new capillaries by the formation of vascular buds not only provides the basis for the growth of blood vessels during normal development, it is also an essential process in the adult in the adjustment of the peripheral circulation to varying functional demands, and in the vascular reactions to inflammatory stimuli. There is evidence to show that, under normal conditions, new capillaries are continually being formed in this way while others may be retracted and absorbed, so that the capillary pattern in any part of the body is plastic and capable of alteration from time to time in response to changes in the immediate environment.[1]

Vascular buds, it should be noted, can only be formed from capillaries— when a vessel acquires muscular or collagenous tissue in its wall so that it becomes converted into an arteriole or a venule, it loses this property. The formation of endothelial sprouts is preceded by active cell division in the endothelium. The tip of the bud is composed of a solid strand of endothelial protoplasm in which at first the endothelial cells appear to form a syncytium with no definable cell boundaries. As the bud extends, it becomes canalized and the endothelial cells become separately defined. Adjacent buds effect connexions with each other and establish in this way a continuous network. It has been established that in the rabbit's ear new capillary buds grow in length at the rate of 200–600 μ a day. Newly formed capillaries can become converted into larger vessels by the deposition in their walls of muscle cells or collagenous fibres. The latter are derived from fibroblasts, and the former apparently from the fibroblast-like 'cells of Rouget' which lie in contact with the capillary endothelium (see p. 188).

With the possible exception of the dorsal segmental branches of the aortae (which appear to sprout out as separately defined vessels), the principal vascular channels in the embryo, including even the heart and the aortae, are differentiated from a diffuse primordial capillary network in which at first no sign of the future vascular pattern is evident. The manner in which this pattern 'crystallizes' out of the homogeneous plexus has been followed in studies of the development of the limb vessels in the fore-limb of the pig.[2] When the fore-limb bud first appears, it is vascularized by

[1] E. R. Clark and E. L. Clark, 'Observations on living preformed blood vessels in the rabbit's ear', *Amer. Journ. Anat.* **49**, 1932.
[2] H. H. Woollard, 'The development of the principal arterial stems in the fore-limb of the pig', *Contrib. to Embryology*, Carnegie Inst. **14**, 1922.

five segmental arteries from the dorsal aorta (the 5th to the 9th), which contribute equally to a common capillary net. As development proceeds, the seventh segmental artery becomes increasingly dominant, while the remaining segmental arteries dwindle and finally disappear altogether. Along the central axis of the limb bud, i.e. in the direction naturally followed by the circulating blood which now enters it through the seventh segmental artery, a main axial artery begins to differentiate. The component elements of the capillary net along this line undergo enlargement and form a retiform line of channels, which soon stands out as a definite pathway in the midst of the generalized capillary net. This axial strand becomes increasingly isolated from the capillary net by the disappearance of many of its direct connexions with the latter, and it is converted into a single main stem partly by the progressive enlargement of certain component parts of the leash and partly by their mergence with each other. Coincidentally, marginal veins become differentiated at the pre-axial and post-axial borders of the limb bud for the return circulation (Fig. 76).

Three stages may be recognized in the development of main arterial stems, (1) the diffuse capillary net, (2) the retiform stage, and (3) the formation of a definitive single stem. The first stage probably depends on purely morphogenetic factors, being determined by the inherent property of certain cells to form blood vessels and blood cells, while the various patterns which the main vessels ultimately assume in the limbs (i.e. stages 2 and 3) are determined mainly by mechanical (haemodynamic) influences working during ontogeny.

In the early human embryo, the circulatory system is relatively simple and symmetrical in its arrangement. Two ventral aortae (which emerge from the truncus arteriosus of the heart tube) are connected with the dorsal aortae by a series of aortic arches encircling the pharynx on either side. The dorsal aortae are continued back into the umbilical arteries which carry impure blood to the placenta. On their way caudalwards, they give off a series of segmental arteries to the body wall of the embryo, and vitelline branches to the yolk sac. Blood is returned to the heart by paired anterior and posterior cardinal veins which receive segmental tributaries from the body wall of the embryo, by vitelline veins from the yolk sac, and by umbilical veins conveying oxygenated blood from the placenta.

This basic pattern is soon disturbed by a series of changes which appear in some cases to be quite revolutionary in nature. The aortic arch system becomes completely remodelled in adaptation to the differentiation of the neck region characteristic of higher vertebrates and in relation to the development of the pulmonary apparatus; segmental arteries in many cases disappear, or longitudinal intersegmental anastomoses which link them together become enlarged to form main stem arteries such as the vertebral artery and the internal mammary artery which run in a longitudinal direction; the cardinal veins become converted, by obliteration of some parts of their extent and the addition of other venous elements, into an asymmetrical system including an approximately median inferior vena cava and a single superior vena cava; the termination of the umbilical and vitelline veins is invaded by growing masses of liver tissue in the initial formation of the portal venous circulation; and axial arteries in the limbs are in large part replaced by the secondary development of other channels. It will be appreciated that these upheavals affecting the vascular system during ontogeny provide opportunities for the appearance of vascular anomalies owing to the possibility of arrested development at one particular stage in embryological history.

Immediately after birth, further abrupt modifications occur in adaptation to the commencement of pulmonary respiration and digestive functions.

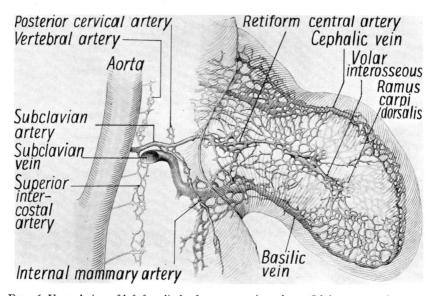

Fig. 76. Ventral view of left fore-limb of a 12-mm. pig embryo. Of the segmental arteries which initially supply the fore-limb, only the stem of the seventh remains to form the subclavian artery. The main vessels in the limb are becoming differentiated from a diffuse capillary network as retiform channels. (From H. H. Woollard.)

The right side of the heart now becomes completely shut off from the left by the obliteration of a foramen (*foramen ovale*) in the interatrial septum; the ductus arteriosus—through which the blood during intra-uterine life is short-circuited from the pulmonary artery into the aorta—becomes obliterated; the umbilical arteries and veins which no longer transmit blood become sealed off as functionless fibrous cords; and the ductus venosus— a short-circuiting channel which allows the blood from the alimentary tract to 'by-pass' the liver—is also closed down.

Attempts have been made to explain these post-natal changes on the basis of direct mechanical influences, such as those associated with movements of the diaphragm, and with the abrupt fall in pressure in the pulmonary circulation which occurs on the first respiratory movements. It is probable, however, that mechanical factors play no more than a very minor role in processes of this kind. In the case of the ductus arteriosus, at least, there is evidence that its definitive *anatomical* obliteration is preceded by a marked proliferation of its lining endothelium, i.e. that it is the result of an active growth process and not merely passive degeneration. On the other hand, the initial *functional* closure of its lumen occurs very rapidly—probably within a few minutes of birth,[1] and this is undoubtedly the direct result of a reflex mechanism leading to contraction of its muscular wall. In contrast with the adjacent vessels with which it connects (whose walls are composed predominantly of elastic tissue), the tunica media of the ductus arteriosus contains a relatively large proportion of plain muscle.

8. VASCULAR PATTERNS

We have noted that the main blood vessels in the embryo become established as a differentiation from a diffuse capillary network. In the latter some channels become larger and more clearly defined, while others may disappear altogether. We have now to consider in more detail what factors determine that certain vascular patterns should develop with a fair degree of constancy from a homogeneous network in individuals of the same or different species. These factors may be divided into two main categories—morphogenetic and mechanical—and, in order to distinguish between them, it is necessary to inquire in more detail how far the formation of a definitive vascular channel depends on an inherent developmental tendency which may be expressed by the tissues even in the absence of a normal circulation, and how far it is conditioned by the rate of flow and the pressure of blood in certain component parts of the primordial network.

Clearly the initial development of blood vessels which takes place before the blood begins to circulate in them must depend rather on general hereditary principles. It has also been shown, by observations on cultures of the skin and subcutaneous tissue of chick embryos,[2] that capillary vessels can sprout and form plexuses *in vitro* in the absence of any circulation. Even after the time when the circulation has usually commenced, the further differentiation of vascular channels may proceed to some extent independently of mechanical factors acting through the circulation. This problem was studied experimentally by removing the heart from chick embryos.[3] It was found that under these abnormal conditions they remained alive for seven or eight days. By the time the circulation begins, the chick embryo has a complete system of blood vessels, and it is interesting to note that after removal of the heart the further development of the vascular system is not immediately stopped. New capillaries are added at the usual rate in the area vasculosa on the surface of the yolk sac, and a large anterior vitelline vein is formed in the normal manner. The differentiation of main vascular channels, however, is apparently very limited. Thus, the omphalomesenteric arteries, which should put in an appearance during the developmental stages in which the experiments were carried out, fail to develop.

[1] A. E. Barclay and K. J. Franklin, 'The time of functional closure of the *foramen ovale* in the lamb', *Journ. Phys.* **94**, 1938.

[2] W. H. Lewis, 'The outgrowth of endothelium and capillaries in tissue culture', *Bull. Johns Hopkins Hosp.* **48**, 1931.

[3] W. B. Chapman, 'The effect of the heart beat upon the development of the vascular system in the chick', *Amer. Journ. Anat.* **23**, 1918.

Similar observations on the vascular development in the tails of acardiac tadpoles have shown that in this case the blood vessels form an irregular and indifferent network with no tendency to the differentiation of arteries and veins—in marked contrast to normal tadpoles.[1] It appears, indeed, that the differentiation of an indifferent capillary network into definite arteries and veins depends very largely on the transmission of the circulating blood in certain particular directions. We are thus able to recognize in the ontogeny of the vascular system the three stages which were postulated many years ago by Roux.[2] First there is the stage in which the formation of vessels depends entirely on genetic factors, then follows a transitional stage in which hereditary formation is gradually supplanted by adaptational factors, and in the final stage the differentiation of vessels is controlled by the haemodynamic factors of the circulation.

The dominance of morphogenetic factors in determining the development of vascular patterns is emphasized by the appearance in the early embryo of a primitive arrangement of arterial and venous channels which is characteristic of the adult form in lower vertebrates. The ancient system of aortic arches, cardinal veins, and other embryonic vessels requires to be almost completely remodelled during human development in order to provide a vascular pattern which is functionally adapted to the requirements of the human body. Yet gross vascular anomalies sometimes occur as the result of the persistence of an aortic arch or of a part of a cardinal vein which normally disappears. These cases may be regarded as evidence of the morphological individuality of specific blood vessels which expresses itself even at the expense of functional adaptation. Indeed, some blood vessels give the appearance of being so conservative that they persist in spite of the fact that they may be compelled, by the displacement during development of the organs which they supply, to follow a course which seems mechanically inconvenient.

In this connexion the blood supplies of the kidney and testis provide an instructive contrast. The definitive kidney is developed from the metanephros in the sacral region, and while in this position it gains its blood supply from the iliac arteries. Subsequently it migrates in a cranial direction to reach the upper lumbar region, and, in doing so, it progressively changes its blood supply at different levels until it finally receives its main artery from the upper end of the abdominal aorta. Here we have an example of functional adaptation which enables the kidney to receive its blood supply by the shortest and most direct route. It may be noted, incidentally, that one of the early arteries of supply to the kidney occasionally persists to form an 'aberrant renal artery', a possibility of some clinical importance since such an anomalous vessel may actually lead (by mechanical pressure) to an obstruction of the ureter.

The testis, like the kidney, also migrates during embryological development, but in this case it starts in the lower thoracic and upper lumbar

[1] E. R. Clark, 'Studies in the growth of blood vessels in the tail of the frog larva', *Amer. Journ. Anat.* **23**, 1918.
[2] W. Roux, 'Über die Verzweigungen der Blutgefäße des Menschen', *Zeitschr. f. Naturwiss.* **12**, 1878.

region and wanders down into the scrotum. Unlike the kidney it does not change its blood supply as it descends. Consequently the main vessel of supply—the testicular artery—becomes drawn out into an exceedingly long channel which has its origin from the upper end of the abdominal aorta and requires to pursue a rather complicated course in order to reach its objective. It remains possible, however, that the curious course of the testicular artery may have some functional significance, for even when it approaches the testis in the scrotum it does not immediately enter it. On the contrary, it often undergoes a most complex ravelling, and in some animals (e.g. the rabbit) it may encircle the testis twice before giving off any branches into its substance. In marsupials, again, the artery may break up into as many as a hundred separate branches which reunite again on reaching the testis, a remarkable example of a 'rete mirabile' (see p. 218). All these complicated vascular patterns of the testicular artery appear to serve a thermo-regulatory function, in association with the fact that in most mammals the temperature of the testis is several degrees lower than the body temperature generally (presumably a requirement for effective spermatogenesis).[1]

The development of the main arteries of the limbs provides a further illustration of the relation of morphogenetic to adaptational factors in the establishment of vascular patterns. In the most primitive mammals (e.g. *Ornithorhynchus*) the main vessel of supply in the forearm is the anterior interosseous artery. In marsupials and most other mammals, this is replaced by the median artery. Only in man and other Primates do the radial and ulnar arteries become well defined and constant. These three stages in the differentiation of the forearm vessels are broadly repeated in human embryology. It may be surmised that at different stages in human development different mechanical conditions prevail in the circulation of the forearm which determine the change-over from one main vessel of supply to another, giving rise to what appears to be a true recapitulation of phylogenetic evolution. On occasion, however, the earlier stages may persist to form what are often regarded as atavistic anomalies, and it is not uncommon to find a large median artery with much reduced or absent radial and ulnar arteries, even though the rest of the anatomy of the forearm is quite normal. It seems that, in these cases, morphogenetic factors have had a stronger influence than adaptational requirements in determining the definitive vascular pattern.

In the lower extremity, the main vessel of supply or 'axis artery' is primitively derived from the sciatic artery, which emerges from the pelvis in the gluteal region (Fig. 77). This arrangement is found in reptiles and also in the early stages of human development. The axis artery runs down the back of the thigh, passes deep to the popliteus muscle behind the knee joint, and continues down the back of the leg to reach the sole of the foot. In the later stages of human development (as in adult mammals generally), the artery is replaced above by the femoral artery in front of the thigh, and

[1] R. G. Harrison, 'The comparative anatomy of the blood supply of the mammalian testis', *Proc. Zool. Soc.* **119**, 1949; R. G. Harrison and J. S. Weiner, 'Vascular patterns of the mammalian testis and their functional significance', *Journ. Exp. Biol.* **26**, 1949.

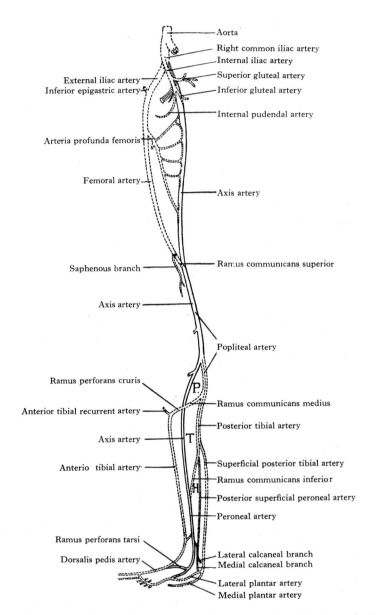

FIG. 77. A diagram showing the development of the main arteries of the lower limb in man. *P* shows the position of the popliteus muscle. The original main axis artery is indicated by unbroken lines: the arteries which become subsequently differentiated to replace it are shown by dotted lines. (From H. D. Senior.)

below by the anterior and posterior tibial arteries.[1] It is also replaced partly by a vessel which passes superficial instead of deep to the popliteus muscle. As with the forearm, this change-over is perhaps correlated with a change in the circulatory conditions in the limb. Here again, however, developmental arrest may occur with the persistence of an embryonic vascular channel. As a very rare abnormality of this kind, for example, the popliteal artery may be found passing deep to the popliteus muscle—a position which appears to be most disadvantageous from the mechanical point of view.

The inherent potentiality of blood vessels as morphological entities for growth and proliferation independently of surrounding tissues and without regard to functional requirements is illustrated, on the pathological side, by the occasional formation of tumours composed almost entirely of a tangled skein of vessels. Such a tumour, which is termed a *haemangioma*, is probably always congenital. It is not the result of the dilatation of preformed capillaries, but of a definite proliferative differentiation of abnormal vessels.

The mechanical factors which determine the later differentiation of arteries or veins now require to be considered. Many years ago they were discussed in detail by Thoma[2] who, as a result of his study of the developmental mechanics of the vascular system, formulated three postulates. These are (1) that the increase in size of the lumen of a blood vessel is directly related to the rate of blood flow, (2) that growth in thickness of a vessel wall is directly dependent on the tension to which it is subjected, this in turn being related to the diameter of the vessel and to the pressure of blood within it, and (3) that an increase of blood pressure above a certain limit provides a direct stimulus for the formation of new capillaries.

It has already been pointed out that the earliest stages in the development of blood vessels occur independently of an actual circulation, and it is clear, therefore, that Thoma's postulates cannot be applied here. In later stages, however, they are certainly applicable, at least in part.

The relation of the diameter of a blood vessel to the rate of blood flow has been studied in some detail by Hughes in the area vasculosa of the chick embryo.[3] He found that such a correlation does in fact exist, though it is not a simple mathematical relationship. His observations on the mitosis of endothelial cells in a growing vessel have suggested how a vessel may increase its diameter in response to an increased flow of blood through it. He concluded that 'at those points along the artery where the fluid friction generated by the circulation was greatest, excessive tensions would be set up within the wall, which would be manifested by nuclear elongation within the endothelium. When cell division took place, the directional increase in area of the endothelium under tension consequent on the orientation of the mitotic spindles would tend to relieve these tensions.'

The increase in the diameter of a vessel and in the thickness of its walls,

[1] H. D. Senior, 'The development of the arteries of the human lower extremity', *Amer. Journ. Anat.* **25**, 1919.

[2] R. Thoma, *Untersuchungen über die Histogenese und Histomechanik des Gefäß-systems*, 1893.

[3] A. F. W. Hughes, 'Studies on the area vasculosa of the embryo chick', *Journ. Anat.* **70**, 1935-6, and **72**, 1937-8.

in relation to increased blood pressure and rate of blood flow, is adequately demonstrated by the changes which occur in the establishment of a collateral circulation. If a main vessel is ligatured, the increased flow of blood which is deflected through collateral anastomotic channels leads to their rapid dilatation, and this proceeds until they are capable of transmitting the normal volume of blood to the part concerned. At the same time, the walls of the collateral vessels increase in thickness. The details of this process have been studied experimentally in the mesenteric vessels of the rat.[1] Interruption of one of these vessels is immediately followed by an increased blood volume in adjacent capillary vessels. One result of this is an exuberant growth of new capillaries which form a collateral network alongside the obliterated channel. Muscle or connective tissue is added to the walls of capillaries which are now called upon to transmit an additional volume of blood, so that they become converted into arterioles or venules. The walls of arteries involved in the establishment of a collateral circulation not only become thicker, but, since the proliferation of muscle tissue occurs in more than one plane, they also grow in length and become tortuous.

The increased thickness in the walls of arteries which occurs in senility is probably in part a direct reaction to raised blood pressure, though there is evidence that it is also partly a degenerative change related to defective nutrition of the tunica media. It has further been observed that the walls of veins become thickened in cases of raised pressure due to chronic obstruction of the venous circulation. The response of a vessel wall to increased pressure has also been demonstrated experimentally by transplanting a length of vein into the path of an artery, this being followed by a great increase in the thickness of the tunica media of the transplanted vessel.[2]

It has been demonstrated that, in the rabbit's ear, the vascular pattern shows a remarkable lability, new capillaries sprouting out and old capillaries undergoing retraction as the result of every slight change in their environment.[3] In this way, the capillary pattern may be completely remodelled in a relatively short space of time. Capillaries may be transformed temporarily into arterioles or venules in response to some local demand, only to revert to capillaries again when the demand no longer exists. In one interesting series of observations a venule was seen to become transformed within two months into a leash of capillary vessels, following a diminution in the local circulation; one month later, as the result of an increase in blood flow, the capillary plexus became transformed into an arteriole.[4] Apart from the size of a vessel and the thickness of its wall, there is some evidence that its structure may also depend on haemodynamic factors. In this connexion reference may be made to a report on the anatomy of an acardiac foetus

[1] H. B. Weyrauch and C. F. de Garis, 'Normal and interrupted vascular patterns in the intestinal mesentery of the rat. An experimental study in collateral circulation', *Amer. Journ. Anat.* **61**, 1937.

[2] B. Fischer and V. Schmieden, 'Experimentelle Untersuchungen über die funktionelle Anpassung der Gefäßwand', *Frankfurt. Zeitschr. Path.* **3**, 1909.

[3] E. R. Clark and E. L. Clark, 'Observations on living preformed blood vessels in the rabbit's ear', *Amer. Journ. Anat.* **49**, 1932.

[4] E. R. Clark and E. L. Clark, 'Microscopic observations on the extra-endothelial cells of living mammalian blood vessels', ibid. **66**, 1940.

attached to a normal twin, the heart of the latter being responsible for the circulation in both. In this anomalous specimen, all the arteries of the acardiac twin were physiologically the peripheral arteries of the normal twin, and it was found that the aorta and other large arteries (whose walls are normally composed mainly of elastic tissue) had the muscular walls characteristic of peripheral arteries.[1] However, a mechanical explanation is not always applicable to vascular structure, since in some vessels abrupt changes in structure may occur along their course. It has already been mentioned that the ductus arteriosus has muscular walls, in sharp contrast to the elastic walls of the neighbouring large vessels; this is evidently a structural adaptation which anticipates the functional demand for the immediate closure of the ductus arteriosus at birth.

As regards the formation of new capillaries in the adult, there is little experimental evidence that this is related specifically to blood pressure. On the other hand, it has been suggested that it may be related to the amount of material which passes through the endothelial wall of the parent vessels, or, in other words, to the metabolic activity of the endothelial cells. The vascular response shown to an inflammatory focus by the sprouting of new capillary vessels in the immediate neighbourhood suggests that this proliferation is in fact dependent on the increased metabolic activity of the tissues involved, which must also involve an increased metabolic activity of the endothelial cells of the capillary vessels already in existence. Such a vascular response is somewhat dramatically shown in the transparent and normally avascular cornea of the eye when it receives a slight injury. Very rapidly a leash of new capillaries may grow in from the margin of the cornea to the site of the injury. When the latter is completely healed, and the tissue metabolism of the cornea is reduced to its normal level, the newly formed vessels often undergo retraction and absorption, and finally may disappear altogether. It seems clear that the absorption of the vessels must also be related to metabolic rather than to haemodynamic factors, since it is not preceded by an interruption of the blood flow from the vessels at the margin of the cornea.

The relation of increase in capillary density to tissue activity is particularly well illustrated in the brain. It has been shown that in the new-born rat the vascularity of the grey matter is relatively poorly developed, and in many cases is no richer than that of the white matter.[2] Within a few days of birth, however, the capillary density in the cortex and in the centres of the brain stem increases rapidly, in close relation to the development of neural activity in these regions. It appears from these observations that functional activity in the central nervous system requires a much greater vascular richness than does metabolic activity associated with growth processes.

Retia mirabilia. It sometimes happens that an arterial stem quite abruptly terminates in a whole series of minute vessels. Such a formation

[1] A. Benninghof and R. Spanner, 'Das Gefäß-system eines Acardiers', *Morph. Jahrb.* **61**, 1929.
[2] E. H. Craigie, 'Changes in vascularity in the brain stem and cerebellum of the albino rat between birth and maturity', *Journ. Comp. Neur.* **38**, 1924–5; 'Postnatal changes in vascularity in the cerebral cortex of the male albino rat', ibid. **39**, 1925.

is termed a *rete mirabile*. In the so-called 'bipolar' type, the branches rejoin to form single stems again. Examples of this are seen in the glomerular vessels in the kidney, in the arteries at the base of the brain in many ungulates, and also in the testicular artery of marsupials (p. 213). In the first case, the arrangement provides a mechanism whereby the rate of flow and the pressure in the glomerular vessels can be controlled by vaso-constriction of either the vasa afferentia or the vasa efferentia, or of both simultaneously. The carotid arterial network at the base of the brain has been shown to assist the flow from the venous sinuses closely related to it. Such retia are found in grazing ungulates in which the venous return from the dependent head tends to be impeded gravitationally.[1]

In the 'unipolar' type of rete mirabile, the leash of small branches con-tinues directly into the ordinary capillary bed. A vascular pattern of this kind is found in the limb arteries of certain mammals, e.g. sloths and lemurs, but it is difficult to assign any particular functional significance to such cases. It is also developed to an exaggerated degree in the intercostal vessels of some aquatic mammals such as whales and porpoises. Here it has been suggested that the replacement of single main stems by multiple small arteries allows for the accommodation of a greater volume of blood,[2] and provides particularly for the temporary storage of oxygenated blood which can be drawn upon during prolonged submersion in the water. How-ever, it has been made clear that the vessels are not capacious enough to serve such a function. In slow moving animals such as sloths and lemurs the arterial plexus (which is intermingled with an equivalent venous plexus of small vessels) appears to facilitate the venous return by the impulse of its rhythmic pulsations. On the other hand, the thoracic retia of Cetacea probably plays a part in equalizing the external pressure and that in the lungs when the animals dive to some depth.[3]

9. THE INNERVATION OF BLOOD VESSELS

Blood vessels are almost everywhere accompanied by a delicate plexus of fine nerve fibres, medullated and non-medullated, which ramify in the tunica adventitia (Fig. 78). This plexus contains vasomotor and also afferent or sensory fibres. The vasomotor fibres are of sympathetic origin, and they are, at least predominantly, vasoconstrictor in function. The presence generally of specific vasodilator fibres to the arterial system has also been suspected, but there is as yet no certain evidence for this. They certainly exist in individual nerves of a parasympathetic nature, such as the chorda tympani branch of the seventh cranial nerve, and there is indirect evidence for the existence of sympathetic vasodilators in the limb vessels of

[1] C. H. Barnett and C. D. Marsden, 'Functions of the mammalian carotid rete mirabile', *Nature*, **191**, 1961.
[2] According to an old observation by P. Bert, the total blood volume of a porpoise is double that of a dog of the same body weight. More recently, the relatively large blood volume in seals has been attested by R. J. Harrison and J. D. W. Tomlinson (*Proc. Zool. Soc. Lond.* **126**, 1956).
[3] C. H. Barnett, R. J. Harrison, and J. D. W. Tomlinson, 'Variations in the venous system of mammals', *Biol. Rev.* **33**, 1958.

man.[1] The vasomotor supply to the coronary and pulmonary arteries also contains a predominance of vasodilator fibres. In a general review of the subject, Burn has called attention to an interesting species difference in regard to dilator mechanisms in mammalian muscle. In the muscle arteries of the dog and hare which are capable of sustained muscular exertion, there is an abundant sympathetic vasodilator supply, but in the monkey and rabbit which become more rapidly fatigued, it appears to be absent.[2]

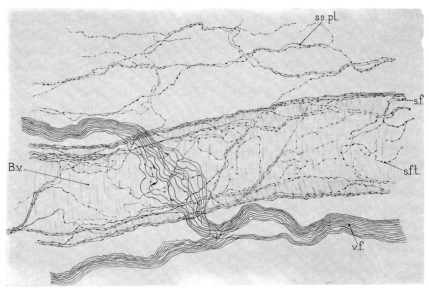

FIG. 78. Portion of the subserous coat of the small intestine of a dog, stained with methylene blue to show the sympathetic nerve fibres passing to a small blood vessel (*B.v.*). *ss.pl.* subserous plexus of sympathetic fibres. *s.f.* sympathetic fibres. *v.f.* vagal fibres. *s.f.t.* sympathetic fibres terminating in the vessel wall. × 240 (approx.). (From C. J. Hill.)

From the practical point of view, the vasoconstrictor fibres of the perivascular plexus are of considerable importance, since they provide the nervous pathway through which the local circulation to different tissues may be regulated. The route followed by vasoconstrictor impulses to reach the vessels which they control has, however, been a matter of controversy in the past, though it was known that they leave the spinal cord by the anterior roots of the spinal nerves, and that they have their cell station in ganglia of the sympathetic chain.

It can be demonstrated readily enough by ordinary methods of dissection that there are two main sources of vascular innervation for the arteries of the limbs—proximally the main trunks receive fibres directly from ganglia of the sympathetic chain, and distally they and their branches receive further accessions from peripheral nerves which lie in close topographical

[1] T. Lewis and G. W. Pickering, 'Vasodilation in the limbs, with evidence for sympathetic vasodilator nerves in man', *Heart*, 16, 1939; T. J. Fatheree and E. V. Allen, 'Sympathetic vasodilator fibres in the upper and lower extremities', *Arch. Int. Med.* 62, 1938.
[2] J. H. Burn, 'Sympathetic vasodilators', *Physiol. Rev.* 18, 1938.

relation to them. All these branches appear to contribute to the formation of the perivascular plexus in the tunica adventitia. In the past there was some uncertainty whether the vasoconstrictor fibres reach the vessel proximally and are thence carried down as a continuous network in the tunica adventitia to the terminal branches, or whether they run in the vascular branches which are supplied at intervals by the peripheral nerves. The former arrangement was once held to be the case and, on the basis of this conception, the operation of 'periarterial sympathectomy' was devised for the treatment of certain pathological conditions associated with spasm of the peripheral vessels. This operation involves stripping off the tunica adventitia with its contained plexus from a short section of the main artery with the object of producing peripheral vasodilatation. Experimental stimulation of the perivascular nerve plexus proximally was also said to produce constriction of the terminal arterioles in the area supplied by the main vessel of a limb, but the obvious inference from this observation was not upheld by many other investigators.

Langley[1] brought forward experimental evidence in favour of the view that the vasoconstrictor fibres probably reach the vessels which they supply by way of the peripheral nerves. Trotter and Davies[2] had previously noted that, in peripheral nerve injuries in man, the anaesthetic areas of the skin were flushed and hotter than normal areas; they concluded, therefore, that vasoconstrictor fibres had been severed in the corresponding cutaneous nerves. These observations were confirmed by Woollard and Phillips,[3] who mapped out cutaneous areas of flushing and, with the aid of a thermopile, areas of raised temperature, after blocking peripheral nerves by the local injection of novocaine. This technique also indicated the fact that vasoconstrictor fibres run in the trunks of peripheral nerves, and afforded no evidence that they extend all along the main stems of arteries.

These results were finally established experimentally by Gilding,[4] who studied the anatomical distribution of vasoconstrictor fibres in cats by noting the coloration of the skin when a diffusible dye (bromo-phenol blue) had been injected into the blood stream. On previous stimulation of the stellate ganglion the resulting vasoconstriction in the fore-limb was shown by the fact that the tissues remained relatively unstained. When a peripheral nerve was cut, the cutaneous area or muscular tissue which it supplied became coloured, as the result, evidently, of the interruption of vasoconstrictor impulses. It was concluded from these experiments that the sympathetic fibres in the limbs follow sensory and motor nerves to the periphery, branching from them in the immediate vicinity of the small blood vessels which they innervate. On the other hand, no evidence was found that the periarterial nerve plexus conducts impulses to the periphery. However, it should be noted that in splanchnic arteries, which supply

[1] J. N. Langley, 'The vascular dilatation caused by the sympathetic and the course of vasomotor nerves', *Journ. Phys.* **58**, 1923.

[2] W. Trotter and H. M. Davies, 'Experimental studies in the innervation of the skin', ibid. **38**, 1909.

[3] H. H. Woollard and R. Phillips, 'The distribution of sympathetic fibres in the extremities', *Journ. Anat.* **67**, 1932–3.

[4] H. P. Gilding, 'The course of the vaso-constrictor nerves to the periphery', *Journ. Phys.* **74**, 1932.

visceral structures, the vascular innervation is certainly effected by peri-vascular nerve plexuses which extend along them. Herein, splanchnic arteries contrast with somatic arteries supplying the body wall and the limbs.

From a study of the vessels of the hind-limb in experimental animals, it appears that the perivascular plexus derived directly from lumbar sym-pathetic ganglia extends down only as far as the upper part of the thigh, its exact extent varying in different species. Farther distally it is formed by branches from peripheral nerves. In regard to the intrinsic details of the plexus, the latter is composed mainly of non-medullated fibres in the proximal part of the main arteries, but in the distal part there is an in-creasing predominance of fine medullated fibres. Finally, in the terminal arterioles non-medullated fibres are again in the majority. From the plexus of non-medullated fibres in the tunica adventitia, a secondary plexus in the muscular tissue of the tunica media is formed. Here, motor terminals are given off which end in pericellular nets round the individual muscle fibres.

The medullated fibres appear to be largely sensory in function, terminat-ing in varicose and thickened endings in the tunica adventitia, or sometimes in Pacinian corpuscles. From these sensory fibres, collaterals may be given off to end in the adjacent tissues, providing an anatomical basis for the axon reflexes which are believed to be responsible for the local vasodilata-tion which is produced by a local stimulus. This reaction is the result of antidromic impulses carried by sensory nerves (see p. 358).

The capillary bed is pervaded with fine sympathetic fibres, but there is no secure evidence that any of these terminate actually on the endothelial wall of the capillary vessels.[1] It may be mentioned that some anatomists have described in detail a rich and intricate terminal reticulum of neuro-fibrillae in the walls of blood vessels having the most intimate relation to muscle fibres and endothelial cells, and even being in cytoplasmic conti-nuity with them. According to these accounts, indeed, the 'terminal reticu-lum' brings every single cell in the vessel wall into direct relation with the sympathetic nervous system. However, a critical analysis of such observa-tions by Nonidez[2] has shown that they have been made on the basis of a silver staining technique which is not entirely selective for nerve fibres. It seems certain, in fact, that, in the literature of this subject, reticular connec-tive-tissue fibres have not infrequently been mistaken for nerve fibres.

It has been noted that arteriovenous anastomoses are particularly well supplied with sympathetic nerves, and that these anastomoses are most abundant in the skin at the distal end of the extremities. In correlation with this fact vascular branches from peripheral nerves in the limbs are most numerous in the hand and foot, and they are also more numerous in relation to superficial as compared with deep vessels.

The new growth of nerve fibres along newly formed blood vessels has been observed in transparent chambers in rabbits' ears.[3] By staining the

[1] J. W. Millen, 'Observations on the innervation of blood vessels', *Journ. Anat.* 82, 1948.
[2] J. F. Nonidez, 'Observations on the innervation of the blood vessels', *Anat. Anz.* 82, 1936.
[3] E. R. Clark, E. L. Clark, and R. G. Williams, 'The new growth of nerves and the establishment of nerve-controlled contraction of newly formed arterioles', *Amer. Journ. Anat.* 55, 1934.

fibres with methylene blue injected intravitally, it has been possible to follow their extension into the muscular walls of recently developed arterioles, and to note that this coincides with the establishment of a nervous control of the reactions of these vessels. The rate of growth of vasomotor fibres was found under these conditions to amount to 70μ a day.

The question of the sensitivity of blood vessels, i.e. whether local stimulation of the vessels can give rise to conscious sensation, has been the subject of some discussion. Ligature of some arteries certainly gives rise to a sensation of pain, and this may be ascribed to the mechanical irritation of nerve fibres in the vessel wall. On the other hand, the possibility cannot be excluded that sensory somatic nerve fibres in the immediate neighbourhood of the blood vessel may be involved in the process of ligature. Pathological conditions of the vascular system associated with spasm of peripheral vessels may be exceedingly painful, and the pain can sometimes be relieved by excision of the appropriate sympathetic ganglia. This has given rise to the suggestion that painful impulses are mediated by the sympathetic fibres which innervate blood vessels. But the pain may be really due to ischaemia and the accumulation of irritant substances in the tissues, and its relief is perhaps the result of the vasodilatation which allows of their rapid removal. However, direct histological studies have shown that fine nerve fibres are abundantly present in the tunica adventitia of blood vessels even when the degeneration of all sympathetic fibres has been brought about by the previous destruction of the sympathetic ganglia from which they are derived.[1] Moreover, these residual nerve fibres are morphologically similar to those which elsewhere are considered to mediate the transmission of painful impulses. It is no doubt these fibres in the tunica adventitia which are responsible for the pain when an artery (and, to a less extent, a vein) is punctured. There is no certainty that the endothelium of a blood vessel is sensitive to pain. Indeed, at least so far as veins are concerned, the evidence is to the contrary. The introduction of sclerosing fluids into a vein (e.g. in the treatment of varicose veins) may cause momentary pain, but the latter is felt only after a delay of about 20 seconds, during which the fluid has presumably diffused through the endothelium to reach the 'pain nerve-fibres' in the outer coats of the vessel wall.

Arteries supplying different kinds of tissue are not all controlled to the same degree by the sympathetic nervous system. Those that are distributed to muscles, for example, have a much poorer innervation than cutaneous vessels. That the former are certainly influenced by vasoconstrictor impulses, however, is shown by Gilding's experiments which have already been referred to (*vide supra*, p. 220), for it was found that, in the case of a muscle receiving a double nerve supply, section of one of these nerves is followed by a sharply defined staining of one-half of the muscle.

There has been some lack of agreement regarding the extent of vasomotor control of arteries in the central nervous system. Much of the earlier experimental work on this problem had led to negative results. One of the difficulties in the investigation of the reactions of the cerebral arteries is

[1] H. H. Woollard, G. Weddell, and J. A. Harpman, 'Observations on the neuro-histological basis of cutaneous pain', *Journ. Anat.* **74**, 1940.

the fact that they are very readily influenced passively by changes of blood pressure in other parts of the body, so that it is not easy to isolate the direct effects of sympathetic stimulation on their calibre. From the anatomical point of view, it is difficult to accept the conclusion that they are not controlled in any way by vasomotor impulses, since it has been demonstrated by many competent anatomists that they are certainly innervated by autonomic fibres, though the latter are less numerous than in the case of most arteries elsewhere in the body.

Nerve fibres run in the tunica adventitia of the main vessels on the surface of the brain, and though they have been actually followed along the smaller branches into the substance of the brain and spinal cord, they probably do not extend very far along these.[1] Fine medullated fibres from the adventitial plexus have been found to terminate in relation to muscle fibres, and coarser fibres—sometimes medullated—end in the tunica adventitia. In correlation with this anatomical mechanism, evidence has accumulated from experimental work which indicates that cerebral vessels are, in fact, under control of the autonomic nervous system, though to a much slighter degree than cutaneous vessels. Forbes and Wolff[2] studied the problem by making a trephine hole in the skulls of experimental animals and fitting in a glass window. The blood vessels on the surface of the brain could thus be observed directly with a microscope, while still preserving the normal conditions under which the intracranial circulation takes place in a closed chamber. By means of photographic records, they demonstrated a slight constriction of the pial arteries on stimulation of the cervical sympathetic, and also on the direct application of epinephrine, while vasodilatation occured on stimulation of the vagus. They found, also, that vasoconstriction of the pial vessels, unlike those of the skin, is readily overcome by a general rise in systemic blood pressure. Similar results have been obtained with the choroid arteries in the ventricles of the brain.[3]

Finally, it has been observed that stimulation of the sympathetic in a completely isolated head, which is perfused with blood from an intact animal, is followed by a constriction of the vessels on the surface of the brain.[4] This experiment eliminates the effects on the cerebral circulation of changes in the general systemic circulation, and for this reason is of somewhat crucial importance.

In a general review of the problem of vasomotor control of cerebral vessels, Forbes and Cobb[5] conclude that, while vasoconstrictor nerves are certainly present, they are distributed unequally and they are only about one-tenth as effective as the vasoconstrictor nerves in the skin.

Special sensory mechanisms of blood vessels. Although it is

[1] S. Clark, 'Innervation of the blood vessels of the medulla and spinal cord', *Journ. Comp. Neur.* **48**, 1929; W. Penfield, 'Intracerebral vascular nerves', *Arch. Neur. & Psych.* **27**, 1932.
[2] H. S. Forbes and H. G. Wolff, 'The vasomotor control of cerebral vessels', ibid. **19**, 1928.
[3] T. J. Putnam and E. Ask-Upmark, 'Microscopic observations on the living choroid plexus and ependyma of the cat', ibid. **32**, 1934.
[4] J. L. Pool, H. S. Forbes, and G. I. Mason, 'Effect of stimulation of the sympathetic nerves on the pial vessels in the isolated head', ibid.
[5] H. S. Forbes and S. S. Cobb, 'Vasomotor control of cerebral vessels', *Brain*, **61**, 1938.

probable that all arteries and veins are supplied with afferent nerves through which the circulatory system as a whole can be influenced to some degree by reflex responses to local vascular stimuli, in certain vessels the sensory nerve endings are elaborated to form specialized receptors of

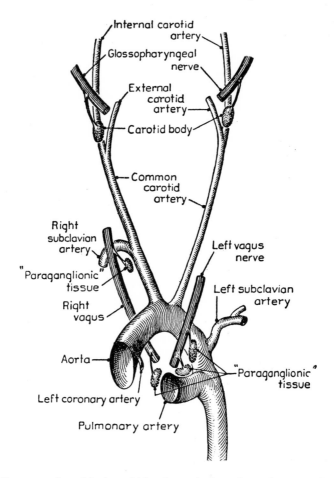

FIG. 79. Reconstruction of the branchial arch arteries in a 26-mm. human embryo to show the positions in which 'paraganglionic' tissue can be found. ×30. (From J. D. Boyd.)

considerable functional importance. It has been shown that, in lower vertebrates, such a nervous apparatus is present in each of the branchial arch arteries. These *pressor-receptor* mechanisms respond to an increase in the blood pressure, and provide for an adjustment of the circulation by leading to an inhibition of the heart's action and general vasodilatation. In man, those arteries which represent persistent parts of the aortic arches are also provided with pressor receptors. The latter are found, therefore, at the commencement of the internal carotid artery (3rd arch), the aorta and right subclavian artery (4th arch), and the pulmonary artery (6th arch) (Fig. 79).

We may refer briefly first to the pressor-receptor mechanism of the internal carotid artery. At its commencement, this vessel commonly shows a bulbous dilatation—the *carotid sinus*—in immediate relation to which is a small 'glandular' structure called the *carotid body*. The latter is a few millimetres in diameter and consists of groups of polygonal epithelioid cells[1] permeated with an abundant plexus of sinusoidal vessels. At one time the carotid body was believed to be composed of chromaffin tissue similar to that found in the adrenal medulla and in relation to sympathetic ganglia generally. It has therefore been frequently termed a 'paraganglion'. More recent histological studies, however, have shown that true chromaffin cells, if present at all, are few and scattered. In its embryological origin the carotid body is derived from the mesoderm of the third pharyngeal arch.

The carotid sinus and carotid body are both richly innervated from the glossopharyngeal nerve (i.e. the nerve of the third branchial arch) by a branch called the *carotid nerve*,[2] and both have been shown experimentally to play an important part in carotid sinus reflexes. In the carotid sinus itself there is a rich plexus of sensory nerve endings in the tunica adventitia, which is here relatively thick. The tunica media of the carotid sinus, on the other hand, is thinned out, and this permits the adventitial receptor mechanism to respond more readily to slight changes of blood pressure. Boyd[3] has pointed out that the local swelling of the internal carotid artery is also a functional adaptation to the same end, for it can be shown by theoretical computation that an increase in pressure will cause a greater change in the tension of a vessel in a dilated segment than in a narrower portion—in other words, changes in blood pressure are amplified at the carotid sinus. While the carotid sinus itself is a pressor-receptor mechanism, responding to changes in the pressure of the blood, the carotid body alongside it is a *chemo-receptor*, responding to changes in the chemical constitution of the blood. In relation to this, the carotid body is richly supplied with sinusoidal blood vessels, and nerve fibres can be demonstrated to terminate in their walls as well as among the epithelioid cells. This dissociation of mechanical and chemical sensitivity has been determined experimentally by interrupting the nerve supply of the carotid sinus and the carotid body independently of each other.[4]

'Paraganglia' quite similar to the carotid body have been found in relation to pressor-receptor mechanisms in the arch of the aorta, the pulmonary artery and the origin of the right subclavian artery.[5] They are

[1] According to Schumacher (*Zeitschr. f. mikroskop.-anat. Forsch.* **43**, 1938) the cells of the carotid body (and other bodies of a similar nature) are really epithelioid muscle cells in the tunica media of extremely tortuous arteriovenous anastomoses.

[2] F. de Castro, 'Sur la structure et l'innervation du sinus carotidien', *Trav. de Lab. de recherches biol. de l'Univer. Madrid*, **25**, 1928.

[3] J. D. Boyd, 'Observations on the nerve supply and anatomy of the human carotid sinus', *Anat. Anz.* **84**, 1937.

[4] J. Bouckaert, L. Dautrebande, and C. Heymans, 'Dissociation anatomo-physiologique des deux sensibilités du sinus carotidien', *Ann. de Phys.* **7**, 1931.

[5] J. F. Nonidez, 'The aortic (depressor) nerve and its associated epithelioid body, the glomus aorticum', *Amer. Journ. Anat.* **57**, 1935; 'Observations on the blood supply and the innervation of the aortic paraganglion in the cat', *Journ. Anat.* **70**, 1936.

innervated by branches of the vagus nerve (which also terminate in pressor-receptor endings in the walls of the main vessels), and usually receive their blood supply directly from the arteries with which they are associated. The 'paraganglion' related to the pulmonary artery, therefore, is supplied with 'venous' blood. By analogy with the carotid body it is to be presumed that these 'paraganglia' are also chemo-receptors.

Receptor mechanisms have also been described in the walls of the large veins close to their entry into the heart—that is to say, in the intraperi-cardial portions of the venae cavae and pulmonary veins. Here, in contrast to the adventitial position of arterial pressor receptors, the sensory nerves terminate in sub-endothelial endings and in perimuscular arborizations. This renders them particularly sensitive to slight changes in the venous blood pressure. These venous receptors provide the anatomical basis for Bainbridge's reflex, as the result of which a rise in the venous pressure leads to cardiac acceleration—a response which has been shown experi-mentally to be abolished by section of the vagus nerves.

IX

BLOOD

So far as its structural composition is concerned, blood consists of a fluid plasma in which are suspended cells or corpuscles of various types. The nature of the cells will be considered briefly, with special reference to their embryological origin and their morphological relationship to other types of cell in the body. These problems have led to the accumulation of a considerable bulk of literature—an expression of the fact that their solution is a matter of great difficulty. In organized tissues the transition from one type of cell to another can often be established by direct study of ordinary stained sections in which the contiguity of intermediate forms is evident. Where cellular elements are isolated and unorganized, as in blood, their morphological relationships are far less easy to determine with certainty.[1]

1. BLOOD CORPUSCLES

There are two main types of blood corpuscle, red and white. The former —which are in the great majority—are concerned with the transport of oxygen from the lungs to all parts of the body, and the conveyance back to the lungs of carbon dioxide. The latter are broadly concerned with protective functions.

Red blood corpuscles, or erythrocytes (Fig. 80 a, a'). In the circulating blood of adult mammals, the red cells have lost their nuclei. They have the simple form of flattened bi-concave disks, and in structure they seem to consist of little more than a protoplasmic envelope containing a solution of haemoglobin mixed with a structureless colloidal matrix. The envelope is highly elastic and forms a semi-permeable plasma membrane. Until it became demonstrable by electron microscopy, its existence was inferred from the changes in shape which a red corpuscle shows when the osmotic pressure of the surrounding medium is changed. Each corpuscle in man is on the average 7–8μ in diameter (as measured in ordinary film preparations). Their size shows a certain degree of variation in different mammals, but this is not necessarily correlated with body size.[2] In addition to their non-nucleated character, the red corpuscles of mammals, in comparison with those of lower vertebrates, are round instead of oval. A curious exception to this generalization is found in the family *Camelidae*, in which

---

[1] For an authoritative discussion on the relationships and developmental origin of blood corpuscles, see the relevant chapters in M. M. Wintrobe, *Clinical Hematology*, 4th edn., London, 1956.

[2] The smallest mammalian erythrocytes are said to occur in the musk-deer, where they are about $2 \cdot 5 \mu$ or less in diameter. The contrast in size of the red corpuscles in vertebrates generally is made very evident by a comparison of their volume. This ranges from $20 \mu^3$ in the goat to $13,800 \mu^3$ in the urodele amphibian *Apmhiuma*; yet the amount of haemoglobin in a unit volume of blood in these two creatures is about the same (M. M. Wintrobe, op. cit.).

the cells are oval in form. In human blood the number of erythrocytes in one cubic millimetre is normally about 5,000,000 and on the average slightly more in men than women. In conditions of relative anoxia, for example at high altitudes, reduced oxygen intake in respiration is compensated by an increased density of erythrocytes in the blood.

As regards their immediate origin, red corpuscles are derived by linear descent from a series of intermediate cellular elements which are collectively termed *normoblasts*.[1] The earliest definite precursor in this series is called a *pronormoblast*; it has basophilic cytoplasm and a large pale nucleus with distinct nucleoli. The pronormoblast, in its turn, is derived from a cell of generalized type—the *haemocytoblast* (see p. 234). In the course of its further development, the nucleoli of the pronormoblast disappear and the nuclear chromatin becomes more dense; this leads to the differentiation of a *basophilic normoblast*. Haemoglobin now begins to appear in the cytoplasm immediately surrounding the nucleus, and as a result the cell becomes less basophilic and more acidophilic. This stage of development is represented by the *polychromatic normoblast*. Lastly, the nucleus becomes smaller and more pyknotic, and with the final development of the erythrocyte it is lost altogether. The disappearance of the nucleus in the formation of an erythrocyte is probably the result of its extrusion as a whole, though some observers have described its disintegration and dispersal within the cytoplasm of the cell. In the early embryo all red corpuscles are nucleated, and even at birth there may be an occasional normoblast in the circulating blood. In the normal adult, however, nucleated red corpuscles are only to be found in blood-forming or haemopoietic tissues; their appearance abnormally in the blood stream is associated with pathological conditions involving excessive blood destruction or excessive blood formation. There is some evidence, it is interesting to note, that under normal conditions about five days are required for the maturation of red corpuscles from haemocytoblasts.[2] Once in the circulation, corpuscles are commonly reckoned to have a mean life-span of about 120 days (as shown by 'tagging' them with radioactive isotopes). It follows from this that rather less than one per cent of the blood is replaced every day. However, by recording the length of time during which corpuscles (of different but compatible groups) remain in the circulation of the recipient of a blood transfusion, some authorities have been led to conclude that the life-span may normally be a good deal shorter.[3] But clearly such a situation is abnormal and any inferences based on it must be liable to error.

Immature red corpuscles which have recently been formed from normoblasts show a reticular structure which can be demonstrated by basic dyes,

---

[1] The nomenclature of developing red corpuscles has in the past been the source of some confusion. It is now generally agreed that while the term 'erythroblast' applies to any nucleated red cell whether normal or pathological, the term 'normoblast' should be limited to the nucleated precursors in a normal developmental series. The term 'megaloblast' is a pathological form of normoblast which, in certain types of anaemia, leads on to the formation of abnormal red cells called 'macrocytes'.

[2] For an account of erythropoiesis, see J. V. Dacie and J. C. White, 'Erythropoiesis with particular reference to its study by biopsy of human bone marrow', *Journ. Clin. Path.* **2**, 1949.

[3] E. B. Krumbhaar, 'The Erythrocyte' in *Special Cytology*, Vol. 2, ed. by E. V. Cowdry, London, 1963.

and are hence termed *reticulocytes*. They are more numerous in the blood circulating through the bone marrow (the site of active blood-cell formation in the adult) than elsewhere in the circulation, and they become quite numerous in the general blood stream following haemorrhage or blood destruction. A reticulocyte count, indeed, provides an index of the activity of blood-forming or haemopoietic tissues.

It has been found that mesenchymal tissue with haemopoietic potentialities will form red blood corpuscles when cultivated *in vitro*. This has been shown by Murray[1] in cultures of the blastoderm of the chick embryo, in which large numbers of haemoglobin-containing erythroblasts are produced. Vascular endothelium also becomes differentiated, and the experiments support the conception that in the embryo the endothelial cells are formed from indifferent haemangioblast cells whose fate is determined by the superficial position in which they find themselves in the blood islands (see p. 206).

White blood corpuscles. Several types of white corpuscle are normally present in the circulating blood. They are all nucleated cells, they all possess amoeboid properties to a greater or lesser degree, and they are probably all concerned to some extent with the protection of the body against bacterial infection and toxic absorption. In certain pathological conditions the relative proportion of white corpuscles in the blood may show considerable variation, hence an estimation of their numbers is often of considerable aid in clinical diagnosis. Normally there are from 5,000 to 10,000 white corpuscles in one cubic millimetre of blood, but following vigorous exercise the number may rise to 25,000. An increase is also a concomitant of mental anxiety, exposure to extreme cold, and other stresses.

For descriptive convenience white corpuscles can be broadly divided into granular and agranular types, though some anatomists would not be prepared to recognize these as natural groups. The former comprises those cells (granulocytes) in which the cytoplasm constantly contains obvious granules which can be selectively stained with acid or basic dyes. It includes polymorphonuclear or neutrophil leucocytes, eosinophil leucocytes, and basophil leucocytes. The cytoplasm of agranular white corpuscles stains evenly with basic dyes, though even here a faintly granular appearance can be demonstrated in some types of cell with special staining methods. Agranulocytes include lymphocytes and large monocytes.

Polymorphonuclear leucocytes (Fig. 80 *d*). These cells—which make up approximately 55–60 per cent of the total white corpuscles—are about 10–12 μ in diameter. They are characterized by their lobulated nucleus, which is often divided into three or more parts interconnected by fine threads of chromatin, and by containing in their cytoplasm a sprinkling of very fine granules. The latter stain a nondescript pink colour with eosin-methylene-blue mixture, and it is for this reason that these leucocytes are also termed *neutrophils*.[2] Polymorphs are highly amoeboid and phagocytic,

---

[1] P. D. F. Murray, 'The development *in vitro* of the blood of the early chick embryo', *Proc. Roy. Soc.* B, **111**, 1932.

[2] The staining reactions of the granules in polymorphonuclear leucocytes vary to some

and they have a marked capacity for ingesting and destroying bacteria. They respond particularly to infections by pyogenic organisms, and under such conditions they appear in large numbers in the circulating blood, giving rise to a characteristic *leucocytosis*. In local infections, also, they emigrate freely from the blood vessels into inflamed tissues and become actively engaged in destroying the infective agent.

Eosinophil leucocytes (Fig. 80 *e*). These corpuscles are somewhat similar to polymorphs in their size and nuclear conformation. Their cytoplasm is filled with coarse, highly refractile granules which stain brilliantly with acid dyes, e.g. eosin. Normally they comprise not more than 1 to 3 per cent of the total white blood corpuscles. Their number increases, giving rise to an *eosinophilia*, in certain pathological conditions such as parasitic infestations, allergic states, and some types of skin disease. The significance of this eosinophilia and the functions of the eosinophil leucocyte are still somewhat uncertain, but it may be noted that antibody-antigen reactions such as occur in allergic conditions are accompanied by the release of histamine, and there is some evidence that eosinophils liberate or absorb this substance.

Basophil leucocytes. These are similar to eosinophil corpuscles, except that they are somewhat smaller, and their granules stain selectively with basic dyes. They are very few in number, forming $\frac{1}{2}$–1 per cent of all white corpuscles, and their function is quite obscure. Their basophil granules, like those of mast cells, contain heparin, and they are also rich in histamine.

It is agreed that all the granulocytes are specialized cells, representing end products of cellular differentiation, and incapable of being transformed into other cell types. In relation to this, they never undergo cell division. Their immediate origin is also not disputed. In blood-forming or haemopoietic tissues (e.g. bone marrow), large granular cells of a primitive type are found which are called *myelocytes* (Fig. 81 *d*, *e*). They have relatively large rounded or reniform nuclei, and they are also differentiated into three groups according to the staining reaction of their granules—neutrophil, eosinophil, and basophil. Complete series of transitional forms linking up these myelocytes with different types of granulocyte can be readily demonstrated.

Lymphocytes (Fig. 80 *b*, Fig. 91). Of the agranular white corpuscles, lymphocytes are by far the most numerous, forming about 25–30 per cent of the total white count. They are small rounded cells of about the same size as red corpuscles (though larger cells are also occasionally seen), and they are almost completely filled with a darkly staining round nucleus which leaves only a narrow rim of surrounding cytoplasm. They have been observed to show a slight degree of active amoeboid movement, the nucleus being characteristically placed at the advancing edge of the cell. There is no good evidence that they can ingest foreign particles by phagocytosis— if they do have a limited capacity to do so, it is quite negligible in

extent in different species of animals. In some mammals no granules are demonstrable; in others they are larger and more conspicuous than in man and, in the rabbit for example, the cells may at first sight be mistaken for eosinophil leucocytes.

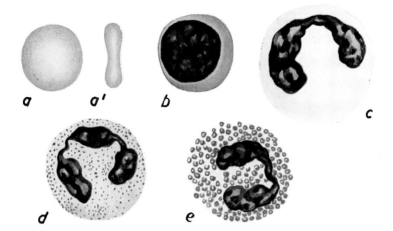

Fig. 80. Blood corpuscles. *a*. Red blood corpuscle seen in surface view and *a'* in profile. *b*. Small lymphocyte. *c*. Monocyte. *d*. Polymorphonuclear leucocyte. *e*. Eosinophil leucocyte. Magnification ×2500 approx.

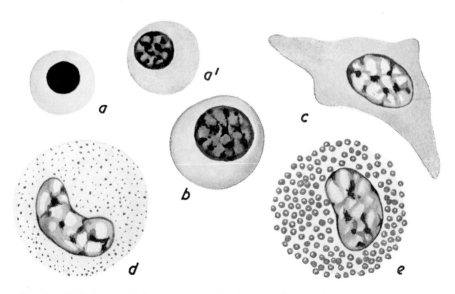

Fig. 81. Cells from which the mature blood corpuscles are derived. (Compare with Fig. 80.) *a*. Nucleated red blood corpuscle, and *a'* its immediate precursor, normoblast. *b*. Lymphoblast, the precursor of lymphocytes. *c*. Reticular cell, which gives rise to monocytes. *d*. Neutrophil myelocyte. *e*. Eosinophil myelocyte. Magnification ×2500 approx.

comparison with the phagocytic power of polymorphs. They respond particularly to infections of a chronic type (e.g. tuberculosis), and in such conditions their accumulation in the blood may give rise to a *lymphocytosis*. As will be seen later (p. 256), lymphocytes are in a process of continual re-circulation, passing out from lymphoid tissues and being filtered back again into the latter from blood capillaries or post-capillary venules. The life-span of a lymphocyte may be surprisingly long—a matter of weeks or months, but it is probably very variable.[1]

Monocytes (Fig. 80 c). Up to 7 per cent of the white blood corpuscles are made up of large basophil agranulocytes, 10–12 μ or sometimes more in diameter, in which the nucleus is oval or reniform and eccentrically placed. These are the monocytes (also called mononuclear cells), and from their appearance in ordinary stained preparations they are not always easy to distinguish from the larger lymphocytes. Much controversy has been de-voted to the relation of monocytes to other types of corpuscle, and to their origin and functional significance. Their general similarity to large lym-phocytes has suggested that they have a lymphocytic derivation, and the occasional occurrence of 'transitional cells' has suggested a relation with granulocytes. By some anatomists they are regarded as circulating elements of the macrophage system (see p. 55), on the grounds that they show the same reaction to vital dyes which are introduced into the blood stream. They have therefore been termed 'blood histiocytes', and it is now gener-ally accepted that there is no clear distinction between monocytes and histiocytes. They have pronounced phagocytic properties in relation to relatively coarse particles and bacteria, as well as to the products of damaged tissues or foreign material introduced into the body. The ultimate derivation of monocytes from reticular cells in blood-forming and lymphoid tissue (as well as from the endothelial cells of sinusoidal vessels) seems also to be fairly well established. On the other hand, there seems to be no proof that (as some have supposed) mature lymphocytes, once formed, can transform into monocytes, or that the latter can become converted into other types of blood cell. In other words, it seems probable that monocytes represent one of the end products of the diverging lines of differentiation which characterize haemopoiesis. This conception is con-sonant with the general opinion of haematologists that cells which are once set free in the circulating blood do not, as a rule, undergo any further signi-ficant transformations.

In a previous chapter (see p. 185) mention was made of the fact that white corpuscles can emigrate from capillary and post-capillary vessels into the surrounding tissues. The mode of penetration of the endothelial lining of a blood vessel has now been precisely determined by electron micro-scopical studies,[2] and it shows interesting contrasts between different types of corpuscle. Thus, while neutrophils and monocytes push their way out *between* adjacent endothelial cells, lymphocytes are transferred *through* the cytoplasm of individual endothelial cells (see Fig. 82).

[1] J. L. Gowans and D. D. McGregor, 'The production of lymphocytes by lymphoid tissue', *Cell Proliferation*, Blackwell Scient. Publ., 1963.
[2] V. T. Marchesi and J. L. Gowans, 'The migration of lymphocytes through the endo-thelium of venules in lymph nodes', *Proc. Roy. Soc.* **159**, 1964.

Blood platelets. Besides the typical corpuscular elements, there are found in blood minute particles of cytoplasm about 2 μ in diameter (but very variable in size), the nature of which has in the past been a matter of considerable controversy. They are termed *blood platelets*, and although their number is difficult to compute with any accuracy, there are estimated to be about 300,000 in a cubic millimetre. They contain fine granules which

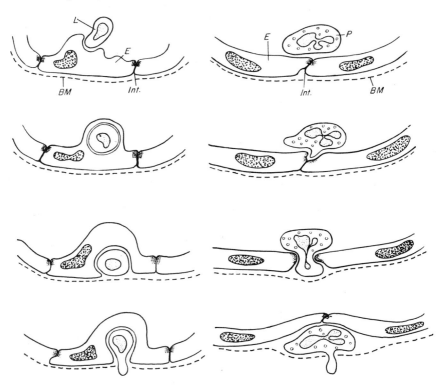

FIG. 82. Diagram illustrating the emigration through the endothelial lining of a blood vessel of a lymphocyte (on the left), and a neutrophil (on the right). Lymphocytes (*L*) penetrate the cytoplasm of the endothelial cells (*E*), while neutrophils (*P*) push their way through the intercellular junctions (*Int.*) of adjacent endothelial cells. Both these types of cell then penetrate the basement membrane (*BM*) to reach the surrounding tissues. (From V. T. Marchesi and J. L. Gowans, *Proc. Roy. Soc.*, **159**, 1964.)

tend to group themselves in the centre of each platelet so as to form what appears to be a nucleus, but this appearance is illusory.

Blood platelets play an important part in the process of coagulation. They rapidly collect together in clumps in freshly drawn blood and then undergo disintegration. This results in the liberation of a factor essential for the production of thromboplastin, which activates prothrombin and leads to a conversion of fibrinogen to fibrin.[1] Because of the essential part which platelets thus play in the production of a clot, they have sometimes

[1] D. E. Bergsagel, 'Viscous metamorphosis of platelets', *Brit. Journ. Haemat.* **2**, 1956.

been referred to as *thrombocytes* (a term which was originally coined on the mistaken assumption that they are true cellular elements). Recent studies have also shown that platelets are carriers of a number of pharmacologically active agents, including histamine, serotonin, and adenosine triphosphate, the release of which at the site of local injury may have important physiological effects.

The origin of blood platelets from giant cells (*megakaryocytes*) which are found in red bone-marrow is now well established (*vide infra*, p. 237). These cells have been described as giving out pseudopodial processes which push their way into the lumen of sinusoidal blood vessels through the lining endothelium and then break off into small fragments which are set free in the circulation as blood platelets. The latter appear to have a life-span of about nine days in the circulation.[1]

2. THE GENEALOGY OF BLOOD CORPUSCLES

In the last section, the immediate origin of the various corpuscular elements of the blood was briefly indicated. The problem of their genealogy now requires consideration in further detail.

It has already been made apparent that the task of elucidating the morphological relationships of different types of blood corpuscle is not easy. Haematologists have in the past been faced with the same sort of problem as evolutionists in their attempts at constructing a genealogical tree. In the absence of direct records of the transformation of one type into another, they have had to decide how far resemblances between one cell and another betoken a real affinity, and how far the various cell types can be linked up satisfactorily in a naturally graded series by the study of cells which seem to be intermediate in form. With the further development of the tissue-culture technique, and particularly the method of recording changes in living cells by micro-cinematography, it is probable that many of these difficult cytological problems will be solved. Indeed, there has already come to hand evidence of this kind which tends to confirm or alter some of the earlier conceptions regarding the developmental potentialities of blood cells.

We have noted that, in the early embryo, white and red corpuscles arise in the first instance from angioblastic cells which become differentiated from primitive mesenchyme. In this sense, of course, all types of corpuscle no doubt have a common ancestry. It is in regard to the origin of corpuscles during postnatal life, however, that differences of opinion become conspicuous.

It may be said that orthodox teaching in the past gave three sources for the derivation of blood cells. Erythrocytes and granulocytes were believed to originate, mainly in the bone marrow, from two different embryonic types of cell, while lymphocytes were assumed to be formed in germinal centres in lymphoid tissue from the active division of more generalized cells called lymphoblasts. This is the polyphyletic conception of blood-cell

[1] C. H. W. Leeksma and J. A. Cohen, 'Determination of the life span of human blood platelets', *Journ. Clin. Invest.* **35**, 1956.

development. In direct opposition to this was the monophyletic idea, according to which all the cellular elements arise from a primitive amoeboid and basophil stem-cell called a *haemocytoblast*, the latter being in structure similar to, and by some considered even to be identical with, the lymphocyte. This cell, again, was said to give rise severally to erythroblasts, monocytes, myelocytes, and histiocytes. The main difficulty in the way of accepting this account of haemopoiesis was the identification of the primitive and generalized haemocytoblastic cell with the lymphocyte. Under normal conditions in the adult at least, lymphocytes seem to be fairly stable elements, separated morphologically and in their functional properties from other types of blood corpuscle, though there is some evidence that they are capable of being transformed into larger cells similar to, and perhaps identical with, lymphoblasts, and that these cells are able to proliferate by cell division (see p. 252).[1] Otherwise they have commonly been regarded as specialized end products of cell differentiation, rather than generalized and plastic cells with wide potencies for development into diverse forms. Indeed, this view is now well established. In any case, it may be noted that lymphoid tissue, which contains large aggregations of lymphocytes and which appears to be the site of their production, is rather clearly segregated from myeloid tissue where red corpuscles and granulocytes are formed, and an abnormal activity of the one tissue bears no obvious relation to that of the other. Pathological conditions associated with excessive destruction and production of red corpuscles may result in hypertrophy and other changes in bone marrow while lymphoid tissue appears to remain unaffected, and over-activity of lymphoid tissue with ensuing lymphocytosis may proceed independently of obvious disturbances of haemopoietic tissues. These observations have provided further evidence against the suggestion that lymphoid tissue has any relation to the production of blood corpuscles other than lymphocytes. Although the results of recent research work has led to the complete abandonment of the idea that lymphocytes play any part in haemopoiesis, it has seemed desirable to make mention of it, partly because this conception still finds lingering consideration in some currently read textbooks, and partly because it has a certain interest in the history of attempts to solve a hitherto unsolved problem, that is, although many millions of lymphocytes enter the blood stream daily from the lymphatic system, yet the lymphocytic content of the blood remains relatively constant (see p. 255).

It is not possible in this brief account to do more than indicate the general problems of haemopoiesis. Whatever their *ultimate* origin and their potentialities for further differentiation, the *immediate* origin of blood corpuscles in the adult may be summarized as follows. Red corpuscles are produced from primitive nucleated forerunners in red marrow, which lose their nuclei before being set free in the circulation. Granulocytes (i.e. polymorphonuclear, eosinophil, and basophil leucocytes) arise from the granular myelocytes which are actively produced in large numbers only in red marrow. Lymphocytes are formed by the proliferation and differentiation

[1] K. Carstairs, 'The human small lymphocyte: its possible pluripotential quality', *Lancet*, 21 April, 1962.

of precursor cells in some of the lymphoid organs of the body. These precursor cells are probably those termed lymphoblasts. Finally, the immediate precursors of monocytes are immature cells called monoblasts which, in turn, are differentiated from extravascular reticular cells of the bone marrow and spleen (and probably of lymphoid tissue generally), and also from the sinusoidal endothelium of the spleen, liver, and bone marrow.

3. HAEMOPOIETIC TISSUES

In the embryo, blood formation may occur in a variety of places in the body; we have already noted that this process begins in the primary mesoderm of the yolk sac, the body stalk, and chorionic villi (p. 207).

The placenta provides a centre for the active development of blood corpuscles during early embryonic life, but later this centre shifts to the liver and spleen. In mid-foetal life (and to some extent almost up to the time of birth), the liver is actively concerned with haemopoiesis, and here masses of erythroblasts and myelocytes are formed in the mesenchymal tissue immediately outside the vascular endothelium. It is for this reason, indeed, that the liver is relatively so much larger in the foetus than it is in postnatal life. In the foetal spleen, also, both red and white corpuscles are actively produced during the later months of intra-uterine life. Apart from the fact that lymphocytes and large mononuclear cells continue to be produced throughout life in lymphoid tissue, towards the end of foetal life the main centres of haemopoietic activity shift once more and become localized in the red marrow of bones. Marrow already participates in this function to some degree during foetal life, but except for the last two months its blood-forming activity is quite secondary to that of the liver and spleen.

In a new-born infant the bony cancellous tissue in all parts of the skeleton is occupied by red marrow, but at this age it has not acquired the dark-red colour characteristic of red marrow in the adult. By the age of 7, fatty tissue begins to appear in the middle of the shafts of the limb bones, replacing the haemopoietic tissue of red marrow, and at 12 it becomes evident macroscopically as a patch of yellow marrow. The latter gradually extends in both directions along the shaft until it reaches the extremities. Even in the adult, however, a small amount of red marrow persists in the region of the metaphysis at the proximal ends of the humerus and femur. In the epiphyses of the long bones (as well as in the bones of the carpus and tarsus), the red marrow is also superseded by yellow marrow, a process which is usually complete by the age of 20 except for the occasional persistence of a few small patches of red marrow.

In the cancellous tissue of the ribs, vertebrae, sternum, skull bones, and os innominatum, red marrow persists as actively haemopoietic tissue throughout life. It is important to recognize, also, that, in the adult, yellow marrow may again be replaced by red marrow under exceptional circumstances, as in certain types of anaemia in which excessive blood destruction requires to be compensated for by increased haemopoiesis.[1] It is also an interesting fact that, in the case of severe destruction of the bone marrow (e.g. in the condition of myelofibrosis), the liver and spleen may revert to the haemopoietic functions of their foetal career.

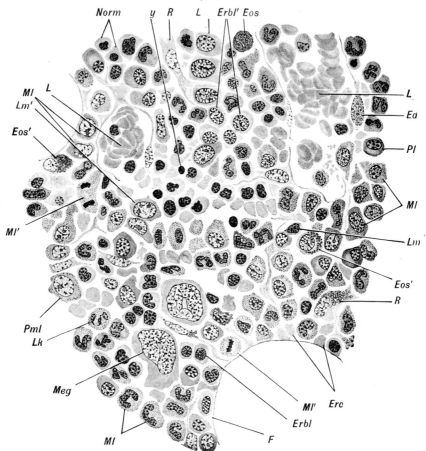

FIG. 83. Section of human bone marrow. *L.* Venous sinus. *Ed.* endothelial cell. *Pl.* Plasma cell. *Ml.* Neutrophil myelocyte. *Lm.* Lymphocyte. *Eos'.* Eosinophil myelocyte. *R.* Reticular cell. *Erc.* Red blood corpuscle. *Ml'.* Neutrophil myelocyte undergoing mitotic division. *Erbl.* Basophilic normoblast. *F.* Fat cell. *Meg.* Megakaryocyte. *Lk.* Polymorphonuclear leucocyte. *Lm'.* Large lymphocyte. *Norm.* Normoblast. *y.* Free normoblast nucleus. *Eos.* Eosinophil leucocyte. *Pml.* Neutrophil promyelocyte. (From A. Maximow and W. Bloom.)

In structure, red marrow contains a loose stroma which is rather similar to the reticular meshwork of lymphoid tissue. That is to say, it is formed of an interlacing network of reticular (argentophil) fibres entangled in which are reticular cells. The latter have phagocytic properties, and they can be mobilized and transformed into freely amoeboid macrophages.

[1] It is interesting to note that myeloid tissue may become differentiated in association with atypical areas of calcification—e.g. in calcified cartilages of the larynx, and in calcareous deposits in the walls of arteries.

The stroma is permeated with sinusoidal vessels lined by endothelial cells whose outlines are often indistinct and which, like the Kupffer cells of the hepatic sinusoids, can also become converted into active macrophages. It has been supposed that the vascular endothelium in red marrow is deficient, so that the vessels open out directly into tissue spaces. However, perfusion experiments suggest that the sinusoidal capillaries are not permanently in direct communication with the marrow parenchyma; on the other hand, as already noted, they may become temporarily open vessels when maturing red corpuscles push their way through the endothelial lining to reach their lumen.[1] In the extravascular tissue spaces can be seen a great variety of blood corpuscles showing all degrees of development— nucleated and mature red cells; myeloblasts, myelocytes, and granulocytes; lymphocytes and mononuclear cells (Fig. 83). In addition, conspicuous giant cells are also present. These are called *megakaryocytes*.[2] They reach a diameter of 40 μ, and are distinguished by possessing a large lobulated nucleus which often takes on a characteristic annular form. As already noted (p. 233) they give origin to the blood platelets, and this they do by fragmentation of their cytoplasm.

[1] C. K. Drinker, K. R. Drinker, and C. C. Lund, 'The circulation in mammalian bone marrow', *Amer. Journ. Physiol.* **62**, 1922.

[2] Megakaryocytes are also found in the spleen, and in relation to the blood vessels of the lung.

X

LYMPHATIC TISSUES

THE lymphatic system, in its narrower sense, consists of a system of fine absorptive vessels permeating most of the tissues of the body and emptying their contents ultimately into the venous system. Along the course of these vessels are groups of lymph nodes (or lymph 'glands') which are small compact bodies composed in the main of masses of lymphocytes entangled in a reticulum of connective-tissue fibrils. Localized masses of lymphoid tissue of different types, closely related in structure to these lymph nodes, are also found in other parts of the body, such as the palatine tonsils, spleen, thymus, and scattered lymphoid follicles embedded in mucous membranes. The functional significance of these aggregations of lymphocytic cells is still in doubt. The function of the lymphatic vessels, however, is fundamentally one of absorption from the connective-tissue spaces of the body, and the role they play is of extreme importance in the consideration of the defence of the body against infection and of the spread of deleterious material from one part of the body to another.

Lymphatic vessels form peripherally a rich meshwork of capillaries which unite into larger trunks until the main vessels are reached. They convey a colourless fluid called lymph which is ultimately poured by way of the thoracic duct (and the right lymphatic duct) into the veins at the base of the neck. During its passage from the periphery to the thoracic duct, the lymph is usually filtered through one or more series of lymphatic nodes.

Even the main lymphatic vessel of the body, the thoracic duct, is in size only comparable to quite a small vein; hence the lymphatic vascular system as a whole is not very evident in gross dissection. Although lymphatic vessels were well known to the ancient anatomists in the third century B.C., it was not until about 2,000 years later that they were 'rediscovered' and described in some detail. In 1627 Aselli (Professor of Anatomy at Pavia) noted them in the mesentery of the intestine of a 'well-nourished dog'.[1] His attention was drawn to them by reason of the fact that the mesenteric vessels convey from the small intestine a large proportion of the fat which is absorbed there, and after a meal rich in fats they beome filled with a milky fluid called 'chyle' which serves to bring them in evidence to naked-eye examination.

1. LYMPHATIC CAPILLARIES

The peripheral lymphatic capillary plexuses are very similar to those of the blood vascular system. In general they lie somewhat deeper than the blood capillaries beneath epithelial surfaces, and the individual vessels, besides being larger, are more irregular in contour, showing dilatations

[1] G. Aselli, *De Lactibus*, Milan, 1627.

and constrictions in their course. As in blood capillaries, the endothelium consists of flattened pavement cells with wavy margins forming interlocking junctions which can be demonstrated by treatment with silver salts (Fig. 84). The endothelium is in direct contact with the surrounding tissues without a perithelial covering.[1]

For many years there was considerable doubt on the question whether or not lymphatic capillaries open freely into the connective-tissue spaces in which they lie. It now appears reasonably certain that they form a closed system, everywhere lined by a complete and intact endothelium, the terminal capillaries ending in a plexus, or blindly in pointed diverticula.

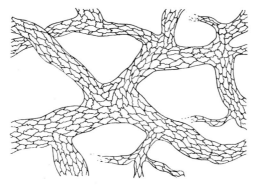

FIG. 84. Drawing of a lymphatic capillary plexus (from the nasal mucous membrane of a rabbit).

Openings or 'stomata' had been described forming gaps between adjacent endothelial cells, but even these openings have now been shown to be artefacts produced by imperfect histological technique. It is evident, therefore, that absorption into the lumen of the lymphatic capillaries must take place through the endothelial cells or between the junctional margins of these cells. The accumulation of experimental evidence shows that both routes are used. In general, it may be said that lymphatic vessels provide the principal route for the absorption of particulate matter and colloid material from connective-tissue spaces, while the blood capillaries, in so far as they play a significant part in absorption from tissue spaces, take up the soluble crystalloids. This, however, is by no means a sharp distinction. As the molecular size of the substance increases, there is a progressive shift from blood to lymph absorption. When a molecular size of the dimensions of serum albumen is reached, absorption by blood capillaries practically ceases and the lymphatic route is alone used. Consequently in cases of lymphatic obstruction in which the tissues become oedematous, they become filled with fluid containing a high proportion of protein.[2]

[1] J. M. Yoffey and F. C. Courtice, *Lymphatics, Lymph, and Lymphoid Tissue*, London, 1956. I. Rusznyák, M. Földi, and G. Szabo, *Lymphatics and Lymph Circulation*, Pergamon Press, 1960. Readers are referred to these monographs for comprehensive accounts of the lymphatic system as a whole.

[2] In their studies of the absorption of toxins from the tissues, Barnes and Trueta (*Lancet*, May 1941) found that viper venom and diphtheria and tetanus toxins, all with a

We must now consider the mechanism of absorption by lymphatic capillaries. This has been studied experimentally, and also by the observation of living capillaries in tissues suitable for anatomical study *in vivo*. As regards lower vertebrates, such observations have been made on the vessels in the transparent tails of amphibian larvae, while mammalian lymphatics have been studied by means of transparent chambers in the ears of rabbits (see p. 6).[1] These technical methods have provided the opportunity for watching the growth and repair of lymphatic vessels, and their reactions under normal and abnormal conditions to mechanical, chemical, and other stimuli.

As already indicated, lymphatic vessels are largely concerned with the absorption of colloid material and extravascular protein material which is not normally taken up by blood capillaries, suggesting that lymphatic endothelium is considerably more permeable than the endothelium of blood vessels. However, the permeability of lymphatics may rapidly change from time to time, for it is markedly increased by mechanical irritation, chemical stimulation, and rise of temperature. It has also been noted that lymphatic capillaries are concerned with the removal of insoluble particles from connective-tissue spaces. A common method of absorption of such particulate material is its phagocytosis by extravascular phagocytes (leucocytes or macrophages) which then gain entrance to the lumen of the vessel by actively penetrating the endothelium with its ingested material. This, however, is by no means the only method of absorption.

In amphibian larvae it has been observed that if a foreign particle (such as a blood corpuscle or a carmine particle) is placed in the connective tissue alongside a lymphatic vessel, a small sprout grows out of the vessel, extending as a diverticulum until its endothelial wall comes into contact with the particle (Fig. 86). The latter then passes through the endothelial cells (which apparently undergo a softening in consistency) until it reaches the lumen, when the diverticulum retracts once more and the capillary assumes its original shape. Although such temporary outgrowths from lymphatic vessels have not been seen to occur in mammalian tissues, the process of absorption through the intact endothelium appears to be otherwise identical. This process, it seems, involves some kind of phagocytosis on the part of the endothelial cells. The latter do not show all the characteristic properties associated with true phagocytes, since, although they may in certain circumstances take up colloidal dyes like the phagocytic cells of the macrophage system, they do not become freely motile. Nevertheless, these observations indicate that lymphatic endothelial cells have the property of

molecular weight exceeding 20,000, are absorbed almost entirely by lymphatic vessels, whereas cobra venom, with a molecular weight below 5,000, is absorbed with equal facility by blood vessels. More recently, the absorption of plasma protein labelled with radioactive iodine has been studied both in animals and in human subjects, and has been shown to be removed from subcutaneous tissues by the lymphatic vessels; only when the latter route is interrupted is there any appreciable absorption into capillary blood vessels (G. W. Taylor *et al.*, 'Lymphatic circulation studied with radioactive plasma protein', *Brit. Med. Journ.* 19 Jan., 1957).

[1] E. R. Clark, 'Observations on living growing lymphatics in the tail of frog larvae', *Anat. Rec.* 3, 1909; E. R. Clark and E. L. Clark, 'The fate of extruded erythrocytes; their removal by lymphatic capillaries and tissue phagocytes', *Amer. Journ. Anat.* 38, 1926; E. R. Clark and E. L. Clark, 'Observations on the new growth of lymphatic vessels as seen in transparent chambers in the rabbit's ear', ibid. 51, 1932.

transporting in some way particulate material from without into the lumen of the vessel. It has been argued that this mechanism of absorption will not

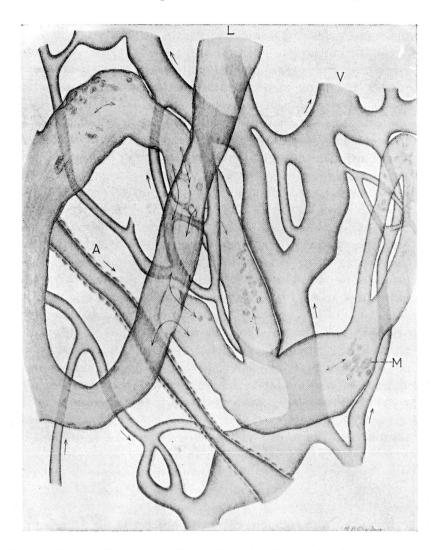

FIG. 85. Camera-lucida drawing of lymphatic and blood-vessel plexuses in a transparent chamber in a rabbit's ear. The two sets of vessels are shown intertwining but not anastomosing. *L.* Lymphatic vessel. *A.* Artery. *V.* Vein. *M.* Macrophage. × 260. (From E. R. Clark and E. L. Clark.)

altogether account for the rapidity with which, in some conditions, absorption by lymphatic vessels may take place.

That temporary ruptures of the lymphatic endothelium may occur, resulting in an impairment of the vessel wall which allows free entry of material into the lumen from outside, has also been shown.[1] Very slight trauma may produce such a rupture, and the opening may persist for

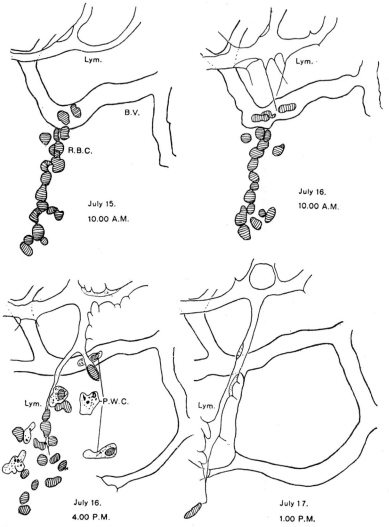

FIG. 86. Series of records showing the growth of a lymphatic capillary towards an extravasation of blood, and the picking up of red blood cells by the newly formed lymphatic sprout and by wandering cells. *Lym.* Lymphatic vessel. *B.V.* Blood vessel. *R.B.C.* Red blood cells. *P.W.C.* Pigmented wandering cells. (From E. R. Clark and E. L. Clark.)

several days. Such artificial openings (though to be regarded as exceptional) are liable to occur in conditions such as inflammation, when free fluid accumulates in the connective-tissue spaces. Hence they provide a means whereby living bacteria may find their way directly into lymphatic vessels.

[1] E. R. Clark and E. L. Clark, 'Observations on living mammalian lymphatic capillaries —their relation to blood vessels', *Amer. Journ. Anat.* **60**, no. 2, 1937.

It has been found that particulate material (such as blood corpuscles and fat globules) or soluble material (as shown by the injection of soluble dyes into the blood stream) can pass *directly* from blood vessels into immediately adjacent lymphatic vessels, penetrating the intact endothelial walls where these come into contact. This phenomenon may explain the presence of red blood corpuscles which are sometimes to be observed in lymphatic vessels. It may also be of considerable significance in those tissues (e.g. the liver) where lymphatic and blood capillaries lie in very close relation with each other. According to Clark and Clark, 'it is probable that much of the fluid in the lymphatics draining the liver—which is generally considered to constitute a large proportion of the lymph entering the thoracic duct—represents a direct "leakage" from the perilobular portal vessels to the accompanying lymphatics and that it has not been in contact with the cells of the liver lobule'. This statement has important implications in regard to the observation that if the pressure in the lobular blood capillaries of the liver is raised by compressing the vena cava just above the diaphragm, there is an immediate outpouring of highly proteinized lymph from the thoracic duct.

In acutely inflamed and oedematous tissues the lymph flow is often considerably accelerated and the lymphatic capillaries become markedly dilated. The mechanism of this dilatation has been studied by Pullinger and Florey.[1] They suggest that it is a passive process dependent upon the fact that the collagen fibrils of the surrounding connective tissue are directly adherent to the endothelial wall of the lymph vessels. With the swelling of the connective tissue in oedema the increased tension of these fibrils exerts a traction on the vessel wall which widely opens the lumen. The dilatation is thus quite distinct from that of blood capillaries in inflamed tissues, for in the latter case it is caused by the increased internal pressure of the circulating blood.

Direct and indirect observations have shown that the flow of lymph in peripheral lymphatics is accelerated by movement and also by raising the blood capillary pressure. As a practical corollary of these observations it may be noted that the best method of increasing the lymph drainage of any part of the body is to produce a temporary venous congestion (e.g. by applying local pressure to the veins from the part) and to follow this up by active movement or massage. This is important in view of the fact that chronic inflammatory conditions sometimes lead to a local stagnation of the lymph flow.

2. COLLECTING LYMPHATIC VESSELS

The lymph from the peripheral plexuses passes into the collecting channels of the lymphatic system. These vessels are similar to small venules in their endothelial lining, and in possessing valves. The latter, however, are much more numerous than in veins.[2] They are always paired, being formed by a reduplication of the endothelium between the layers of which

[1] B. D. Pullinger and H. W. Florey, 'Some observations on the structure and functions of lymphatics', *Brit. Journ. Exp. Path.* **6**, 1935.

[2] Valves are said to be absent in the lymphatic vessels of cold-blooded vertebrates in which the lymph circulation is directed by rhythmically contractile 'lymph hearts'.

is a thin support of delicate connective tissue. The valves direct the flow of lymph, and in vessels which radiate in different directions from an area towards various groups of lymph nodes they are disposed accordingly. Lymphatic vessels in the body frequently show local dilatations immediately proximal to each pair of valves, which give the vessels a characteristic beaded or knotted appearance.

Three layers can be distinguished in the walls of the large lymphatic vessels, tunica intima, media, and adventitia, and in their relative development and consitution they are very similar to those of small veins. The tunica intima consists of the endothelium with a thin sub-endothelial coat of connective tissue, the tunica media contains circular and oblique muscle fibres mingled with connective tissue and some sparse elastic fibres, while the tunica adventitia (which is the thickest) consists mainly of a condensed interlacing network of connective-tissue fibres (collagenous and elastic) with some plain muscle fibres. The plain muscle in the walls of lymphatic vessels is usually more abundant immediately proximal to the valves, and in this position it probably plays a part in the propulsion of the lymph by active contraction.

Lymphatics have been observed to contract on sympathetic stimulation. Contraction in response to electrical and mechanical stimuli of lymphatic capillaries without any muscle fibres in their wall has also been observed in some animals, from which it appears that the endothelium itself may have contractile properties.

Lymphatic vessels are innervated by very fine non-medullated fibres forming a network in the tunica adventitia. These nerves are believed to be predominantly motor, since they are mainly found in vessels with muscle fibres in the tunica media. However, fine terminals may extend as far as the tunica intima where endings of a receptor type have been described. Rhythmic contractions of lymphatic vessels have been observed in certain mammals (e.g. rat and guinea-pig), and also in human patients in the course of surgical operations, but it is not certain whether this contractility is dependent on nervous impulses.[1] Besides a nerve supply, the walls of the larger lymphatics are vascularized by fine blood vessels.

The thoracic duct. The lymph from the greater part of the body is eventually poured into the main lymph vessel—the *thoracic duct*, which commences in an elongated sac, the *cisterna chyli*. In man, the thoracic duct is about 18 inches long and rather varicose in appearance, its diameter being on the average about 2 or 3 mm. Extending up through the thorax in close contact with the bodies of the thoracic vertebrae, it ultimately reaches the base of the neck on the left side where it opens into the commencement of the left innominate vein, i.e. at the junction of the left internal jugular and subclavian veins. The duct has a structure similar to a vein of comparable size, except that in the tunica media the plain muscle tissue is rather more predominant.

[1] H. M. Carleton and H. Florey, 'The mammalian lacteal; its histological structure in relation to its physiological properties', *Proc. Roy. Soc.* B, **102**, 1927; J. B. Kinmonth and G. W. Taylor, 'Spontaneous rhythm in contractility in human lymphatics', *Journ. Physiol.* **133**, 1956.

It is important to note that the opening of the thoracic duct into the innominate vein is not the only route whereby lymph reaches the blood stream. On the right side of the neck is the *right lymphatic duct* which, by the union of the right jugular trunk, subclavian trunk, and mediastinal trunk, conveys lymph from the right side of the head and neck, the right

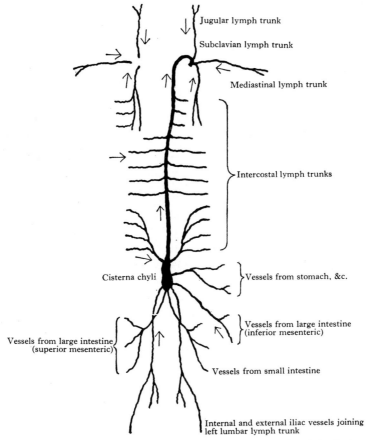

Jugular lymph trunk

Subclavian lymph trunk

Mediastinal lymph trunk

Intercostal lymph trunks

Cisterna chyli

Vessels from stomach, &c.

Vessels from large intestine (inferior mesenteric)

Vessels from large intestine (superior mesenteric)

Vessels from small intestine

Internal and external iliac vessels joining left lumbar lymph trunk

FIG. 87. Diagram showing the main lymphatic drainage vessels of the body. (From Cunningham.)

upper extremity and the right side of the thorax, to the right innominate vein. Quite commonly, these three trunks open separately into the right internal jugular, subclavian, and innominate veins. It is also to be noted that in some animals—e.g. the cat and dog[1]—the thoracic duct may be in open communication with the azygos vein in the thorax. In the rat lymphatico-venous communications have been found with the inferior vena cava and the portal vein.[2] In certain South American monkeys, again, the

[1] L. W. Freeman, 'Lymphatic pathways from the intestine in the dog', *Anat. Rec.* **82**, 1942.
[2] T. T. Job, 'Lymphatico-venous communications in the rat', *Amer. Journ. Anat.* **24**, 1918.

lymphatic vessels from the abdominal viscera and the lower limbs open constantly into the renal veins or into the inferior vena cava at this level.[1] A similar communication has been observed in some species of squirrel. Such variations are explicable on the embryological origin of lymphatic vessels as outgrowths from veins. The communicating channels of these outgrowths usually become obliterated except in the case of the thoracic and right lymphatic ducts at the base of the neck. Nevertheless, the recognition of their occasional persistence in other regions—either as individual anomalies or specific variations—is of obvious importance in the experimental investigation of the factors controlling the circulation of the lymph.[2]

The thoracic duct is well provided with valves. However, its entry into the innominate vein is not guarded very efficiently, and there is frequently observed a regurgitation of venous blood into the upper part of the duct.

The cisterna chyli is to be regarded as the sole survivor of a series of lymph sacs which are formed embryologically in the earliest development of the lymphatic system (*vide infra*, p. 263). In man it is an elongated dilatation of the lower end of the thoracic duct, measuring about 2 or 3 inches in length and $\frac{1}{4}$ inch in width, and situated on the anterior surface of the first and second lumbar vertebrae in the posterior abdominal wall. It receives lymphatic trunks draining the lower limbs, the pelvis, most of the abdominal viscera, and the lower part of the thorax (Fig. 87). It is extremely variable in its form and may be partially or completely replaced by a plexus of lymphatic vessels intermingled with lymph nodes.

3. THE GENERAL DISTRIBUTION OF LYMPHATIC VESSELS

Lymphatic vessels are absent in those tissues which are devoid of a blood supply—e.g. hyaline cartilage and epidermis. They have also not been certainly observed in the bone marrow or the red pulp of the spleen. In the central nervous system they are entirely absent—indeed they have never been observed inside the skull or the vertebral canal. It should be noted, in this connexion, that perivascular channels and other spaces in connective tissue have sometimes been referred to as 'lymph spaces'. This designation is quite erroneous, for they form no part of the lymphatic vascular system and do not contain lymph.[3] Similarly, serous cavities such as the peritoneal and pleural sacs, as well as bursal and joint cavities, were at one time included by some authors under the category of the lymphatic system.

[1] C. F. Silvester, 'On the presence of permanent communications between the lymphatic and venous systems at the level of the renal veins in South American monkeys', *Amer. Journ. Anat.* **12**, 1912.

[2] For an account of the variations in the lymphatico-venous connexions in different mammals, see McClure and Silvester, *Anat. Rec.* **2**, 1909. It may be noted that direct lymphatico-venous connexions have been described as of normal occurrence in the thyroid gland. Their existence, however, is not proved.

[3] The use of the term 'lymph' for tissue fluid has also led to some confusion in the past. By definition, the term can only be applied to fluids identical in composition with the contents of the lymphatic vessels.

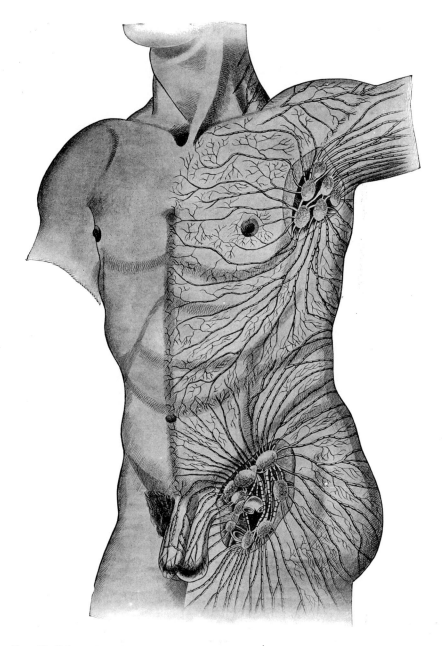

FIG. 88. Schematic representation of the superficial lymphatic vessels of the anterior surface of the trunk, showing their termination in axillary and inguinal groups of lymph nodes. (From Cunningham after Sappey.)

However, such cavities are developed independently of, and in a manner different from, lymphatic vessels, and they have no direct connexion with the latter (although the older anatomists supposed this to be the case).

Where lymphatics are present they usually run in the looser areas of connective tissue. Hence their distribution is determined to a large extent by fascial planes, and they also show a tendency to run close alongside blood vessels in the loose tissue of the perivascular spaces. In the limbs and the body wall, the superficial lymphatics of the integument are separated rather sharply from the deep lymphatics by the deep fascia, particularly where the latter is dense and aponeurotic as in the lower extremity. Anastomoses between the superficial and deep sets of vessels are practically insignificant, and the two unite only proximally where the main collecting trunks of the superficial set pass through gaps in the deep fascia to enter the underlying tissues. The recognition of this fact has formed the basis of the treatment of certain types of lymphatic obstruction such as elephantiasis. In this tropical disease, the superficial lymphatics of the lower extremity are sometimes blocked at the top of the thigh through the invasion of the vessels by the filaria parasite. One operative procedure which has been devised to deal with this condition (but unfortunately not with unqualified success) is the removal of a long strip of the deep fascia so as to allow the establishment of free anastomoses between the obstructed superficial lymphatics and the deep lymphatics, thus providing alternative routes for the drainage of the skin.

The abdominal and thoracic viscera are richly supplied with lymphatics and, here again, the peripheral vessels are divided into two sets, a submucous and a subserous plexus. These have been commonly regarded as fairly distinct with very sparse anastomotic connexions. However, studies on the lymphatics of the stomach have shown that a mutual anastomosis between the two sets is considerably more free than has been generally realized.[1]

The collecting trunks of the lymphatic system usually run in close relation to the larger blood vessels, especially veins. Indeed, the course followed by lymphatic vessels draining any particular region of the body can almost always be inferred from a consideration of the corresponding venous drainage. This intimate relation between the main lymphatic and venous channels is perhaps to be expected on the basis of the initial derivation of lymphatic vessels from the venous system in the embryo.

4. METHODS OF INVESTIGATING THE ANATOMY OF LYMPHATIC PATHS

The clinical importance of a knowledge of lymph drainage routes has led to extensive anatomical studies in order to map out these paths. Direct dissection is of little use for this purpose, since lymphatic vessels are so tenuous and difficult to see. It is therefore necessary to display them by experimental methods, or to infer their disposition by tracing the

---

[1] J. H. Gray, 'The lymphatics of the stomach', *Journ. Anat.* **71**, 1937.

destination in lymph nodes of pathological or experimentally injected material from any particular part of the body. The clinical observation of the position of lymph nodes involved in the spread of inflammatory conditions or of malignant tumours from different organs or tissues has led to the accumulation of very valuable data regarding drainage routes in man. Local injections of soluble dyes or supensions of particulate material (such as carmine or bacteria), and their subsequent detection by histological examination in lymph nodes, have provided similar information with regard to lower mammals.

Direct injection of lymphatic vessels allows the visual demonstration of drainage routes. Peripheral lymphatics may be shown up by perfusion of the blood-vascular system with silver nitrate, the lymphatic vessels being then injected directly with material such as mercury. This is an old method that was used by Sappey in 1874, and in many anatomical museums admirable preparations of this type may be seen today. The common method now employed for injecting lymphatic vessels is to insert a hypodermic needle in the loose connective tissue of the region to be investigated and inject blindly. The injection materials used are various, but we may here mention Gerota's injection mass (which contains Prussian Blue), aqueous solutions of dyes such as Berlin Blue, and Indian ink. By using steady pressure on the syringe, and varying the position of the needle point, the injection can usually be persuaded to enter lymphatic vessels. The flow of the material is aided by light massage of the part, and in this way the vessels may be filled up to the lymph nodes in which they terminate. This method is used on dead tissues. It was originally employed by Jamieson and Dobson on the human foetus for an extensive investigation of the lymphatic system in man,[1] the results of which have provided the basis for many of the diagrams in modern textbooks of human anatomy.

Lymphatic routes may also be demonstrated in living animals experimentally by the local deposition in the tissues of soluble dyes or Indian ink. Within a short time the material is taken up into the vessels, and the lymph nodes in which they terminate can be identified by their staining. Potassium ferrocyanide can be used in the same way. In this case the blood-vascular system is washed through with normal saline immediately after death in order to remove the solution from the blood capillaries, and the tissues are then placed in weak hydrochloric acid, which leads to the deposition of Prussian Blue in the lymphatic vessels and lymph nodes which have taken up the ferrocyanide.

It is possible to inject lymphatics in a reverse direction in spite of the presence of valves. This is done by first ligaturing the main ducts during life, which results in a distension of the peripheral tributaries and a resulting incompetence of their delicate valves. It is suggested that in certain pathological conditions a retrograde flow of lymph may similarly occur as the result of blockage of the main channels and that bacteria may thus be carried along the vessels in a distal direction.

The demonstration of lymphatic vessels by radiography has provided

[1] J. K. Jamieson and J. F. Dobson, 'On the injection of lymphatics by Prussian Blue', *Journ. Anat.* **45**, 1910–11.

opportunities for the study not only of the direction of lymph flow in the living body, but of the rate of flow and the factors which influence it.[1] For example, by using lipiodol or thorotrast for injecting the peripheral lymphatics in animals it is possible to follow the flow of lymph, the formation of collateral paths in the case of obstruction of the usual channels, and the regeneration of lymphatic vessels. From such observations, it appears very doubtful whether a steady and continuous flow of lymph occurs in the peripheral vessels during rest. On the other hand, the radiographic shadows alter their appearance very rapidly if active or passive movement of the part is allowed. In recent years the visualization of lymphatic vessels by the injection of dyes and contrast media (lymphangiography) has been extensively applied for the purpose of clinical diagnosis.[2]

5. LYMPH NODES

In the course of the main lymphatic vessels in certain regions, and usually occurring in fairly well-defined groups, are compact bodies formed of lymphoid tissue and called lymph nodes, or lymph 'glands'. The latter term, which is still in common usage, is somewhat misleading, for it may be taken to imply secretory functions which they do not possess.

Lymph nodes are flattened oval, or rounded, structures with a smooth surface. In man they vary in diameter from a few millimetres to as much as an inch or so. Commonly they are somewhat bean-shaped, with a hilum. An efferent lymphatic vessel (usually single) leaves the hilum of each lymph node, while a number of afferent vessels, carrying lymph into the node, enter it along the convex border. In these vessels the valves are disposed so as to direct the lymph through the lymph node in a proximal direction. Normally the lymph drained by all lymphatic vessels from any part of the body passes through one or more lymph nodes in this way before it is finally collected into the thoracic or right lymphatic duct and poured into the blood stream. Lymph nodes are particularly numerous at the proximal ends of the limbs, in the neck, the pelvis, the mesenteries of the abdominal viscera, the posterior abdominal wall, and the mediastinal region of the thorax. As we shall see, their precise location and the areas from which they receive lymphatic vessels are of great importance from the clinical point of view.

In structure, a lymph node consists essentially of masses of lymphocytes enmeshed in a network of fine connective tissue and enclosed altogether in a firm, smooth capsule (Figs. 89 and 90). The capsule is formed of fibrous tissue in which there is a certain amount of plain muscle. The latter is particularly evident in certain mammals such as the cat, and it is probable that its contraction may help to accelerate the flow of lymph from the afferent to the efferent vessels. From the deep aspect of the capsule irregular trabeculae of fibrous tissue (evident in some mammalian species but inconspicuous in man) extend into the substance of the lymph node,

[1] S. Funaoka et al., 'Die Röntgenographie des Lymphgefäßes', Arb. a. d. Anat. Inst. d. Kaiserl. Univ. Kyoto, 1930.
[2] M. H. Gough, E. J. Guiney, and J. B. Kinmonth, 'Lymphangiography: new techniques and uses', Brit. Med. Journ. 1, 1963.

partially dividing it into a number of loculated compartments, and, by splitting up into finer and finer processes which connect up with each other, form a fine reticulum in the centre of the node which provides a supporting framework for the lymphoid tissue. The peripheral loculi which are separated by the main trabeculae, and the central reticulum, are filled with dense collections of lymphocytes. In the loculi these form rounded and rather circumscribed masses which are termed *lymphoid follicles* or *nodules*.

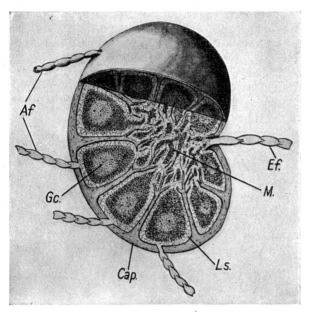

FIG. 89. Diagram illustrating schematically the essential structure of a lymph node. The node is enclosed in a capsule (*Cap.*) from which trabeculae penetrate into its substance. In relation to the capsule and the trabeculae are seen the lymph sinuses (*Ls.*) Germinal centres (*Gc.*) are shown in the middle of the lymph follicles. Afferent lymphatic vessels (*Af.*) enter the node along its convex surface, while an efferent vessel leaves at the hilum. The centre of the node or medulla (*M.*) is occupied by medullary cords of lymphoid tissue.

They comprise the so-called cortex of the lymph node, the central and somewhat more diffuse tissue comprising the 'medulla'. In the medulla the lymphocytes tend to be arranged in anastomosing cords of cells. A clear-cut distinction between cortex and medulla, however, is usually quite absent.

Between the cortical lymphoid follicles and the capsule with its septal processes are zones of comparatively loose tissue which form a series of lymph channels or *sinuses*. The latter send extensions from the cortex into the medulla along the finer ramifications of the main septa. They ultimately reach the hilum of the node, thus forming a continuous series of channels linking up the afferent with the efferent lymphatic vessels. The lymph channel immediately under the capsule is termed the *marginal sinus*, and into it the afferent lymphatics open after piercing the capsule obliquely. The lymph sinuses are bridged across by fine trabeculae of connective-tissue fibres which are clothed by branched reticular cells with highly phagocytic properties.

In the centre of each lymph nodule in the cortex the lymphoid tissue is more loosely arranged, forming, in stained sections, a rounded clear area. These central areas are termed *germinal centres*, for they are the site of a very active cell proliferation containing cells somewhat larger than lymphocytes. These are the lymphoblasts, many of which can be seen in different

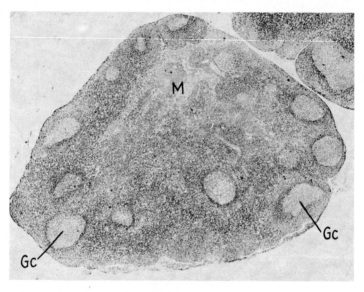

FIG. 90. Section through a small lymph node, stained with haematoxylin and eosin. Scattered round the periphery of the node are seen a number of circular germinal centres (*Gc*), embedded in the dense cortex. The looser tissue of the medulla is seen in the centre (*M*). The capsule of this small node is extremely thin. × 24.

phases of mitotic division. Surrounding each germinal centre is a dense zone of lymphocytes. The germinal centres show a cyclic rhythm of proliferative activity. They are absent in the foetus, and also disappear in old age. They are also absent in 'germ-free' animals, that is to say, animals which have been born and bred under completely sterile conditions. If, however, the animals are infected with pathogenic organisms or an allergic reaction of some sort is induced, germinal centres make an appearance. Until recently it was supposed that the centres are essentially or entirely concerned with the production of young lymphocytes to replenish the older cells as they come to the end of their life cycle. While this supposition is perhaps partly true, there is good reason to infer that they are also sites of formation of plasma cells whose main function, as we have seen (p. 52), appears to be the elaboration of antibodies. It remains possible, and even probable, that both plasma cells and lymphocytes are the maturation products of lymphoblasts, and that they are therefore closely related cellular types, but this is a conjecture that still remains without final proof. The lymphocytes which are so densely aggregated in lymph nodes find their way into the lymph channels and so gain entry into the efferent lymphatic vessel of the node. It is believed by some authorities that they may also enter directly the blood capillaries which vascularize the medulla.

The function of lymph nodes and vessels. From their anatomical disposition it is clear that lymphatic vessels are primarily absorbent in function. As already noted, it is possible under certain conditions for ab-

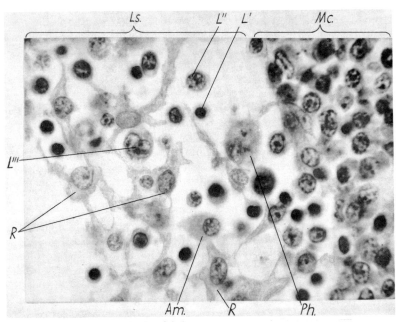

FIG. 91. Microphotograph of a portion of a section through a lymph node, showing a lymph sinus (*Ls.*) and part of a medullary cord (*Mc.*). A number of small lymphocytes (*L'*) are seen, as well as lymphoblasts (*L''*) and a large lymphocyte (*L'''*). These cells are held in a network of reticular cells (*R*). A wandering, amoeboid cell (*Am.*) is also seen. *Ph.* Macrophage with débris in the cytoplasm. × 850 approx. (Photograph by Dr. K. C. Richardson.)

sorption of material from the tissue spaces to be carried out by the blood capillaries. This they are able to do in virtue of the difference between the osmotic pressure of the blood plasma and that of the tissue fluid, but under conditions of intense tissue activity this difference in osmotic pressure is more than counterbalanced by the rise in 'hydrostatic pressure' within the dilated capillaries. Were it not for the lymphatic vessels, the tissues would then become filled with exudate and a condition of oedema would result. In addition, as already described, the lymphatic vessels—as the result of the high degree of selective absorption shown by their lining endothelium —are able more readily to remove from the tissue spaces colloid material and particulate material (such as cellular débris and foreign matter).

The mesenteric lymphatic vessels are particularly important for transporting fat which has been absorbed through the intestinal mucosa. It is estimated that practically all the ingested fat absorbed during digestion is conveyed to the blood stream by way of the thoracic duct. In each of the villous processes of the small intestine there is a central lymphatic vessel or 'lacteal'[1] which takes up particles of fat from the sub-epithelial tissue spaces. A small proportion of the fat may be transported into the lumen of the central lacteal by phagocytic cells, though this method of absorption is probably of little importance.[2]

For a long time one of the main functions of lymph nodes has been inferred from their microscopic anatomy to be the formation of lymphocytes. To what extent this is so now remains doubtful, for as we have noted above, plasma cells are almost certainly produced in the nodes and their relationship to lymphocytes is not securely known.

A secondary function of the lymph nodes (and of extreme importance clinically) is protective against the entrance of noxious material from tissue spaces into the blood circulation. They provide a series of very closely meshed filters through which the lymph must percolate before it reaches the thoracic duct and the blood stream. In its slow passage through the intricate labyrinth of lymph sinuses in the nodes, any deleterious material will tend to be taken up by the phagocytic reticular cells which line the sinuses. Thus, dust particles which gain entrance to the lungs in inhaled air are caught up by the lymph nodes at the hilum of the lung, and in town-dwellers who live in a smoke-laden atmosphere these nodes soon become black with carbon deposit. Living micro-organisms which gain entrance to the 'tissues may also be filtered off in the same way by the lymph nodes which drain the area. By this mechanism an infection is often localized to a considerable extent. It may lead to an enlarged, tender, and inflamed lymph node (lymphadenitis) which perhaps becomes the site of abscess formation, but it minimizes the possibility of a general septicaemia that might result from the entrance of bacteria into the blood stream.

Some writers have protested against the emphasis which is usually laid on the protective functions of the lymphatic vascular system and the lymph nodes. Far from guarding the body by limiting and localizing infection, they argue, the lymphatic vessels provide dangerous channels for the rapid spread of infectious material. But if it be granted that the vessels provide a necessary mechanism for the absorption under normal conditions of extravascular proteins and other material in the tissue spaces, the value of the lymph nodes as filters for clearing the lymph flow of noxious material becomes quite evident.

In acute local infections of the subcutaneous tissue, vessels which drain the affected area may themselves be inflamed (lymphangitis), and become visible as a series of red streaks in the skin extending up towards lymph nodes. This appearance depends on the fact that the endothelium of the collecting trunks is permeable to certain chemical substances (produced in the inflammatory process) which induce a local hyperaemia in the blood capillaries immediately surrounding the vessels. It has been shown that

---

[1] The lymphatic vessels draining the small intestine are sometimes called lacteals because, following a meal rich in fat, they can be seen to contain a 'milky' fluid.
[2] E. H. Leach, 'The role of leucocytes in fat absorption', *Journ. Phys.* **93**, 1938.

adrenalin and histamin injected into the peripheral lymphatics can readily escape from the collecting vessels in this way.

It is well known that lymphatic vessels provide routes facilitating the extension of certain types of malignant neoplasm from the primary focus. On reaching a lymph node the cancerous cells may be held up temporarily; their proliferation in the node then leads to the formation of secondary growths.

A consideration of these facts serves to emphasize the importance to the clinician of an accurate knowledge of the topographical anatomy of the various groups of lymph nodes in the body, and of the precise territories which they drain. In a patient with an enlarged and inflamed node, the primary focus of infection must be sought for in the appropriate areas. In the case of the operative treatment of many types of cancer, the surgeon is not content with the removal of the primary neoplasm, but he must also excise, as extensively as possible, the lymph nodes and vessels which immediately drain the site of the growth in order to obviate, if he can, the subsequent appearance of a secondary growth.

6. THE ANATOMY OF LYMPH

The fluid basis of lymph is a plasma very similar to blood plasma, with similar proteins and salts. The protein composition shows a certain degree of variation depending on the part of the body from which it has been collected, and upon tissue activity. In the thoracic duct, it usually contains emulsified fat which is particularly evident after a meal rich in fat.

The cellular content of lymph consists almost entirely of lymphocytes. These range in number in the thoracic duct of man up to 5,000 or more per cubic mm. They are much more numerous in the efferent than in the afferent vessels of lymph nodes, and they can be considerably increased by massage over groups of nodes.[1] Yoffey[2] has shown that in the dog the lymphocytic content of the thoracic duct lymph varies partly with age (being more numerous in younger animals) and partly with hourly fluctuations. On the average, however, over 200,000,000 lymphocytes are poured into the blood stream every hour, enough to replace the entire lymphocytic content of the blood twice daily. Yet the lymphocytic content of the blood remains constant, except for slight hourly fluctuations. In some animals such as the rabbit the number of lymphocytes entering the blood circulation by the thoracic duct is sufficient to replace those already in the blood as much as five times daily, and it appears that, on the average, these cells remain in the blood for only about five hours. The problem is, therefore, what happens to them? For it was generally assumed that the great majority were newly formed lymphocytes produced by the proliferative activity of the germinal centres of the lymph nodes, though it was admitted that some might be cells which, after circulation in the blood stream, had been redeposited in lymph nodes. In this connexion it must be noted that, according to

[1] F. N. Haynes and M. E. Field, 'The cell content of dog lymph', *Amer. Journ. Physiol.* **97**, 1931.
[2] J. M. Yoffey, 'Variations in lymphocytic production', *Journ. Anat.* **70**, 1935–6.

the experimental observations of Sanders and Florey,[1] the operative removal in rats of most of the lymphoid tissue of the body (including the
lymph nodes, spleen, and thymus) is not followed by any marked fall in
the number of lymphocytes in the blood. But it was found that this
extensive lymphadenectomy is followed by the compensatory development
of new lymphoid tissue in other sites such as the liver and the lungs, and
that these new formations show histological evidence of excessive lympho-
cytogenic activity.

If it be true that, in the dog, as many as 200,000,000 lymphocytes may be
produced hourly by the activity of lymphoid tissues and poured into the
blood stream, it must be presumed that, in order to preserve a balance, they
are continually being destroyed somewhere, or become converted into other
types of cell, or that there is a continuous circulation (through the blood
stream back again to lymphoid tissue) on a very much more extensive scale
than had previously been envisaged. A suggestion that they are conveyed
to the alimentary canal, there to be discharged into the lumen and destroyed, was negatived by a study of the effect of removal of the intestinal
tract on the lymphocyte content of the blood.[2] We have already noted the
possibility, at one time entertained, that lymphocytes provide the basis for
the elaboration in haemopoietic tissues of other types of white corpuscle
and also red corpuscles, a possibility which has now been finally excluded.
Some observations, it is true, had suggested that, in tissue cultures,
lymphocytes can become directly transformed into granular leucocytes and
myelocytes.[3] Similar evidence had been adduced to support the contention
that lymphocytes may undergo a metaplastic transformation into monocytes and macrophages. These conclusions, however, were criticized on the
grounds that the cultures were not pure. Lewis[4] found, on the contrary,
that in cultures of lymph-node tissue lymphocytes and monocytes are
always quite distinct in morphology and in their mode of locomotion, and
he was never able to follow the transformation of one into the other. He
concluded that they are probably independent strains of cells. Continuous
observation of individual lymphocytes in transparent chambers in the rabbit's ear has also provided no evidence that they change into other cells.[5]

Recent experimental studies by Gowans, Knight, and Marchesi have at
last settled the problem of what was termed 'the mystery of the vanishing
lymphocyte', for they have demonstrated quite clearly, by labelling injected
lymphocytes with isotopes and by electron microscopical investigations,
that a very high proportion of the lymphocytes circulating in the blood are
rapidly returned to lymph nodes, Peyer's patches, and the spleen—probably,
indeed, to all lymphoid tissue with the exception of the thymus. This they
do by penetrating the post-capillary venules for the endothelial cells of

[1] A. G. Sanders and H. W. Florey, 'The effects of the removal of lymphoid tissue',
Brit. Journ. Exp. Path. 21, 1940.
[2] J. M. Yoffey, 'Enterectomy and the blood lymphocytes', Journ. Anat. 76, 1942.
[3] W. Bloom, 'Transformation of lymphocytes into granulocytes in vitro', Anat. Rec. 69,
1937.
[4] W. H. Lewis, 'Lymphocytes and monocytes in tissue cultures of lymph nodes', ibid.
70, 1938 (abstract).
[5] R. H. Ebert, A. G. Sanders, and H. W. Florey, 'Observations on lymphocytes in
chambers in the rabbit's ear', Brit. Journ. Exp. Path. 21, 1940.

which they appear to have a specific affinity. They then recirculate back into peripheral lymphatics, once more to be returned to the blood stream by way of lymphatico-venous connexions (of which the thoracic duct and the right lymphatic duct in man are the only main channels).[1]

Among the functions which have been ascribed to lymphocytes are the elaboration or storage of enzymes (proteases), trophic functions, or the conveyance of some growth-promoting substance which is required by actively growing and dividing cells. In this connexion it may be noted that they tend to accumulate in areas of repair, in certain regions of highly active cell proliferation (e.g. bone marrow) and also around malignant cells. This last association raises the very important question of the role of lymphocytes in relation to cancerous growths. Tissue culture experiments have suggested that they have a special affinity for malignant cells, and also for cells generally which are in process of mitosis.[2] If this is so, it may be that by supplying some essential growth-promoting substance lymphocytes actually operate to the disadvantage of the victim of malignant disease. On the other hand, the fact that, in their initial spread from the primary growth, malignant cells may be temporarily 'held up' in lymph nodes no doubt offers some degree of protection against a fatal outcome of the disease if surgical measures are promptly taken. Since fully differentiated small lymphocytes consist of little more than nuclei with a minimal amount of cytoplasm, and since the latter contains none of the cytoplasmic elements related to processes of active synthesis, e.g. endoplasmic reticula and Golgi apparatus, it is unlikely that they are themselves capable of elaborating enzymes and antibodies. But it remains possible that they exert trophic functions by supplying break-down products of their own disintegration— for example, nucleoprotein and amino-acids.[3]

Besides lymphocytes, it is not uncommon to find a few red blood corpuscles in lymph. Particularly is this the case with lymph draining inflamed or congested tissues, and, as already mentioned, it may be the result of a direct leakage from a blood capillary lying in immediate contact with a lymphatic capillary. Occasionally, also, polymorphonuclear leucocytes or macrophages are found.

The lymph flow. Some of the factors which determine and control the flow of lymph have already been mentioned, i.e. the muscle tissue in the walls of lymphatic vessels and in the capsule of lymph nodes, the contractile properties of the lymphatic endothelium (at least in some animals), and the capillary blood pressure. Other factors are general muscular activity, respiratory movements, and arterial pulsation. With regard to the first, it is clear that intermittent contraction of muscles will compress lymphatics in the neighbourhood and thus serve to propel the lymph in the direction determined by the valves in the vessels. It has been observed that the lymph

[1] J. L. Gowans and J. Knight, 'The route of re-circulation of lymphocytes in the rat', *Proc. Roy. Soc.* **159**, 1964; V. T. Marchesi and J. L. Gowans, 'The migration of lymphocytes through the endothelium of venules in lymph nodes: an electron microscope study', ibid.

[2] J. G. Humble, W. H. Jayne, and R. J. V. Pulvertaft, 'Biological interaction between lymphocytes and other cells', *Brit. Journ. Haemat.* **2**, 1956.

[3] J. F. Loutit, 'Immunological and trophic functions of lymphocytes', *Lancet*, 24 Nov. 1962.

flow from an immobile limb in an animal is practically negligible, but it becomes very active as soon as the limb is moved. Thus, if a limb is completely immobilized (e.g. in a Plaster of Paris casing) its lymphatic drainage can be almost entirely arrested and the absorption of toxic material thereby prevented. This procedure is sometimes used in the treatment of septic wounds and it is said to lessen the possibility of an ensuing toxaemia.[1] The effect of arterial pulsation on the lymphatic circulation generally must be relatively small, but there is little doubt that, where an artery and a lymphatic vessel lie in immediate relationship to each other, it must influence the lymph flow locally to some extent. Respiratory movements play a considerable part in propelling the lymph from the abdominal part of the thoracic duct to its thoracic part, while the negative pressure in the great veins at the base of the neck determines the flow of lymph into them from the terminal part of the thoracic and right lymphatic ducts.

7. GROWTH AND REPAIR OF LYMPHATIC VESSELS

The capacity of lymphatic vessels for regeneration has been abundantly demonstrated by direct observation on living vessels in mammals and lower vertebrates. Their marked proliferation in association with inflammatory changes has also been observed and recorded by micro-injection of the vessels in mice.[2] This proliferation is clearly an adaptive response to cope with the accumulation of fluid and metabolites in the tissue spaces resulting from increased cellular activity and hyperaemia.

Direct observation of living vessels in transparent chambers in rabbits' ears[3] shows that, in the progressive vascularization of newly formed tissue, lymphatic vessels grow in later than blood vessels, and they are always formed by the direct extension of pre-existing vessels. From the latter, sprouts grow out by the mitotic division of the endothelial cells. These sprouts are at first solid, but the lumen soon extends into them. They branch and anastomose with each other, and though they show a tendency to take up a position alongside blood vessels (where the connective tissue is looser), they show no affinity at all for the endothelium of the blood-vascular system. The vascular pattern of growing lymphatics is determined to a considerable extent by the arrangement of the connective tissue in which they lie. Where this tissue is dense, the extension of the vessels is limited or impeded. This observation is consistent with the established fact that, where much scar tissue is formed at the site of an injury, the re-establishment of interrupted lymphatic drainage paths is considerably delayed.

The rapidity with which interrupted lymphatic paths can be re-established under favourable conditions of healing was demonstrated many years

[1] J. M. Barnes and J. Trueta, 'Absorption of bacteria, toxins, and snake venoms from the tissues', *Lancet*, May 1941.
[2] B. D. Pullinger and H. W. Florey, 'Proliferation of lymphatics in inflammation', *Journ. Path. & Bact.* 45, 1937.
[3] E. R. Clark and E. L. Clark, 'Observations on living mammalian lymphatic capillaries', *Amer. Journ. Anat.* 60, 1937. A number of previous papers by these authors on the same subject in the *American Journal of Anatomy* and the *Anatomical Record* should also be consulted.

ago by Reichert.[1] In his experiments all the tissues of the hind-limb in a dog were completely transected with the exception of the main artery and vein and the femur. This operation is followed by a transient oedema of the limb, but in a few days the lymphatic channels may be repaired and efficient lymph drainage established. So far as the regeneration of new lymphatic channels has been studied up to the present, this process seems to be limited to the formation of capillary vessels. The regeneration of larger vessels with valves has not been directly observed. The regeneration of cutaneous lymphatics in the rabbit has been studied by Gray.[2] Lymphatic vessels were interrupted by a skin incision, and their subsequent growth followed by radiographic examination after thorotrast injection. Capillary anastomoses across the line of the incision were established after three weeks. Transplantation of short lengths of lymphatic vessels was also performed successfully, with the establishment of their normal functions as conducting channels.

Lymph nodes show a partial regenerative capacity. That is to say, if the greater part of a node is excised, the remaining fragment can form the basis for its reconstruction to the original size. If, however, the node is entirely removed, it is not replaced by the development of a new node except in young animals.[3]

8. THE SPLEEN

The spleen comprises the largest single mass of lymphoid tissue in the human body. The main substance of this organ is a red pulp consisting of a rich plexus of venous sinuses embedded in a loose meshwork of connective tissue. The meshwork contains abundant red corpuscles, white corpuscles, monocytes (sometimes called *splenocytes*), and large phagocytic cells of the reticulo-endothelial system. The latter can occasionally be seen in microscopical preparations ingesting and destroying fragmentary and apparently effete red corpuscles. Penetrating the red pulp are branches of the splenic artery, the terminal ramifications of which are ensheathed in globular masses of lymphocytes. These masses are similar in their structure to lymph follicles elsewhere, and in the spleen they are termed *Malpighian corpuscles*. They are formed by the infiltration of the perivascular reticulum and the tunica adventitia of the vessel wall with lymphocytes.

The Malpighian corpuscles may contain germinal centres which are the site of active production of lymphocytes; the latter migrate into the surrounding red pulp, eventually finding their way into the venous sinuses. Hence the blood leaving the spleen by way of the splenic vein may contain more lymphocytes than the splenic artery. The splenic blood is also comparatively rich in macrophage cells which are derived from the endothelium lining the sinusoidal vessels. There are no lymphatic vessels in the red pulp of the spleen (though they have been demonstrated in its fibrous capsule). The spleen appears to be similar to a lymph node in that it provides for the

[1] F. K. Reichert, 'Regeneration of lymphatics', *Arch. Surg.* **13**, 1926.
[2] J. H. Gray, 'Studies on the regeneration of lymphatic vessels', *Journ. Anat.* **74**, 1940.
[3] W. J. Furuta, 'An experimental study of lymph node regeneration in rabbits', *Amer. Journ. Anat.* **80**, 1947.

storage of lymphocytes and possibly aids in their production, but this is of quite minor importance and perhaps only incidental to its other functions, such as its capacity for removing foreign matter in the blood by means of its reticular cells. It has been mentioned above that red corpuscles may sometimes be seen to be in process of disintegration in the reticular cells. This has given rise to the idea that the spleen is mainly concerned with the destruction of red corpuscles. Probably, however, this only occurs to any significant extent in pathological conditions, for there is good evidence that normally the spleen is actually concerned with prolonging the life of the corpuscles by providing them with temporary shelter from certain ionic changes to which they are exposed in the circulation.[1]

Another important function which can be ascribed to the spleen, on the basis of Barcroft's studies, is that of storing blood.[2] When necessary it can, by virtue of the plain muscle in its capsule, expel its contents into the circulation. This occurs during haemorrhage and also during exercise, and the same reaction has been shown to take place in a rarefied atmosphere, e.g. at high altitudes. The spleen thus plays an important part in circulating functional haemoglobin according to the needs of the body. In relation to this, it may be noted that the size of the normal spleen after death is often much smaller than it is during life.

9. HAEMAL NODES

Haemal nodes are composed mainly of lymphoid tissue intermixed with which are numerous red corpuscles. They resemble the spleen in their general cellular matrix and in their arrangement of venous sinuses, and they are devoid of afferent and efferent lymphatic vessels. Haemal nodes are said to be particularly numerous in certain groups of mammals (particularly ruminants) where they are situated in the posterior abdominal wall alongside the aorta, and also in relationship to the kidneys and spleen. They have been described in man, though some authorities doubt whether they are of normal occurrence here. But they certainly exist in the human body as the so-called 'accessory spleens', that is to say, small nodules of splenic tissue often to be found in the peritoneal pedicle of the spleen. Indeed, it is of some importance to note the existence of these nodules, for following splenectomy they may enlarge to a degree as to suggest that the spleen can in course of time become completely regenerated.

Haemolymph nodes have also been described, with a structure somewhat intermediate between lymph nodes and haemal nodes. These, however, may represent transitional stages in the transformation of lymph nodes into haemal nodes, involving the gradual disappearance of the lymphatic vessels which are originally connected with them. Such a transformation is said to occur when the spleen has been removed.[3] It should be noted that in typical lymph nodes red corpuscles may occasionally be found in the lymph sinuses.

[1] J. G. Stephens, 'Prolongation of red cell life by the spleen', *Journ. Physiol.* 95, 1939.
[2] See J. Barcroft, 'Some recent work on the functions of the spleen', *Lancet*, 1, 1926.
[3] J. Loesch and J. L. Witts, 'Extirpation and exclusion of the spleen', *Journ. Metab. Res.* 6, 1924.

10. THE EPITHELIO-LYMPHOID SYSTEM

Developed in intimate relation with the epithelial lining of the alimentary tract are accumulations of lymphoid tissue, in some cases forming rather large circumscribed masses, and in others appearing as scattered solitary follicles. This lymphoid tissue includes the palatine tonsils, the pharyngeal tonsils (adenoids), the thymus, and circumscribed aggregations of lymph follicles in the mucous membrane of the alimentary tract, particularly in the lower part of the small intestine (Peyer's patches), and the appendix and large intestine.

The significance of this epithelio-lymphoid system is a matter of conjecture. It is mostly composed of lymph follicles similar to those found in the cortex of lymph nodes, often containing germinal centres. There is evidence that, in many cases, the lymphocytes from these follicles reach the epithelial surface with which they are in relation and there undergo destruction, though it is possible that a certain proportion gain entrance to lymphatic vessels and so add to lymphocyte production in lymphoid tissue elsewhere. Those that push their way through the overlying epithelium have been supposed to play some part in the protection of the tissues from bacterial invasion. This is the role which was at one time commonly ascribed to the lymphoid masses which 'guard' the entrance from the mouth cavity to the alimentary tract (e.g. the palatine tonsils). However, it is now well established that the phagocytic activities associated with lymphoid tissue are not carried out by the lymphocytes but by the reticulo-endothelial cells which help to form the reticulum in which the lymphocytes are entangled. Experimental observations *in vivo* indicate that lymphocytes probably do not normally act as phagocytes, or only to a very limited extent. It has been found, for instance, that if minute quantities of different foreign substances are injected into the extravascular tissues of a tadpole's tail, while polymorphonuclear leucocytes are actively phagocytic towards all of them, lymphocytes show no positive response.[1]

The lymphocytes of the epithelio-lymphoid system appear morphologically to be identical with those of the lymph-node system. But the possibility that they are not physiologically equivalent is suggested by the fact that, in certain pathological affections of the lymphatic system, lymph nodes may be extensively affected while the lymphoid tissue of the epithelio-lymphoid system is not obviously involved. The origin of the lymphocytes in the latter is still in dispute. Some authorities believe that they are primarily produced elsewhere (e.g. in the lymph nodes), and migrate to take up their position in the wall of the alimentary tract. In this connexion the observation of F. T. Lewis may be noted, that in the development of the human embryo lymph nodes are differentiated before there appears to be any true lymphoid tissue elsewhere in the body.[2]

[1] E. R. Clark, E. L. Clark, and R. O. Rex, 'Observations on polymorphonuclear leucocytes in the living animal', *Amer. Journ. Anat.* **59**, 1936.

[2] F. T. Lewis, 'The first lymph glands in rabbit and human embryos', *Anat. Rec.* **3**, 1909.

S

The development of epithelio-lymphoid bodies is preceded by a local, characteristic proliferation of the neighbouring entodermal cells. These invade the subjacent tissue in a remarkable way, becoming intimately intermingled with mesodermal cells, with no basement membrane to form a limiting boundary. The process is associated with a marked proliferation of the mesodermal cells, and lymphoid tissue actually appears first in immediate relation to the columns of invading entodermal cells. It has thus been suggested that the latter, by some process of induction, bring about a local differentiation of lymphocytes from undifferentiated connective-tissue cells, at least in the tonsil and the thymus.[1] The tonsil is developed in relation to the pharyngeal entoderm. Its absence in certain mammals (e.g. rodents) suggests that it is not a morphological entity necessarily associated with a local region of the branchial arch system (as, for instance, is the case with the thyroid gland), but the result of the local metabolic influences of the entoderm.[2]

Also developed in relation to the pharyngeal entoderm, but in this case formed from bilateral diverticula associated with the third pharyngeal pouches, is the thymus. Endocrine functions have been ascribed to this organ following work by Rowntree and others[3] suggesting that the thymus elaborates some specific hormone which plays a part in regulating the growth of the body as a whole, but their conclusions have not received confirmation. That the thymus has other than lymphoid functions is indicated by the fact that it really contains two distinct types of tissue—lymphoid and epithelial. The lymphoid tissue forms the cortex of the gland, and consists of densely packed lymphocytes, but, unlike a lymph node, there are usually no lymph sinuses or clearly circumscribed germinal centres. However, the mitotic rate among the lymphoid cells is astonishingly high; indeed, according to Miller the mitotic indices of the cells are about seven times higher than those of lymphoid cells elsewhere.[4] What happens to the products of this remarkable lymphocytic proliferation is not known. Possibly they enter blood vessels in the thymus and are transported elsewhere, but there is no certain evidence of this.

In the medulla of the thymus, and also in the cortex (though here they are less conspicuous), are flattened, branching reticular cells, characteristic of the thymus. They appear at first sight to be ordinary connective-tissue elements, but they have been shown definitely to be derived from the pharyngeal entoderm. They have an obviously epithelial character in the embryo, and, if the neighbouring lymphocytes are destroyed by exposure to X-rays in the adult, they lose their reticulo-endothelial appearance and once more assume the characters and disposition of epithelial cells. Moreover, unlike the connective-tissue cells which they resemble, they do not take up and store vital dyes such as trypan blue. If the endocrine functions

[1] The possibility that the lymphocytes may be produced directly from entodermal cells seems to be adequately discounted by the fact that transitional stages between entodermal cells and lymphocytes cannot be certainly detected histologically.

[2] B. F. Kingsbury and W. M. Rogue, 'The development of the palatine tonsil', *Amer. Journ. Anat.* **39**, 1927.

[3] Rowntree *et al.*, 'Further studies on the thymus and pineal glands', *Ann. Int. Med.* **9**, 1935.

[4] F. A. P. Miller, 'Role of the thymus in immunity', *Brit. Med. Journ.* 24 Aug. 1963.

assigned to the thymus should be substantiated, the anatomical substratum of these activities is probably to be found in the reticular cells.

It has been known for many years that the thymus reaches its maximal development and differentiation at puberty. Thereafter it undergoes a slow involution. While at puberty in man the gland weighs about 40 grammes, in old age it shrinks to a quarter this size. Such age changes have suggested a relation between thymus activity and growth processes. However, while experimental removal of the thymus in adult (but not fully grown) animals appears to have no effect on the subsequent growth rate, a similar operation performed on new-born animals leads to death from a wasting disease in a very short time. This is accompanied by a great reduction in the number of lymphocytes in lymphoid tissue generally and in the circulation. It has been inferred from such experiments that the thymus plays an important role in promoting the development of lymphoid tissues generally in the body, and in effecting the differentiation of lymphocytes from their cellular precursors. It has further been demonstrated that, in adult as well as young animals, the thymus is essential for the promotion and maintenance of certain immunological reactions. How these functions of the thymus are mediated is not yet known, but it has been conjectured on the basis of indirect evidence that they may depend on an endocrine factor elaborated by the epithelial reticular cells of the organ.

11. THE DEVELOPMENT OF THE LYMPHATIC SYSTEM

The earliest development of the lymph-vascular system consists of the formation of a series of lymph sacs (Fig. 92). There are five of these, and they all appear in close relationship to large veins. Two jugular sacs are formed alongside the internal jugular veins, two posterior sacs by the iliac veins, and one unpaired sac in the posterior abdominal wall, the cisterna chyli. In mammals, only the cisterna chyli retains its saccular character, and even this may ultimately disappear as a clearly defined sac. In most sub-mammalian vertebrates (e.g. amphibia and reptiles) the jugular and posterior sacs persist and, by acquiring plain muscle tissue in their walls, become specialized to form conspicuous pulsatile structures—the so-called lymph hearts—which play an important part in promoting the lymph circulation by propelling the lymph into adjacent veins. It may be surmised that, with the development of the diaphragm in mammals, and the aid consequently given to the lymph flow by respiratory movements, these lymph hearts are no longer necessary. The corresponding lymph sacs in the mammalian embryo are the site of formation of groups of lymph nodes and become broken up into networks of lymphatic vessels.

The precise origin of the lymph sacs and of lymphatic vessels in the embryo is still in doubt. The opinion was widely held at one time that the lymphatic endothelium arises *in situ* from the differentiation of mesenchymal cells, and that lymph sacs and lymphatic vessels are formed by the confluence of extra-intimal connective-tissue spaces in close relation to the developing veins. On this view, communications between the lymphatic system and veins are formed secondarily. This conception was based on the study of stained sections. It has been cogently argued, however, that such a method is open to fallacy, seeing that it is not only extremely difficult to detect the finer lymphatic vessels in sections, but artificial shrinkage spaces may be mistaken for them.

Micro-injection of embryos has led to the conclusion that lymph sacs and vessels are in direct communication with veins from their earliest

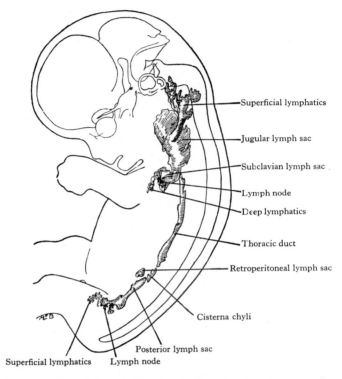

Superficial lymphatics

Jugular lymph sac

Subclavian lymph sac

Lymph node

Deep lymphatics

Thoracic duct

Retroperitoneal lymph sac

Cisterna chyli

Posterior lymph sac

Superficial lymphatics   Lymph node

FIG. 92. Reconstruction of the primitive lymphatic system in a human embryo of two months. × 3. (From Arey, after Sabin in Keibel and Mall.)

appearance, and this technique has provided the main foundation for the belief that the endothelial lining of the lymph-vascular system is initially derived by a direct budding-out of venous endothelium. On this view the lymph sacs are developed as diverticula from the veins in their neighbourhood. In most mammals their connexions with these veins all usually disappear except those of the jugular sacs, which persist to form the terminations of the thoracic and right lymphatic ducts. The peripheral lymphatic vessels are also believed to develop as a direct extension throughout the tissues of the body of endothelial sprouts from the primary lymph sacs. The opponents of this view argue that the micro-injection technique is not adequate for the demonstration of the isolated spaces in the connective tissue which, in their opinion, form continuous vessels by confluence, since, obviously, the injection material used only passes into channels which at the time do communicate with each other.[1]

Although there still remains an element of uncertainty in regard to the primary origin of lymphatic endothelium, there seems no doubt at all that, at least after the formation of the lymph sacs and the main lymphatic vessels, this endothelium acquires a morphological specificity, and henceforth all new vessels are derived by a direct extension of the already existing vessels. This has been shown to be almost certainly the case with the new formation of lymphatic vessels in postnatal life. New lymphatic capillaries are derived from the proliferation of the endothelial cells of older vessels and are probably never formed *in situ* by the differentiation of connective-tissue cells.

Lymph nodes develop in the course of lymphatic vessels already formed and, in the human embryo, are first seen at the 50-mm. stage. At the site of a developing node, the lymphatic vessels take on a rich plexiform arrangement of sinuses. In the interstices of this meshwork a proliferation of mesenchyme occurs with the differentiation of lymphocytes from its cellular elements. Localized accumulations of lymphocytes bulge into the lumen of the lymph sinuses, invaginating the lining endothelium. In this manner a compact solid body is produced, the surrounding connective-tissue becoming condensed to form a capsule. The lymphatic vessels entering and leaving the primary network become the afferent and efferent vessels of the node, while the endothelial lining of the plexiform sinuses appears to form the reticulo-endothelial lining of the lymph channels in the fully developed node. Except possibly for a short time after birth, it is doubtful whether new lymph nodes are formed in postnatal life.

[1] See F. R. Sabin, 'On the origin and development of the lymphatic system from the veins', *Amer. Journ. Anat.* 1, 1902; 'A critical study of the evidence on the development of the lymphatic system', *Anat. Rec.* 5, 1911; 'The origin and development of the lymphatic system', *Johns Hopkins Hospital Report*, 5, 1913. C. F. W. McClure, 'The endothelial problem', *Anat. Rec.* 22, 1921. O. F. Kampmeier, 'The development of the jugular lymph sacs', *Amer. Journ. Anat.* 107, 1960.

XI

MUCOUS MEMBRANES AND GLANDS

IN multicellular organisms of the simplest type the surface epithelial layer of cells has multifarious functions to perform. Not only does it provide a mechanically protective cuticle, and a sensory surface through which the organism receives sensory impressions from the environment, it is also concerned with metabolic processes of nutrition and respiration and must therefore allow for absorption of nutritive material from without, for excretion of waste products, and for gaseous interchange.

In more complex types of animal these functions become allocated to different tracts of epithelium. A portion of the surface layer becomes invaginated to form the lining of an alimentary cavity. It thus becomes sheltered from immediate contact with the external environment and can elaborate its absorptive and excretory functions at the expense of its protective and sensory functions. Moreover, the formation of such a cavity inside the organism allows ingested nutrient material to remain in contact with the absorptive epithelium over a length of time. This, in turn, permits the development of a process whereby nutrient matter which is not ready to hand in an assimilable form may be acted upon by chemical substances and ferments elaborated by the organism itself, so that it can then be absorbed. The organism is thus enabled, for its nutritional requirements, to draw on a much wider range of material which presents itself in the environment.

The process of digestion, which is conditioned on the morphological side by the formation of an alimentary cavity or *enteron*, involves the development of secretory functions by the epithelial cells concerned. Different areas of the epithelium lining the cavity become specialized to form glandular structures concerned in the elaboration of digestive juices. These glandular structures (whose *external* secretions are poured into the alimentary cavity) may also secrete specific chemical substances which are taken up directly into the blood stream as *internal* secretions, and carried to other glands which they serve to activate. An integrating mechanism is thus provided whereby the activity of the different glands involved in the different processes of digestion is co-ordinated.

A further progressive specialization is shown by certain glands which lose altogether their connexion with the alimentary cavity from which they were initially derived, and become entirely glands of internal secretion (or endocrine glands). Broadly speaking, these are concerned with the co-ordination of metabolic processes related to growth and to the maintenance of the physiological equilibrium on which the survival of the individual depends, and (as an extension of this) of all those processes which have immediate or ancillary functions related to the continuance of the species.

In early stages of embryonic differentiation the body of the embryo consists of little more than a primitive type of connective tissue (mesoderm) faced on all aspects by an epithelial formation. The latter is differentiated into an external layer (ectoderm) which covers the outer surface of the embryo, and an internal layer (entoderm) which lines a primitive alimentary cavity. It is from the entoderm that most of the lining membrane of the alimentary tract and its derivatives in the adult is formed. In relation to its absorptive properties this membrane is relatively thin and vascular, and because of its moist surface is commonly called *mucous membrane*. The term is perhaps open to criticism since, although many types of mucous membrane contain actual mucus-producing cells and glands, and are normally covered with a thin film of mucous secretion, others are quite devoid of these structures.

Mucous membranes are not entirely of entodermal origin. The mucous membrane lining the nasal cavity, the mouth, and the lower part of the anal canal is partly derived from ectodermal epithelium, which turns in here from the surface, while the lining of considerable parts of the urogenital tract is differentiated from mesodermal tissue. There appears to be no essentially specific relation between the various morphological types of mucous membrane and the germinal layers from which they are derived in the embryo, so that it is not possible to infer the embryological origin of an epithelium simply by reference to its morphological character in the adult. There is reason to suppose, indeed, that in some cases, e.g. the mucous membrane lining certain parts of the genital tract, an epithelium may have a different developmental origin in different species of mammals.

1. THE EPITHELIUM OF MUCOUS MEMBRANES

In the early embryo the whole length of the alimentary tract is lined by a simple epithelium consisting of a single layer of cuboidal or columnar cells. Where processes of absorption predominate in the adult this type of epithelium persists relatively unaltered, for it forms a thin cellular layer through which material can readily pass in order to reach blood or lymphatic vessels in the sub-epithelial tissues. In other regions the requirements of mechanical protection are met by the development of a many-layered epithelium in which the cells are arranged in a stratified formation.

A recognition of the different types of epithelium covering mucous membranes is not only important for a consideration of their functional implications, but also from the clinico-pathological point of view, since the cytological characteristics of new growths (or neoplasms) which may arise in them are determined by the nature of the cells of which they are composed. We may now briefly review these types of epithelium.

Simple columnar epithelium (Fig. 93 *a*). In this type, which is well represented by the intestinal mucous membrane, the epithelium is composed of a single layer of columnar cells. As might be anticipated on mechanical grounds, the cells, as the result of mutual pressure, are approximately hexagonal in cross-section, so that they fit together side by side, closely packed in a honeycomb mosaic. The cell base rests on a thin

amorphous, hyaline membrane, the basement membrane, which is about
1μ in thickness and provides a supporting matrix, separating the epithelium
from the sub-epithelial tissues. The precise origin of basement membranes
which are to be found in relation to epithelia is not surely known. The free
edges of the intestinal epithelial cells may show a slight marginal differentiation in the form of what appears under the light microscope to be a thin
protoplasmic membrane or cuticle. Moreover, in ordinary preparations
this cuticle often has a striated appearance and may seem to extend as a
continuous membrane over the surface of the epithelium. However, in
many cases the electron microscope has shown that such striated borders
are composed of very minute processes (microvilli) which project vertically
from the free end of each cell and are tightly packed to form a close pile.

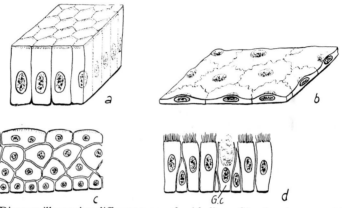

FIG. 93. Diagram illustrating different types of epithelia. *a.* Simple columnar epithelium.
b. Pavement epithelium. *c.* Transitional epithelium. *d.* Columnar ciliated epithelium,
showing also a mucus-secreting (goblet) cell (*G.c.*).

Simple columnar epithelium is found lining the gall bladder and bile
ducts, and the ducts of many different kinds of glands, but it does not
everywhere show a marginal cuticular differentiation. The height of the
cells varies in different regions, and they may be so low as to require
another term for their description—*cubical epithelium.*

Columnar ciliated epithelium (Fig. 93 *d*). In several parts of the
body the mucous membrane is lined by columnar cells each of which is
furnished with fine hair-like protoplasmic processes projecting from its
free border. These *cilia* take their roots deep in the cytoplasm of the cell
and are capable of a rhythmic motility. They 'beat' in a definite direction
and produce wave-like movements on the surface which are sufficiently
strong to carry along minute particles lying in contact with them. The rate
at which particles are transported by ciliary movement no doubt varies from
place to place and from time to time under different conditions. According
to some observations on ciliary activity in the nasal cavity, here it may move
particles through a distance of 6 mm. or more in a minute.[1]

[1] J. P. Schaeffer, 'The mucous membrane of the nasal cavity and the paranasal
sinuses', in *Special Cytology*, ed. by E. V. Cowdry, London, 1963.

Ciliated epithelium is found lining the cavities of the nose, the accessory nasal air sinuses and the nasopharynx, the trachea, bronchi, and bronchioles, and the oviducts and uterine cavity. In some of these regions, e.g. the trachea and bronchi, the epithelium is stratified, for deep to the columnar ciliated cells are several layers of conical and oval cells.

Ciliary action is independent of nervous control—indeed it can continue for a considerable length of time in a portion of mucous membrane which has been excised and kept in saline solution. Chambers and Rényi have shown, also, that if a ciliated cell is damaged without any apparent change in the nucleus, the cilia still continue to beat, but if the nucleus is injured they stop moving at once. As just noted, ciliary action is used for the transport of particulate material over the surface of mucous membranes. In the accessory nasal air sinuses it is apparently responsible for the removal of dust particles and mucous secretion, and here the cilia all beat towards the openings by which the sinuses communicate with the nasal cavity.[1] In the oviducts the cilia are often stated to play a part in conducting the ovum from the ovary to the uterine cavity, but it is improbable that they do so to any significant extent; the rhythmic contraction of the musculature of the tube is the more important factor.

Ciliated cells are commonly regarded as specialized cells. Nevertheless, where they form part of a general secretory epithelium they have been observed occasionally to lose their cilia and assume secretory functions. It is probable, also, that this change is reversible, i.e. that they can again acquire ciliary processes after completing a secretory phase.

The lining epithelium of the ventricles of the brain and the central canal of the spinal cord is also formed of columnar ciliated cells, but here the functional significance of the cilia is quite obscure.

Pavement epithelium (Fig. 93 b). In this type the epithelial cells are flattened into a single layer of thin nucleated plates resembling the endothelial cells of blood vessels. Their margins are usually irregular and are united along crenulated junctions which can be rendered visible by treatment with silver nitrate. Apart from certain regions in the duct systems of the salivary glands and the pancreas, pavement epithelium is found lining the alveoli of the lungs. In the latter, it provides an extremely attenuated membrane which facilitates gaseous interchange between the alveolar air and the blood in the lung capillaries. A similar type of epithelium is found covering the chorionic villi of the mature placenta.

Transitional epithelium (Fig. 93 c). This type of epithelium is termed 'transitional' because morphologically it occupies a position between a simple and stratified epithelium. It is the characteristic lining of the urinary passages, extending from the kidney down the ureter into the bladder and urethra, and is not found elsewhere in the body.

The superficial layer of transitional epithelium consists of somewhat flattened cuboidal cells which quite frequently have two nuclei. The surface zone of these cells shows a structural differentiation to form a vaguely

[1] Hence the rational surgical treatment of a blocked air sinus consists in the opening up and enlarging of the normal apertures rather than in making new openings in different positions.

defined cuticle. The deeper layers (which average three or four in number) consist of oval cells embedded in a slimy cement substance which, when the cells are separated by microdissection, becomes drawn out into delicate adhesive strands.[1] According to one interpretation, it is in virtue of this intercellular slime that the cells of transitional epithelium can readily slip over each other so that, when the cavity which they limit is distended, the deeper layers become thinned out very considerably. However, other histological studies appear to indicate that a reduction in the number of cell layers in a distended bladder is apparent rather than real, being the result of extensive lateral deformation of individual cells.[2] It appears, then, that transitional epithelium is particularly adapted to allow of rapid changes in the distension and contraction of a viscus. It is probable, also, that the superficial cells are modified to serve as a protection against the acidity of the urine with which they are in contact. It is a matter of common know-ledge that in cases where uncontrolled micturition leads to a wetting of the epidermis of the skin the latter soon becomes sodden and excoriated.

Stratified squamous epithelium (Fig. 98). The epidermis of the skin is a stratified squamous epithelium (p. 290), that is to say, it is com-posed of serial layers of cells which become more and more flattened as they approach the surface. They also become progressively converted into a horny material called *keratin*. Epithelium of a similar structure is found covering the mucous membrane which lines the buccal cavity, the lower part of the pharynx, the oesophagus, anal canal, vagina, &c. It is not so thick here, however, and keratin formation is either absent or is much less marked. The adaptive significance of a stratified squamous epithelium is suggested by the fact that simple columnar epithelium may undergo a metaplasia and become converted into stratified epithelium with keratiniza-tion if it is exposed to mechanical irritation.

The amplification of absorptive surfaces. Rapidity of absorption through a membrane is directly related to the surface area exposed to the material to be absorbed. Where absorption occurs, therefore, the area is commonly found to be increased by direct extension, or by the formation of surface irregularities. In the case of the digestive tract this is partly brought about by a lengthening of the whole intestinal canal which (in man) forms a tortuous tube extending over many feet, and partly by the formation of folds and villous processes in the mucous membrane. Thus, in the upper part of the small intestine (where processes of absorption predominate), the mucous membrane is not only richly folded in a circular manner to form the *plicae circulares* of the duodenum and jejunum, it is also covered with minute villi which give it the appearance (as seen under a hand lens) of a velvet pile. In the lower part of the small intestine (the ileum) the villi are also present, but the plicae have almost disappeared. In the large intestine, where absorption is not so active (being normally restricted to the removal of water for the concentration of its contents), both villi and plicae circulares

---

[1] R. Chambers and G. S. Rényi, 'The structure of the cells in tissues as revealed by microdissection', *Amer. Journ. Anat.* **35**, 1925.

[2] J. M. N. Boss, 'The structure and behaviour of transitional epithelium', *Journ. Physiol.* **159**, 1961.

are absent. It has been estimated that (in the cat) the intestinal mucosa may be amplified as much as fifteen times by plications and villous formations.[1] Each villus is covered by a continuous layer of absorptive epithelium, while in its central core are minute blood vessels and a centrally placed lymphatic vessel.[2]

The extension of absorptive membranes by the production of surface irregularities is also illustrated by the complicated chorionic villi of the placenta, and (to a less striking degree) by the formation of villous folds in the synovial membranes of joints (p. 171). The total absorptive area of the human placenta (which in bulk measures about 7 inches in diameter) is approximately 70 square feet.[3]

The epithelium of the alimentary tract. The buccal cavity is everywhere lined by stratified squamous epithelium. The superficial cells here are easily detached, so that isolated squamous cells can be obtained for microscopical examination by gently scraping the mucous membrane of the lips and cheek. Scattered over the stratified epithelium of the tongue are the sensory organs of taste (taste buds), which are particularly numerous in relation to the large vallate papillae. Isolated taste buds may also be present in the mucous membrane of the palate and even over the epiglottis. They are said to have a more extensive distribution in the foetus than in the adult.

The lower part of the pharynx is also lined by squamous stratified epithelium. The mucous membrane of the upper part (nasopharynx), however, is covered by columnar ciliated epithelium (like the nasal cavities), and this extends up the auditory tube as far as the middle-ear cavity.

Squamous epithelium extends down the whole length of the oesophagus. It ceases quite abruptly at the opening into the stomach, where it is replaced by simple columnar epithelium. The junction of the two types of epithelia is usually recognizable to the naked eye as a crenated line. It is of some importance to note that there may be isolated patches of epithelium in the oesophagus which resemble the characteristic mucosal lining of the stomach; these occasional 'outliers' of gastric epithelium possibly form the basis of certain pathological conditions which may affect the oesophagus.[4]

Throughout the length of the gastro-intestinal tract the lining epithelium is of the simple columnar variety. Glandular diverticula form gastric glands of different types in the stomach, Brunner's glands in the duodenum, and the crypts of Lieberkühn in the whole extent of the small and large intestine. In the large intestine these crypts contain a relatively large proportion of mucous cells. At the termination of the alimentary tract, the columnar epithelial lining ceases abruptly at the middle of the anal canal, meeting

[1] H. O. Wood, 'The surface area of the intestinal mucosa in the rat and cat', *Journ. Anat.* **78,** 1944. For methods to be employed in the accurate estimation of the surface area of absorptive membranes in relation to linear measurements, see R. B. Fisher and D. S. Parsons, 'The gradient of mucosal surface area in the small intestine of the rat', ibid. **84,** 1950.

[2] The intestinal villi also contain plain muscular tissue and have been observed to undergo rhythmic contractions (at the rate of six a minute in the dog). It is believed that this effects a kind of pumping action which accelerates absorption.

[3] G. S. Dodds, 'The area of the chorionic villi in the full term placenta', *Anat. Rec.* **24,** 1922.

[4] L. E. Rector and M. L. Connerley, 'Aberrant mucosa in the oesophagus', *Arch. Path.* **31,** 1941.

here the stratified squamous epithelium which turns in from the skin to line the lower half of the canal. The columnar epithelium lining the intestinal canal is covered on its free surface by an apparently continuous cuticular membrane of protein substance. This is the striated border to which reference has already been made (*vide supra*), and since all the products of digestion must pass through it, its structure is clearly of some physiological interest. Examined under the highest powers of the light microscope, it appears to be pierced by minute pores, and during the absorption of fat these may be seen here and there to contain particles of lipoid material which are presumably passing through to reach the cytoplasm of the epithelial cells.[1] Electron microphotographs have now clearly demonstrated that the striated border consists of microvilli packed in close formation, and it is therefore clear that the 'pores' are really the interstices between the microvillous processes. The average diameter of each process is estimated to be 0·08 μ, and since a single cell may bear as many as 3,000 microvilli they must very significantly increase the absorptive surface presented by the epithelium as a whole.[2]

The early development of the mucous membrane of the alimentary canal in the embryo is characterized by a remarkable proliferation of the epithelium. The latter becomes greatly thickened, and this may even lead to a temporary occlusion here and there of the lumen (particularly in the oesophagus and duodenum). Subsequently, an irregular vacuolation appears in the hypertrophied epithelium, forming large spaces which effect a communication with the lumen. In this way the lumen is opened up and re-established. It is possible that certain congenital anomalies of the alimentary tract, such as atresia (a localized narrowing or constriction) or the development of abnormal pouch-like diverticula, may be due to the partial persistence of the foetal proliferation of the epithelial lining.

The epithelium of the respiratory tract. The trachea and larynx are lined by ciliated epithelium. Over the vocal cords, however, this is replaced by stratified squamous epithelium. Ciliated epithelium continues to line the respiratory tract as far distally as the terminal bronchioles. Here it is superseded by a cuboidal non-ciliated epithelium whose component cells become flatter as the alveoli are approached. The latter are lined by an extremely attenuated pavement epithelium, the cytological details of which are very difficult to determine by ordinary histological methods.

The epithelium of the urinary tract. It has already been mentioned that the greater part of the urinary tract is lined by transitional epithelium. This is the case with the pelvis of the kidney, the ureter, the bladder, and most of the urethra. In the uriniferous tubules of the kidney the terminal dilated vesicles (which are invaginated by glomerular tufts of blood capillaries to form Bowman's capsules) are lined by very thin pavement epithelial cells. In the convoluted portion of the tubules the epithelium is of the cuboidal type, the height of the individual cells varying in relation to

[1] J. R. Baker, 'The free border of the intestinal epithelial cell of vertebrates', *Quart. Journ. Micr. Sc.* **84**, 1942; 'The absorption of lipoid by the intestinal epithelium of the mouse', ibid. **92**, 1951.
[2] B. Granger and R. F. Baker, 'Electron microscope investigation of the striated border of intestinal epithelium', *Anat. Rec.* **107**, 1950.

the activity of the kidney. Finally, the collecting tubules are lined by a cuboidal epithelium which gradually becomes columnar in type as the pelvis of the kidney is approached. Just before reaching the latter, however, the cells again assume a flattened shape.

The epithelium of the genital tract. The seminiferous tubules of the testis are lined by germinal epithelium which, in maturity, is the site of active spermatogenesis. The spermatozoa are conveyed by a system of small ductules (vasa efferentia) to an intricately coiled tube—the epididymis. This tube is lined by a columnar ciliated epithelium, the cilia of which are rather long and non-motile. It eventually becomes continuous with the ductus deferens in which the epithelium is of a similar type, but non-ciliated.

In the female, the Fallopian tube (oviduct) is lined by a simple columnar epithelium which is partly ciliated; the cilia 'beat' in the direction of the uterus. The uterine epithelium is also of the columnar ciliated type, but the cilia are not uniformly present and in many mammals they are completely absent. The whole extent of the uterine mucosa is beset with tubular glands which reach down to, and sometimes into, the muscular coat. At the opening of the uterine cervix into the vagina there is an abrupt transition from columnar to stratified squamous epithelium.

2. GLANDULAR MECHANISMS

It has already been noted that the processes of digestion and assimilation involve the production by certain cells of specific chemical substances whose function it is to act on foodstuffs in such a way that they can be absorbed through the wall of the alimentary tract. As an extension of this activity, some cells also elaborate chemical activators which are carried to all parts of the body in the blood stream and control or co-ordinate the activity of other secretory cells. In all but the lowest organisms, these functions are largely carried out by specialized groups of cells which are collected together in the form of *glands*.

The definition of a gland is not altogether easy. All kinds of tissues are concerned in some degree with the elaboration of chemical substances which often exert an influence on other tissues. Muscle cells, when they contract, produce metabolites which cause a dilatation of the neighbouring blood vessels; substances are liberated at the terminals of nerve fibres when they are stimulated and act directly on effector mechanisms such as muscle cells, or on other nerve cells; the epidermis of the skin may be said to secrete scleroproteins, and adipose tissue to secrete fat, and so forth. Indeed, the development of humoral physiology has suggested the conception that all types of cells are concerned with the secretion of some characteristic chemical substance. In most tissues, however, this process is incidental to the main function of the cells. On the other hand, glands may be defined as circumscribed collections of cells which are *specifically differentiated* for the production of certain chemical substances, and the latter are either required for use by other tissues (in which case the process is termed *secretion*) or they are waste products to be eliminated from the body (in which case the process is termed *excretion*).

Glands may be separated into two categories, *exocrine glands*, which discharge their secretory products on to a free surface directly or by ducts, and *endocrine* or *ductless glands*, whose secretion is taken up directly into the circulation. Some authorities have also listed as a third category *cytogenic glands*, whose function is not to elaborate chemical substances but to 'secrete' special types of cells which are carried to other parts of the body. These 'glands' would include lymph nodes, concerned in the production of

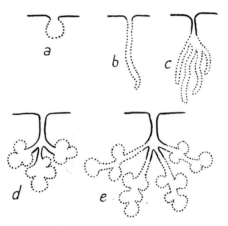

FIG. 94. Diagram illustrating different types of glands. The secretory epithelium is represented by dotted lines. *a*. Simple alveolar gland. *b*. Simple tubular gland. *c*. Compound tubular gland. *d*. Compound alveolar gland. *e*. Compound tubulo-alveolar gland.

lymphocytes and plasma cells, and bone marrow in which red blood corpuscles and granulocytes are differentiated. Such a classification, however, extends the usage of the term 'secretion' beyond its legitimate meaning.

Exocrine glands. Glandular mechanisms are represented in their most elementary form by general secretory epithelia in which no particular structural differentiation occurs. Such, for example, is the epithelial lining of the gut cavity in simple coelenterates, in which all the cells are equally concerned in exocrine secretion. The ependymal epithelium which covers the choroid plexuses in the ventricles of the vertebrate brain is similarly concerned as a whole with the formation of cerebrospinal fluid. A further elaboration is shown by the differentiation of special secretory cells in the epithelium, such as the goblet cells which are scattered over the intestinal mucosa and many other mucous membranes.

Goblet cells are *unicellular glands* and are concerned with the production of mucin (Fig. 93 *d*). It is probable that they can be developed during life by the transformation of indifferent cells of the general epithelium, and that after a phase of active secretion they can again become converted into ordinary epithelial cells. The next stage in the development of glandular mechanisms is reached when individual secretory cells become aggregated to form simple *multicellular glands*. These are exemplified by the small isolated mucous glands scattered over the buccal cavity and lying close to the surface of the mucous membrane (Fig. 94 *a*).

A further development involves the sinking in of localized groups of specialized secretory cells so as to form a pocket or follicle which opens on to the surface by a small ductule. Aggregations of such follicles may extend more deeply, and their connexions with the surface then become drawn out into a series of ducts—e.g. the sublingual salivary gland in the floor of the mouth. An extension of the same process leads to the stage where the ducts from individual follicles or groups of follicles open into a single main duct which may traverse some distance before it reaches the surface of the mucous membrane (e.g. the parotid gland).

The process of the sinking in of groups of secretory cells from the surface to form complex glands provided with ducts serves several purposes. It shelters the secreting cells from the surface; it allows a great extension of the secreting surface by the formation of ramifications of closely packed follicles; and by the formation of a duct of varying length it enables a massive gland to be accommodated in a convenient position (perhaps at some distance from the position of its original development) and permits the outflow of secretory juices to be concentrated on one localized area of the mucosal surface.

According to the degree of their elaboration and the contour of their secretory elements, glands can be classified in several categories. *Simple glands* are unbranched and usually tubular in contour, e.g. the sweat glands of the skin, the uterine glands, and the crypts of Lieberkühn in the small intestine (Fig. 94 *b*). *Compound glands* are formed by the branching to various degrees of the ducts and their subdivisions. This is shown in its simplest expression in the pyloric glands of the stomach and the small mucous glands (glands of Littré) in the urethra. A more complicated variety is the lachrymal gland, which constitutes what is termed a *compound tubular gland* (Fig. 94 *c*). In many glands the terminal secretory portion of each ramification is formed by a sac-like diverticulum lined by the secretory cells, which is called an *alveolus* or *acinus*. Such a gland is called a *compound alveolar gland* (Fig. 94 *d*), and this type is illustrated by the parotid gland. In most cases, however, the actual secretory mechanism includes not only the alveoli themselves but also the terminal tubular canals into which the alveoli open. This constitutes a *compound tubulo-alveolar gland* (Fig. 94 *e*), and is represented by the sublingual gland, the pancreas, prostate, and mammary glands.

One more type of gland requires mention here—the *reticulate type*, in which the secreting cells are arranged in anastomosing cords. While this arrangement is common enough in endocrine glands, it is found in only one exocrine gland, the liver. This organ is developed as an outgrowth from the upper part of the intestinal canal and, by active proliferation, its cells give rise to reticulating masses which invade the venous plexuses formed by the termination of the vitelline veins. In this way the plexuses become broken up into a labyrinthine series of sinusoidal vessels whereby the hepatic cells are brought into the most intimate relationship with the blood circulation.

Endocrine glands. It has been mentioned that, besides glands which pour their secretion on to an epithelial surface by ducts, there are other glands whose secretory products are taken up directly into the blood stream. These internal secretory or endocrine glands are found in different parts of the body and are heterogeneous in their origin and structure. Anatomically they show certain peculiarities in that their embryological derivation often seems to bear no obvious relation to their functional significance, suggesting, from the evolutionary point of view, that they have been developed by the modification of structures which were initially concerned with other functions.

Some endocrine glands arise in the early embryo from diverticula of the alimentary tract. These are the anterior lobe and pars intermedia of the pituitary gland, the thyroid and parathyroid glands, the thymus (at least so far as its epithelial component is concerned), and the islets of Langerhans in the pancreas. Such an origin may be taken to indicate that these were originally external secretory organs opening into the corresponding part of the alimentary tract by ducts, and that the ducts have subsequently been lost with the assumption of internal secretory functions. The thyroid gland, for example, is developed in the embryo from a median diverticulum which grows down into the neck from the floor of the pharynx, the stalk of the diverticulum forming the thyro-glossal duct. In the lowest chordates, as represented by *Amphioxus* and the larval lamprey, the thyro-glossal duct remains in a functional state, and in these forms the gland secretes a mucoid fluid which is poured into the pharynx and which evidently plays a part in preparing the food for digestion. During the metamorphosis of the larval lamprey, this exocrine gland (which is called the *endostyle*) becomes converted into a series of closed sacs similar in structure and secretory function to the thyroid gland of vertebrates generally. In all higher vertebrates the thyro-glossal duct becomes obliterated during embryonic development, though even in man a considerable portion of it may persist (in a non-patent form) as an anomalous variation.

The anterior lobe and pars intermedia of the pituitary gland are developed from a diverticulum in the roof of the primitive mouth (the stomodeum), the parathyroid glands and the thymus from the entodermal lining of the pharynx, and the islets of Langerhans from the diverticular outgrowths which also form the exocrine tissue of the pancreas.

Two endocrine organs, the posterior lobe of the pituitary gland and the pineal body, are derived from the neural tissue of the fore-brain. In their intrinsic anatomy these organs appear to have little in common with typical secretory tissue. Yet the posterior lobe of the pituitary is known to exert a hormonal control over the water balance of the body by regulating renal activity, while there is some evidence (admittedly incomplete), partly based on experimentation and partly on clinical observation, that the pineal may influence growth processes. However, so far as the posterior lobe of the pituitary is concerned, there is a growing accumulation of evidence in favour of the suggestion that its hormonal products, or at any rate their chemical precursors, may in part be elaborated by certain groups of 'neuro-secretory' cells in the hypothalamus and transported thence to the posterior lobe along connecting tracts of nerve fibres.[1] If this is so, these groups of nerve cells should properly be included in the endocrine system (see p. 341). The medulla of the adrenal gland is a specialized derivative of the sympathetic nervous system and is functionally directly related to sympathetic activities.

Of the remaining endocrine glands, the interstitial cells of the ovary and testis and the corpus luteum of the ovary are all derived embryologically from mesodermal tissue which is incorporated in the sex glands, and their functions are predominantly related to those of the reproductive system. The adrenal cortex has its embryological origin in the intermediate cell mass, and its histogenesis is intimately related to that of the urogenital apparatus. It elaborates active substances which are of vital importance for the maintenance and regulation of cell metabolism in health and disease by indirectly controlling the salt and water balance in the body, and it is believed to play a most intimate part in the resistance of the body against injury and infections of various kinds. The hormonal products of the adrenal cortex also have a close functional relationship to the activities of other endocrine glands, particularly the sex glands.

It may be surmised that, in their early evolutionary history, endocrine glands were developed as an extension of the essential basic mechanisms responsible for maintaining the existence and growth of the individual and for promoting the continuance of the species. These basic mechanisms are the glandular epithelial structures concerned with the absorption and assimilation of nutritive material, the primitive neural apparatus controlling the vegetative functions of the body, and the reproductive organs. With the progressive elaboration of metabolic processes in the vertebrates, the endocrine glands have become increasingly specialized, and their coordinating influence on tissue growth and visceral activities more subtly complex. Their functional interrelations have also become exceedingly intricate, so that it often becomes a matter of considerable difficulty to determine whether a particular endocrine gland exerts its influence on the growth and metabolism of other tissues by the direct action of its secretory hormone, or indirectly by disturbing the secretory activities of other endocrine glands.

To the anatomist, the study of endocrinology assumes a greater importance when it is recognized that fundamental processes of growth, tissue differentiation, and modification of anatomical structures are determined and controlled to a very large extent by the interaction of hormones secreted by the endocrine glands.

The influence on processes of growth and differentiation of those endocrine organs which embryologically are derived from the pharyngeal wall suggests that they are related in some way to the organizing regions of the early embryo which are responsible for the induction of structural differentiation and organization in adjacent tissues (see p. 33). The pharyngeal entoderm lies in immediate relation to growth centres which have an inducing action in the organization of the embryo, and the suggestion has been made that it functions as a secondary organizing region in the early stages of

---

[1] See the discussion in G. W. Harris, *Neural Control of the Pituitary Gland*, London, 1955.

development.[1] In the later stages, and after birth, this activity is continued in modified form by the regulatory mechanisms of ductless glands which arise from it. Since, in its origin, the pituitary gland is intimately connected with the primary growth centre of the prechordal region of the embryo, it remains the dominant member of the whole endocrine series in the functional interrelation of hormonal secretions during life. Incidentally, it may be observed that the part which the entoderm of the pharyngeal pouches plays in the genesis of endocrine organs provides an explanation of the persistence of these pouches in the embryological development of the higher vertebrates in which they no longer culminate in the formation of actual gill slits.

3. THE MACROSCOPIC FEATURES OF GLANDS

Seen in gross dissection, most compound glands appear as rather compact structures showing some degree of lobulation and a relatively smooth surface. The lobulation is related to the manner in which a gland develops as a diverticular outgrowth which repeatedly divides and subdivides to form a branching system of ducts. When the ultimate ramifications are bound together in loose connective tissue, the lobulation remains quite apparent, as in the case of the salivary glands and the pancreas. In other glands the primary lobulation may become obscured superficially during development by the close packing of the individual lobules, and is then only revealed by a close inspection of the internal structure, e.g. in the liver or kidney. In the foetus the kidney is quite obviously lobulated in its surface appearance, and occasionally this condition may persist into adult life. In the liver the primary subdivisions of the embryonic hepatic diverticulum are represented in the adult by the main lobes of the organ, but the individual lobules are not evident except on microscopical examination. It may be noted that the lobulation of the liver is somewhat exceptional in that the lobules are differentiated around hepatic venules and not around ductules; the lobules only become well defined by connective-tissue septa shortly after birth.

Most glands are provided with connective-tissue capsules which in some cases may form a relatively dense fascial investment. This has a supporting function and, where it contains an element of unstriped muscle, it may play a part in expressing secretory products.

The details of the shape assumed by a gland are largely a reflection of the pressure of surrounding structures. It should be recognized that glandular tissue (as it occurs in bulk) is exceptionally plastic and therefore readily moulded by external forces. For example, the liver shows on its surface certain impressions for which surrounding structures such as the stomach, oesophagus, inferior vena cava, and right kidney are responsible. Moreover, the impressions formed by hollow viscera vary in appearance from time to time in relation to the distension and emptying of the latter. The precise shape and contour of solid glandular organs in the abdominal cavity as seen in the dissecting-room, therefore, will depend to some extent on the position

[1] H. H. Woollard, 'The potency of the pharyngeal entoderm', *Journ. Anat.* **66**, 1932.

and condition of neighbouring viscera at the time of death. Glandular organs in the neck are also moulded in relation to neighbouring structures. It is customary in topographical anatomy to describe various grooves, eminences, and surfaces on the thyroid and salivary glands; these are merely the impressions of neighbouring blood vessels, muscles, bony elements, &c., in among which the glands are neatly packed.

In the process of their developmental outgrowth, glands may surround adjacent structures (such as vessels and nerves) which then come to be embedded in their substance. For example, the external carotid artery and the branches of the facial nerve run through the middle of the parotid gland, and the pancreas almost encircles the superior mesenteric artery and vein. In a similar manner, the submandibular gland may come to enclose in its substance adjacent lymph nodes, and the parathyroid glands sometimes become secondarily embedded within the thyroid gland. It will be readily understood that such complications may seriously add to the difficulties of experimental and surgical procedures.

4. THE ANATOMY OF SECRETION

The activity of a glandular cell is reflected in its internal anatomy. A secretory phase is usually first indicated by the appearance in the cytoplasm of minute granules which can often be demonstrated by selective staining. In the case of mucus-secreting cells, these granules are composed of a substance called *mucigen*, and they make their first appearance near the free border of the cell. They become rapidly transformed into clear droplets of mucus which run together to form a confluent mass, often distending the cell so that it assumes the characteristic shape of a 'goblet cell'. Finally, the mucus is discharged at the free surface (i.e. the discharging pole) of the cell, and the latter collapses. In the case of serous cells, which elaborate enzymes, the granules are highly refractile and are composed of a substance called *zymogen*. They usually appear first in the middle part of the cell, between the nucleus and the free border. With the progress of glandular activity, they increase in number and, apparently as the result of imbibing water, they also grow larger and stain less intensely. Their accumulation in the cytoplasm frequently displaces the nucleus towards the basal end of the cell. Some of the granules may become converted into conspicuous vacuoles. Finally, the secretion granules and vacuoles become pushed towards the discharging pole of the cell and, on their liberation, the zymogen is converted into an active enzyme (Fig. 95).

Secretory products appear to be formed in close relation to the mitochondrial elements of the cytoplasm, and it had been sugested, therefore, that mitochondria are directly transformed into these products. Detailed cytological studies have shown such a conception to be incorrect; nevertheless it remains certain that mitochondria indirectly play an important part in the secretory processes of the cell through the activity of enzymes concentrated at their surface or on the double membranes which are shown in electron microphotographs to form partial septa in their interior. Another element of the cell which is also concerned in some way with the formation of secretion granules is the Golgi apparatus (see p. 15). This structure undergoes characteristic changes in relation to glandular activity and appears constantly to take up a position between the nucleus and the discharging pole of the cell. The position of the Golgi apparatus has therefore been taken as an indication of the secretory polarity of cells of endocrine glands in which there is no duct system. According to one view, the actual synthesis of secretory products first occurs at the surface of the mitochondria where the intracellular enzymes are mainly located.[1] These products then diffuse out into the surrounding cytoplasm and are concentrated into discrete droplets or granules in the interstices of the Golgi apparatus.

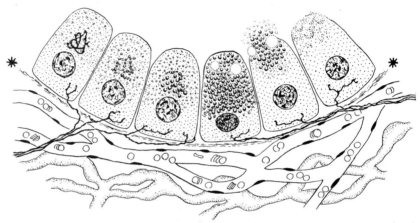

FIG. 95. Diagram illustrating schematically the essential elements of a glandular unit and the cycle of secretory activity in a serous gland. In the cell on the left is shown the appearance of the Golgi apparatus as it can be demonstrated by special staining methods. While the products of secretion are synthesized in the cytoplasm generally, they appear to become segregated in the form of granules in close relation to the Golgi apparatus. The granules increase in size and number, and tend to compress the nucleus towards the basal end of the cell. At the same time some of the granules become converted, by a process of liquefaction, into conspicuous vacuoles. At the height of secretory activity the granules and the contents of the vacuoles are discharged at the free end of the cell. The basal ends of the cells are separated by the basement membrane (*–*) from the blood capillaries and (on a deeper plane) lymphatic capillaries. Secreto-motor nerve fibres are shown penetrating the basement membrane to make contact with the surface of the gland cells (or, according to the view of some histologists which is not generally accepted, actually terminating within the cell).

Whether this is simply a surface phenomenon, the Golgi apparatus acting passively as a conglomeration of condensation membranes, or whether the apparatus is concerned more directly with the formation of the secretory products, seems uncertain. The latter might be expected to be the case if, as generally supposed, the Golgi apparatus represents a focal centre of synthetic processes in the cell.[2]

The final discharge of the secretion at the surface of a cell may be brought about in several ways. In *holocrine* glands the whole cell, having

[1] R. J. Ludford, 'The vital staining of normal and malignant cells', *Proc. Roy. Soc. London*, **103**, 1928.

[2] See R. H. Bowen, 'The cytology of glandular secretion', *Quart. Rev. Biol.* **4**, 1929.

become filled and distended with secretion, undergoes complete disintegration. The cell dies and the secretory products are thereby set free. This extravagant method of secretion is found in the sebaceous glands of the skin. It involves the continuous proliferative activity of basal cells in order to replace those which are destroyed in the process of secretion. In *apocrine* glands, only the free or superficial part of the cell disintegrates. The basal portion with the nucleus remains intact and provides the basis for the regeneration of the whole cell. Such a type of gland is represented by certain types of sweat gland and also (according to some histologists) by the mammary gland. A more economical method of secretion is shown by *epicrine* glands, which constitute most of the glands in the body. In this case the secretory products are discharged on the surface without any disruption of the cell. These different modes of secretory discharge are in part related to the nature of the secretory substance. Viscous substances are not able to pass through the intact cell membrane, so that the latter is necessarily ruptured by their discharge. This is what occurs in apocrine glands. Sebum is evidently too viscous to be released in this way, and so the cells of the sebaceous glands have to disintegrate entirely in order to discharge their secretion.

In some secretory cells a system of intracellular canaliculi has been described, through which the secretion is conducted to the exterior. It seems doubtful, however, whether these are real morphological entities, for the possibility has not been eliminated that they may be artefacts produced during the course of histological preparation of the tissues.

From the above description it will be realized that phases of glandular activity can be followed to some extent by purely anatomical methods. These methods allow a study of the effects of pharmacological agents or nervous stimulation, and the secretory function of cells whose significance is otherwise doubtful may thus be established.

5. THE EMBRYOLOGY OF SECRETION

We have noted that, in most cases, glands are developed as diverticular outgrowths from epithelial surfaces, and that the pedicle of the diverticulum eventually forms the duct of the gland. In general, therefore, the developmental origin of any gland can usually be inferred by a consideration of the duct system in the adult. The opening of the duct usually marks the site of the embryonic outgrowth, a single duct suggests that the gland has been developed from a single diverticulum, while two or more ducts are indicative of multiple diverticula.

As the diverticular process of a gland grows out it commonly branches by successive dichotomy so as ultimately to form a series of ductules which unite evenly in successive stages to form the larger ducts. Exceptions to this arrangement, however, may be found, as in the pancreas where a main longitudinal duct runs the whole length of the gland receiving secondary tributaries as it courses towards the duodenum.

Since the initial diverticular outgrowth forms the duct of the mature gland, it follows that the duct is the primary formation from which the secretory elements of the gland are ultimately differentiated. This developmental potency of the duct epithelium is in many cases retained in the adult for, as we shall see, regeneration and repair of secretory acini can take place by its proliferation.

In the kidney, the development of the main duct (the ureter) occurs independently of the development of the secretory elements of the renal tubules, the two component parts of the system becoming secondarily connected up with each other. Even in this case, however, the renal tubules apparently develop under the influence of the ureteric diverticulum, for it has been shown by embryological and experimental studies that, in some animals, if the ureter fails to develop, differentiation of the embryonic kidney tissue (the metanephric tubules) does not occur.[1]

Little is known of the factors which initiate and control the development of glandular structures in the embryo. By analogy they may be presumed to depend, at least partly, on processes of induction by adjacent tissues. The development of the kidney in response to the influence of the ureteric outgrowth is an example of this. In the case of the endocrine glands it is known by experimental studies on amphibian larvae that the development of the thyroid gland depends on the presence of the developing pituitary gland, and there is abundant evidence to show that the full development of the accessory sexual glands is conditioned by the activity of the appropriate sex hormones.

Embryonic rudiments of glandular tissue can also in some cases undergo a surprising degree of self-differentiation when transplanted away from their normal developmental environment. For example, if the embryonic stomach of a 14-day rat embryo is grafted into the anterior chamber of the eye of an adult animal, it differentiates to form a typical gastric mucosa with pits and glands, and the different types of cellular element also become clearly recognizable. Similar transplants of embryonic intestine form villi and intestinal glands, as well as mucous cells in the epithelium. Moreover, the circular and longitudinal muscle coats of the intestine also become differentiated.[2] Another series of experiments has shown that, if the rudiments of the nephric region of 10- to 12-day rabbit embryos are grafted on to the omentum, they may develop into complete but minute kidneys.[3] Finally, explanted rudiments of salivary glands have been found to undergo a remarkable degree of differentiation in vitro, in some cases growing at almost the normal rate.[4]

The main glands of the alimentary system commence their development at a very early stage. The liver diverticulum first appears in the 2·5-mm. human embryo, and the pancreatic outgrowths are seen in embryos of 3 to 4 mm. The salivary glands commence their formation a little later—at the sixth week of development, while the smaller glands of the intestinal

[1] C. H. Seevers, 'Potencies of the end bud and other caudal levels of the early chick embryo', Anat. Rec. 54, 1932; E. A. Boyden, 'An enquiry into the cause of congenital absence of the kidney', ibid. 52, 1932.

[2] A. Kammeraad, 'Homotransplantation of embryonic and adult gastric and intestinal mucosa to the anterior chamber of the eye', Journ. Exp. Zool. 91, 1942.

[3] A. J. Waterman, 'Growth and differentiation of kidney tissue of the rabbit in omental grafts', Journ. Morph. 67, 1940.

[4] E. Borghese, 'The development in vitro of the submandibular and sublingual glands of Mus musculus', Journ. Anat. 84, 1950.

mucosa do not appear till the third month. Cutaneous glands begin to develop still later—the sweat glands in the fourth month, sebaceous glands and the mammary glands in the fifth month. There is no evidence that the sweat glands are functional before birth. On the other hand, the sebaceous glands are certainly active at this time for they provide an oily protective covering to the skin which, with desquamated epithelium and hair, is called the *vernix caseosa*.

Glandular cells do not reach their full differentiation until their secretory functions are imminent. In some glands this occurs quite early during foetal life.

There is some evidence that the production of digestive enzymes may occur before birth, for Keene and Hewer found that proteolytic ferments are present in the alimentary tract of the human foetus at sixteen weeks, and pancreatic lipase at nineteen weeks; on the other hand, carbohydrate splitting ferments are very rare at twenty-four weeks and are not constant even at full term. However, these results require to be accepted with caution for (as the authors point out) there are many difficulties in the assessment of an investigation such as this which is made on stillborn (and therefore not normal) human foetuses.[1]

The true kidney can probably perform excretory functions in the human embryo as early as the ninth week, though under normal conditions they are certainly of negligible significance. It is known that embryos in which both kidneys have failed to develop may be otherwise quite normal at birth. In the uterus the placenta provides the main excretory apparatus of the foetus and immediately after birth the role is suddenly transferred to the kidneys. In association with this abrupt change there is an equally abrupt modification in the structure of the renal glomerulus.[2] Before birth the glomerulus is covered by a columnar epithelium which has a correspondingly low permeability. Soon after birth the expansion of the glomerulus (associated with an increased blood flow to the kidneys) leads to a rupture of the columnar epithelium which is replaced by an extremely thin layer of pavement cells (the 'visceral' layer of Bowman's capsule). The resulting increase in permeability greatly facilitates the process of filtration at the glomerulus.

As regards endocrine glands, there is good evidence that these also commence to function during intra-uterine life. The pituitary gland shows quite a considerable degree of histological differentiation by the twelfth week, and an extract from a pituitary gland of an eight weeks' human foetus gives the characteristic melanophore reaction. In the thyroid gland, colloid material can be demonstrated in the vesicles at eleven weeks, and the lining cells give the appearance of being actively engaged in secretion. Iodine has been found to be present in the gland in the sixth month. The

[1] M. F. Lucas Keene and E. E. Hewer, 'Glandular activity in the human foetus', *Lancet*, 19 July 1924; 'Digestive enzymes of the human foetus', ibid., 13 April 1929. For a comprehensive account of the physiology of the digestive and other systems of the foetus, see W. F. Windle, *Physiology of the Foetus*, 1940, and C. A. Smith, *The Physiology of the New-born Infant*, 3rd edn., Springfield, Ill., 1959.

[2] P. Gruenwald and H. Popper, 'The histogenesis and physiology of the renal glomerulus in early post-natal life', *J. Urol.* **43**, 1940.

adrenal glands are remarkably large at birth, as the result of the great hypertrophy of a zone of cells in the cortex (sometimes called the 'foetal cortex'). Shortly after birth this zone undergoes rapid regression and finally disappears; its function is still a matter of conjecture. The medulla of the adrenal gland elaborates adrenalin quite early in the embryonic life of many mammals, but in man it has not been found to be present to any significant extent until shortly before birth.

The internal secretions of the sex glands are of considerable developmental importance in the embryo, for under their influence the development and differentiation of the genital apparatus are partly determined and directed. Defects in the development of the genital organs may thus be the result of an abnormal hormonal environment. This is clearly demonstrated in the study of 'freemartin' twins in cattle, that is to say, twins in which the foetal membranes are freely interconnected by vascular anastomoses. If the foetuses are of different sex, the hormone elaborated by the gonads of the male lead to gross disturbances in the development of the genital system of the female. There is evidence to suggest that the testes start to become active in their hormonal secretion as soon as they become histologically differentiated; in the human foetus this occurs at about the seventh week.

6. GROWTH AND REPAIR OF MUCOUS MEMBRANES AND GLANDS

The epithelium covering mucous membranes, like the epidermis of the skin, possesses considerable powers of repair. This is particularly emphasized in the case of the columnar epithelium of the uterus, a large proportion of which is denuded and regenerated in each menstrual cycle. In the intestinal mucosa, mitoses are frequently to be seen in the base and walls of the glandular crypts, and this proliferative activity provides the basis for the regeneration of epithelium and intestinal glands which occurs after local injuries.

The gastric mucosa has similar reparative properties. Ferguson[1] has shown that, if a piece of mucous membrane as large as 25 sq. cm. is removed from the stomach of a dog, it becomes rapidly regenerated (with the formation of new glands), new epithelium extending in from the margin at the rate of 2 mm. a week.[2]

The transitional epithelium of the urinary tract is regenerated from the basal and central layers in which mitotic figures occur regularly.[3]

Even in some of the larger and more specialized glands, considerable regeneration may occur after partial destruction of secretory tissue. In the rat, for example, the liver may be restored to its normal dimensions after

[1] A. N. Ferguson, 'A cytological study of the regeneration of gastric glands following the experimental removal of large areas of mucosa', *Amer. Journ. Anat.* **42**, 1928.

[2] On the other hand, Ferguson found very little regeneration of epithelium and no new gland formation in the rabbit's stomach. This is attributed to the fact that the rabbit's stomach is never empty and therefore never functionally quiescent.

[3] A. Brauer, 'The regeneration of transitional epithelium', *Anat. Rec.* **33**, 1926.

removal of as much as two-thirds of its bulk, regeneration being complete in three weeks.[1] Pancreatic tissue can be formed anew from duct epithelium. Grauer[2] found that if the pancreas of a rabbit is almost entirely extirpated, leaving only a system of branching ducts, it is restored to its normal state in twenty-five days, both acini and islets of Langerhans being regenerated. In another series of experiments by Shaw and Latimer,[3] portions of the pancreatic duct were transplanted into the intestinal wall in wholly or partially pancreatectomized dogs, and in some of these cases new acini and islet tissue were formed from the transplants. Further, new ducts can be formed in the process of regeneration, for if the main pancreatic duct is ligatured, a new duct may grow out from the proximal end and effect a functional union with the distal part of the pancreas.[4] The cortex of the adrenal gland may undergo regeneration in a surprisingly short time. For example, Greep and Deane reported that following 'enucleation' of both adrenals in the adult rat (an operation which leaves behind fragments of the glomerulosal zone of cells), within a month the cortex had been restored to its normal appearance by proliferation of the residual cells.[5]

In relation to the regenerative capacity of glandular tissue, we may mention the compensatory hypertrophy which occurs under certain circumstances. If one kidney is removed, it is found that the remaining kidney enlarges. At first this enlargement is merely a transient pseudo-hypertrophy due to vascular congestion, but this is eventually followed by a true hypertrophy.[6] The glomeruli and renal tubules become enlarged much above their normal size, but they show no increase in number. A similar hypertrophy of a single kidney has been described in a case of congenital absence of the other kidney and ureter in man.[7] That this hypertrophy is conditioned by functional demands *after* birth is indicated by another observation, that in a human foetus showing congenital absence of one kidney the remaining kidney may be normal in size. Experimental studies on young rats have also emphasized the relation between the growth of kidney tissue and the work imposed on it, for not only does the growth rate vary in the direction of the protein content of the diet, but if one kidney is removed and the protein intake increased the remaining kidney may even become larger than the original two kidneys.[8]

Parallel observations have been made on certain endocrine glands. For example, if one adrenal gland is removed in a rat, the remaining gland

[1] R. D. Harkness, 'Regeneration of Liver', *Brit. Med. Bull.* **13**, 1957.

[2] T. P. Grauer, 'Regeneration in the pancreas of the rabbit', *Amer. Journ. Anat.* **38**, 1926.

[3] J. W. Shaw and E. O. Latimer, 'Regeneration of pancreatic tissue from the transplanted pancreatic duct in the dog', *Amer. Journ. Phys.* **76**, 1926.

[4] N. F. Fisher, 'Regeneration of the pancreas from the pancreatic duct', *Journ. Amer. Med. Ass.* **83**, 1924.

[5] R. O. Greep and H. W. Deane, 'Histological, cytochemical and physiological observations on the regeneration of the rat's adrenal gland following enucleation', *Endocrinology*, **45**, 1949.

[6] C. M. Jackson and N. M. Levine, 'Rate and character of the compensatory renal hypertrophy after unilateral nephrectomy in young albino rats', *Anat. Rec.* **41**, 1928.

[7] R. J. Gladstone, 'A note on the post-natal growth of the kidney, thyroid gland and liver', *Journ. Anat.* **52**, 1917-18.

[8] T. Addis and W. Lew, 'The restoration of lost organ tissue', *Journ. Exp. Med.* **71**, 1940.

becomes markedly hypertrophied within a few days.[1] This hypertrophy can be inhibited by administration of the active principle of the adrenal cortex. In the normal animal, also, atrophy of the adrenal glands can be induced by the injection of large doses of cortin.[2] Similarly, partial removal of the thyroid gland is followed by a cell hypertrophy of the residual portion (but with no obvious increase in the number of cells or follicles),[3] and regressive changes in the gland follow the administration of excessive amounts of thyroid extract.[4] It thus appears that the amount of glandular tissue which is formed by growth and differentiation is at least partly related to the functional demands for the active principle which it elaborates.

Perhaps the most striking example of the relation of glandular growth to function is shown by the mammary glands. Up till puberty in the female, these consist of little more than a simplified duct system in which the potentialities for the development of active secretory tissue remain quiescent until it is required. At puberty the gland shows its first marked expansion by the proliferation of the duct system, but, until pregnancy occurs, the actual alveoli of the gland are only developed to a very rudimentary degree. During pregnancy the alveoli become differentiated, and secretory activity commences. Active secretion continues after parturition until weaning occurs, when, as the result of the fact that the secretion is no longer removed by suckling, the mammary gland undergoes a structural involution and once more enters on a quiescent phase. The development and growth of the mammary gland have been shown experimentally to be controlled by a series of hormonal influences, and similar factors are concerned with the growth of the accessory sex glands in the male.

7. THE VASCULAR AND NERVE SUPPLY OF GLANDS

Actively functional glands are richly vascular. In the case of exocrine glands the blood conveys to the cells the material from which the specific secretory substances are elaborated, and in the endocrine glands it also serves to transport the products of secretion to other parts of the body. As might be expected, therefore, the vascularity of endocrine is greater than that of exocrine glands.

In most glands the main vessel of supply enters at a fairly definite point, the hilum, and is distributed along the course of the duct system when this is present. The vascular pattern formed by the terminal capillaries is a reflection of the cellular arrangement in the gland, and unstained sections of glands in which the vessels have been injected will therefore show in outline the disposition of these cells in alveoli, lobules, or columns.

[1] C. A. Winter and F. E. Emery, 'Compensatory adrenal hypertrophy in the rat as influenced by sex, castration, time and thyroidectomy', *Anat. Rec.* **66**, 1936.
[2] D. J. Ingle, G. H. Higgins, and E. C. Kendall, 'Atrophy of the adrenal cortex in the rat produced by the administration of large amounts of cortin', ibid. **71**, 1938.
[3] J. H. Logothetopoulos and I. Doniach, 'Compensatory hypertrophy of the rat thyroid after partial thyroidectomy', *Brit. Journ. Exp. Path.* **36**, 1955.
[4] L. Loeb, R. B. Bassett, and H. Friedman, 'Further investigations concerning the stimulating effect of anterior pituitary gland preparations on the thyroid gland', *Proc. Soc. Exp. Biol. & Med.* **28**, 1930.

In a previous chapter (p. 205), reference has already been made to the presence in some glands of arteriovenous anastomoses which provide a regulating mechanism through which the circulation can be adjusted to secretory activity.

In the case of endocrine glands, the demand for an intimate relationship between the blood and the secretory cells is met by the formation of sinusoidal capillaries. Besides showing local dilatations which permit of a slowing-up of the blood stream, these vessels have extremely thin walls which are composed of little more than a single layer of endothelial cells,

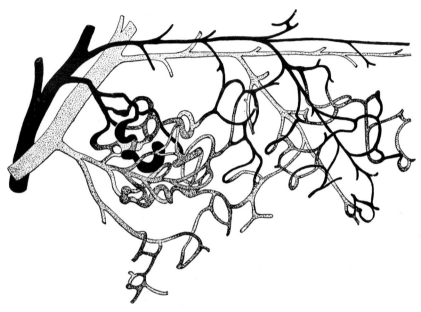

FIG. 96. A schematic drawing of the blood supply of a pancreatic lobule in a mouse. The arteries are black, and the veins stippled. The position of an islet of Langerhans is indicated by the plexus of sinusoidal vessels. (From E. V. Cowdry after Beck and Berg.)

and the latter may be perforated by minute pores. The blood can therefore be brought into the closest proximity with the glandular cells. The contrast between the vascular supply of endocrine and exocrine glands may be illustrated by reference to the pancreas in which endocrine tissue is scattered (in the form of the islets of Langerhans) throughout the acinar formations (Fig. 96).

The pancreas is supplied by several arteries from different sources. These vessels break up to form an interlobular plexus from which blood is distributed to the lobules. According to Wharton,[1] the blood in the larger lobules first passes to the islet tissue, and then forms a terminal capillary plexus among the externally secreting acini. In the islets the vessels are large and sinusoidal in character, being tortuous and lined by

[1] G. K. Wharton, 'The blood supply of the pancreas', *Anat. Rec.* **53**, 1932.

a simple endothelial layer supported only by scanty strands of connective tissue. Some end in blind pouches surrounded by islet cells. The efferent vessels from the islets are smaller in calibre. It is suggested that the wide calibre of the vessels in the islets, the blind pouches, and the small size of the efferent vessels will necessarily produce a slowing of the blood stream and in this way assist in the absorption of the insulin which is elaborated by the islet cells. The capillary plexus related to the acini of the pancreas shows none of these special adaptations; the vessels retain an even calibre and they are separated from the glandular cells by a thin layer of perithelial tissue.

Glands are commonly provided with a rich plexus of lymphatic capillaries which, however, do not have such an intimate relationship with the secretory elements as the blood capillaries. The lymph flow from a gland is increased during secretory activity, but this is evidently secondary to the increased blood flow.

Entering glands in company with blood vessels are always to be found fine nerve plexuses derived from the autonomic nervous system. If these nerves are followed anatomically to their termination, most of them are seen to end in relation to vessel walls; in other words, they are vasomotor nerves and through them the secretory activity of the gland can be modified indirectly by vasoconstriction or vasodilatation.

In the case of some glands, nerve terminals can be traced into direct relationship with secretory cells, thus providing an anatomical demonstration of the existence of *secreto-motor* nerves. Such nerves are found, for example, in the sweat glands, the salivary glands, the pancreas, and the medulla of the adrenals. Physiological observation has in some cases demonstrated the presence of these secreto-motor fibres, but the difficulty in such experiments may be to differentiate between the effects of nerve stimulation which are due to a direct activation of glandular cells and those which are the secondary result of vasodilatation. In the salivary glands this differentiation has been made by observing that the secretory pressure may exceed the intraglandular blood pressure, and that, after the injection of atropin, stimulation of the chorda tympani produces vasodilatation without inducing secretory activity. In the sweat glands, again, it has been found that sympathetic stimulation can induce secretion even in the absence of a cutaneous circulation.

In many glands (such as the thyroid gland, the liver, and the adrenal cortex), repeated experimentation has failed to show any convincing evidence that their secretion is under direct nervous control. In these structures the terminal nerve fibres appear histologically to be all related to blood vessels rather than to the glandular cells themselves.

In its relation to the nervous system, the medulla of the adrenal gland occupies a unique position, for here the glandular cells are morphologically equivalent to the postganglionic neurons of the sympathetic nervous system (see p. 395). They are therefore directly innervated by preganglionic neurons, and stimulation of the splanchnic nerves leads to an outpouring of their secretions, adrenalin and noradrenalin, into the blood stream.

The posterior lobe of the pituitary gland, being derived embryologically from an outgrowth of the fore-brain, is innervated directly from the central nervous system by a tract of fibres (the supra-optico-hypophysial tract) which enters it from the hypothalamus. The functional importance of this tract is indicated by the fact that, if it is interrupted bilaterally, a condition of diabetes insipidus[1] ensues.

[1] Diabetes insipidus is a metabolic disorder characterized by the excretion of excessive quantities of sugar-free urine with a low specific gravity.

XII

SKIN

THE whole of the surface of the body is covered by a layer of skin, which thus provides a protective cuticle for all the underlying tissues. Even the transparent cornea of the eye is overlain by a continuous layer of modified skin. The skin also turns in to line orifices such as the mouth, nostrils, auditory meatus, and anal canal.

Besides its obviously protective value, the skin serves a number of other important functions which are related to the fact that it forms a surface of contact between the organism which it covers and the external environment. For example, it plays an important part in the temperature regulation of the body by surface evaporation of sweat and by the cooling of the blood in its superficial capillary vessels, and, since it is impervious to water, it helps to conserve the moisture of underlying tissues. It also provides an extensive sensory surface richly strewn with nerve endings for the appreciation of general sensory stimuli. There are yet other functions which can be ascribed to the skin, and which will be considered in the later sections of this chapter.

1. THE STRUCTURE OF THE SKIN

The skin consists of a superficial layer of stratified epithelium—the *epidermis*—laid on a foundation of firm connective tissue—the *dermis* or *corium*. Its thickness shows considerable regional variations, ranging in the human body from less than $\frac{1}{10}$ of a millimetre up to 3 or even 4 millimetres. It is generally thicker on extensor than on flexor surfaces, but is thickest on the palms of the hands and the soles of the feet. Since it is well recognized that an area of skin which is exposed to pressure and friction becomes thickened, it may be assumed that the thickness of the skin on the palms and soles is related to these factors, and it is interesting to note, therefore, that the characteristic features of the palmar and plantar skin are already well developed before birth. Some authorities in the past have been tempted to see in this phenomenon an expression of the inheritance of acquired characters, but it has been pointed out that it is obviously an adaptive character which would be susceptible to the influence of natural selection. Thus, it is of definite advantage to the individual that the skin of the sole should be resistant to the effects of pressure and friction as soon as it is put to use.

Epidermis.[1] The stratified epithelium of the epidermis is superficially converted into cornified material which is continually being worn away by usage, and as continually being replaced by proliferation from the deeper strata. Consequently, its layers represent every transition from basal cells

[1] For a discussion of the structure of the epidermis see P. B. Medawar, 'The micro-anatomy of the mammalian epidermis', *Quart. Journ. Micro. Anat.* **94**, 1953.

with well-defined nuclei to superficial flaky débris in which the nuclei and all evidence of cell structure have disappeared (Fig. 98).

In a typical part of the epidermis there can be recognized a number of different strata in which the cells have distinctive anatomical features. From below, the first stratum is the *basal layer* or *layer of Malpighi*. Its cells are mostly polygonal in shape, the deepest tending to a cylindrical columnar form, and the most superficial becoming somewhat flattened.

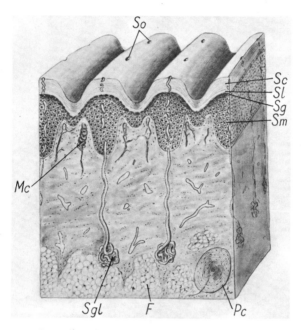

FIG. 97. Diagram illustrating the structure of skin covered with papillary ridges. The sweat glands (*Sgl*) are seen in the deeper part of the dermis and their ducts are shown passing up through the epidermis to open at minute orifices (*So*) along the summits of the papillary ridges. A Meissner's corpuscle (*Mc*) is depicted in one of the dermal papillae, and a Pacinian corpuscle (*Pc*) in the deep part of the dermis. Deep to the dermis is the subcutaneous fat (*F*). The various strata of the epidermis are indicated, str. corneum (*Sc*), str. lucidum (*Sl*), str. granulosum (*Sg*), and the stratum of Malpighi (*Sm*).

Active mitotic proliferation takes place in the deeper layers, the development of new cells leading to a gradual displacement of the older cells towards the surface. Hence this stratum is also called the *stratum germinativum*. Its proliferative activity is intermittent and varies from one area of skin to another at any particular time. As might be expected, it is most marked in those regions of the skin which are most exposed to the effects of pressure and friction. Apart from regional variations, there is also a diurnal rhythm in the mitotic activity of the germinal layer. For example, Cooper found from the study of the epidermis of the prepuce (of infants) removed at circumcision that the activity is greatest between 9 and 10 p.m., and least between 5 and 10 a.m.[1]

[1] Z. K. Cooper, 'Mitotic rhythm in human epidermis', *Journ. Invest. Derm.* **2**, 1939.

The epidermis is quite avascular, and between the cells of the stratum germinativum there are fine intercellular channels which probably allow the transmission of nutrient fluids derived from the capillary blood vessels in the subjacent dermis. These channels are bridged across by delicate protoplasmic threads connecting one cell with another. The stratum germinativum, therefore, appears to be a syncytium of cells. The reality

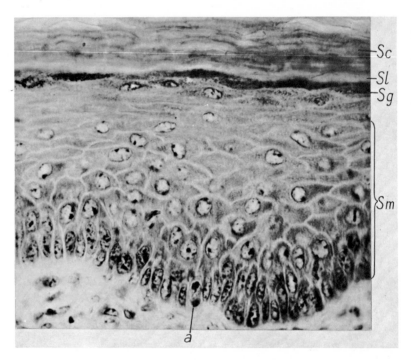

FIG. 98. Microphotograph of the epidermis of a monkey. The various strata of the epidermis are evident; *Sc.* Stratum corneum, *Sl.* Stratum lucidum, *Sg.* Stratum granulosum, and *Sm.* Stratum of Malpighi. In the deepest layer of the stratum of Malpighi can be seen a cell in the telophase of mitotic division (*a*). The fine intercellular bridges between the cells of the stratum of Malpighi can be faintly discerned. × 650 approx. (Photograph lent by Dr. K. C. Richardson.)

of this syncytium was at one time presumed by Chambers and Rényi[1] to have been demonstrated by microdissection of the living cells of human epidermis. They found that the cytoplasm of the cells of the stratum germinativum is particularly tough and resistant (contrasting with the soft and viscid cytoplasm of epithelial cells lining mucous membranes), and the numerous protoplasmic bridges by which they are united require some force to break through. If one cell of a group is deliberately injured, the effects of the injury (resulting in coagulation of the nucleus) rapidly spread to neighbouring cells by means of these connexions. In the cells of the

[1] R. Chambers and G. S. Rényi, 'The physical relationships of the cells in epithelia', *Amer. Journ. Anat.* **35**, 1925.

superficial strata of the epidermis, the protoplasmic connexions are no longer present or they are so brittle that they easily break, so that these cells can readily be separated from each other. From electron microscopy studies it now appears that the intercellular bridges do not betoken an actual syncytial continuity of the cytoplasm of one epidermal cell with its neighbour, for they are interrupted by cellular interfaces which are not detectable with the light microscope. But it is evident that the bridges are strong enough to resist separation at the interfaces when stretched by micromanipulation, so that they may tear through at their attachment to neighbouring cells and thus injure the latter indirectly.

Intracellular fibrils are commonly described and depicted in the cells of the germinal stratum, continuous from cell to cell and reinforcing the intercellular bridges. Microdissection of fresh tissue, however, indicates that these fibrils probably do not exist as anatomical entities. Their appearance is apparently due to tension lines in a structurally homogeneous cytoplasm, produced by traction of the protoplasmic intercellular bridges.

Covering the stratum of Malpighi is a layer two or three cells deep called the *stratum granulosum*. The cells are flattened and contain in their cytoplasm an abundance of granules which stain conspicuously with haematoxylin and other dyes. These granules are composed of a colloid material, and they are believed to represent the first stage in the transformation of the epidermal cells into a horny material—*keratin*.

More superficial is the *stratum lucidum*—so called because it has a clear translucent appearance in stained sections. At this level of their growth towards the surface, the cells have lost their clear-cut outline, the nuclei are becoming indistinct, and the granules of the subjacent stratum have become converted into larger masses of an achromatic substance.

The surface stratum forms the greater part of the whole thickness of the epidermis in many parts of the skin. Because of the horny character of the cellular elements which compose it, it is called the *stratum corneum*. Nuclei are no longer evident in these elements, cell structure has become completely obscured, and from below upwards the flattened remains of cells become gradually converted into cornified flakes. The stratum corneum is traversed by the ducts of sweat glands and by hairs where these are present.

The relative thickness of the various strata of the epidermis of the skin shows some local variations. The stratum lucidum and stratum granulosum may be extremely thin or quite absent. The stratum corneum and stratum germinativum, on the other hand, are always present. These variations naturally raise the question whether the granules so characteristic of the stratum granulosum really represent a stage in the formation of keratin, since it appears that keratinization can proceed without their appearance. This doubt is emphasized by the abrupt transition from the stratum granulosum to the stratum lucidum, as well as the sharp contrast in their staining properties. Further, it may be noted that in the embryo the stratum corneum is developed before the stratum granulosum becomes differentiated, while in the adult there is no constant relation between the thickness of these two strata. In fact, the biochemical processes involved in the cornification of the epidermal cells still remain rather obscure.[1]

Besides the epithelial cells of the epidermis, the latter contains cells of a different category whose importance has only recently been fully appreciated. These are the *dendritic cells*, or *cells of Langerhans*. Described originally in 1868, and for many years regarded as inconstant elements of no great significance, they have now been shown to be abundant in the epidermis of the skin and to play an essential part in pigmentation.[2] Dendritic cells are situated in the basal layer of the epidermis and are provided with slender and elaborately branching processes which spread out along the intercellular spaces. Each cytoplasmic process ends in a flattened expansion which applies itself, in sucker-like fashion, to an epithelial cell (Fig. 101). In this way, almost every cell of the Malpighian layer is capped by the end process of a dendritic cell. The processes of adjacent dendritic cells frequently run into direct confluence with each other, thus forming a syncytial network. In pigmented skin the dendritic cells are laden with pigment granules. They are also present in unpigmented skin, but here they are less conspicuous and may only be detected by special staining methods (such as methylene blue or silver impregnation). They are absent in the modified epidermis of the tongue and cornea. It has been supposed by some histologists that dendritic cells are merely aberrant types of the ordinary basal cells of the epidermis. It now appears, however, that they have a different embryological origin, being probably derivatives of the neural crest (like the pigment cells of amphibia) and secondarily invading the epithelial layers of the epidermis. It is uncertain whether dendritic cells undertake any functions other than those concerned with pigment formation, but it has been suggested that they serve a nutritive role by passing on nutrient substances from the dermis to the cells of the avascular epidermis.

Dermis. The dermis or corium consists of a dense feltwork of connective tissue in which bundles of collagenous fibres predominate, mingled with a certain proportion of elastic tissue in superficial levels. At the surface of contact between the epidermis and the dermis the elastic-tissue element is particularly well marked, and here it plays a part in anchoring down the epidermis to the subjacent dermis. The elastic tissue gives considerable resilience to the skin so that, by stretching, it can adapt itself to the movements of the body. It also explains why a simple skin incision often leads to a gaping wound—the margin of the cut being retracted by elastic tension. In old age the elastic tissue atrophies progressively, and the skin loses much of its elasticity and becomes wrinkled. When the skin becomes stretched beyond a certain limit by rapidly growing tumours, deposition of subcutaneous fat, or some other condition, the deeper layers may undergo a partial cleavage or rupture in irregular lines along which fibrous tissue is deposited to form a kind of cicatrice. This process becomes visible on the surface as a series of white streaks. Such white lines are commonly found

---

[1] Keratinization of epithelial cells is a commonly observed phenomenon in tissue cultures in which the nutrient medium is for some reason inadequate or unsuitable.

[2] R. E. Billingham, 'Dendritic cells', *Journ. Anat.* **82**, 1948; 'Dendritic cells in pigmented human skin', ibid. **83**, 1949.

in the skin of the abdominal wall after it has been stretched during pregnancy and are in this case termed *lineae gravidarum*.

The dermis becomes looser in texture in deeper planes, and the transition to the subcutaneous connective tissue (or superficial fascia) may be practically imperceptible. It contains fine plexuses of blood vessels, lymphatics and nerves, hair follicles, sweat glands, and sebaceous glands. In certain regions, e.g. the nipple, the scrotum, the genital region generally, and occasionally in the skin of the face, it also contains an element of plain muscle. This dermal unstriped musculature is better developed in lower mammals.

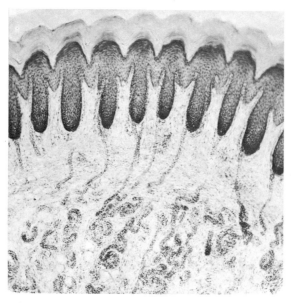

FIG. 99. Microphotograph of a section through the plantar skin of a marmoset. This section shows well the papillary ridges and dermal papillae. It also shows the coiled sweat glands in the subcutaneous tissue, and their ducts running towards the surface. × 60.

Dermal papillae. The boundary between the epidermis and dermis is not evenly horizontal, for the latter projects up at regular intervals in the form of small vascular papillae (Figs. 97 and 99). These papillae evidently provide for a considerable extension of the area of contact between the vascular dermis and the avascular epidermis, and so facilitate the diffusion of nutritive fluids from one to the other. The thicker the epidermis, therefore, the more prominent are the papillae, and in warty growths of the skin (in which the epidermis becomes greatly thickened) they are much exaggerated in size and complexity. In the skin of pachydermatous mammals, also, they are very long and attenuated.[1]

---

[1] This character reaches an extreme development in the horn of the rhinoceros, where the greatly elongated and thread-like papillae give it a fibrous texture, and this has led to the common but erroneous statement that a rhinoceros horn is constructed of thickly matted hairs.

On the palmar surface of the hand and the soles of the feet the dermal papillae are arranged as pairs in rows, each double row lying at the base of a surface elevation of the epidermis. These elevations are termed papillary ridges, and they are arranged in characteristic and distinctive patterns. The dermal papillae are not themselves responsible for elevating the epidermis, for they tend rather to occupy a position underneath the grooves that intervene between the epidermal ridges, and in certain areas of the skin, such as the nail bed, they may be quite well developed while the surface of the overlying epidermis remains smooth.

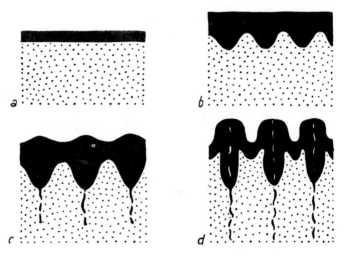

FIG. 100. Diagram of sections through the skin showing the development of papillary ridges in the human embryo. *a.* At the 6th week of development, when the epidermis forms a thin, even layer. *b.* In the 3rd month. At this stage dermal papillae have appeared, while the surface of the epidermis still remains flat. *c.* In the 5th month. The sweat glands are sprouting out as solid buds from the deep aspect of the epidermal thickenings, and the latter at the same time appear on the surface as elevations. *d.* In the 8th month, when the sweat glands have become canalized, and have acquired openings on the surface by the formation of channels through the epidermis.

In the process of embryonic development dermal papillae are formed secondarily at the intervals alternating between regularly spaced thickenings of the epidermis (Fig. 100). These thickenings at first lead to a corrugation of the line of contact between the dermis and epidermis while the surface of the latter is still quite smooth. Later on (in the fifth month in the human foetus), they give rise to down-growths which form the sweat glands, and then, becoming still more accentuated, they appear as surface elevations. The dermal papillae in the palms and soles are further complicated by secondary elevations and depressions and are richly supplied with sensory nerve endings. Papillary ridges provide a means whereby these nerve endings are brought as close to the surface as possible in the intervals between them (thus enhancing tactile sensitivity), while at the same time the thickened epidermal ridges themselves provide a mechanical protection and a gripping surface which is necessary over these cutaneous areas in connexion with their tactile and prehensile functions.[1]

By reference to the details of the papillary ridge patterns of the fingers, the distribution of arches, loops, and whorls, the inter-connexions of adjacent ridges and irregularities in their course, a system of finger-print identification was many years ago devised whereby it is possible to mark the identity of any individual. As is well known, this system is extensively used today for criminological purposes. The finger-print pattern remains unchanged from infancy to old age in each individual, and it is so stable that even if the epidermis on the finger-tips is largely scraped away or burnt off with acid, it is reproduced with identical details when the skin becomes regenerated. Attempts to establish distinctive racial features in the papillary ridge pattern of the fingers and the palm of the hand have shown no very consistent or conspicuous differences, but there are differences in the relative frequency of certain patterns.[2]

If the papillary ridges of the fingers are viewed with a hand lens, a row of minute depressed points will be seen along the centre of each one at intervals of approximately half a millimetre (Fig. 97). They are the orifices of the ducts of the sweat glands. These ducts, running up from the dermis, enter the epidermis between adjacent paired dermal papillae and pursue a spiral course to the surface.

2. FLEXURE LINES

The mobility of the skin over underlying structures depends on its separation from the latter by a zone of loose areolar tissue, and where this is present the skin can easily be plucked up in folds. Elsewhere—as, for instance, over the back of the neck and trunk, the lower part of the nose, the ear, and the palms and soles—it is much more firmly adherent to subjacent tissues. The skin also tends to be bound down to the periosteum covering bony eminences where the latter approach the surface; hence these are sometimes marked on the surface of the body not by elevations, but by dimples and furrows. In many regions it is crossed by permanent lines or creases. These are flexure lines—or 'skin joints'—about which the skin can be easily folded in association with movements of the parts which it covers. A section through a flexure line shows that here the skin is usually somewhat adherent to the underlying deep fascia.

The creases in the palm of the hand have long attracted attention because of the popular superstition that their patterns reveal the past and future history and the character of each individual. These lines are already present at birth, and much the same general pattern is found in the hands of lower Primates (monkeys and apes). The 'line of life' is a flexure line developed in relation to the movements of the thumb in opposition to the rest of the

[1] The interesting suggestion has been made that the thickened epidermal ridges separating the dermal papillae act as amplifying lever mechanisms for the transmission of tactile stimuli to the nerve endings in the papillae (N. Cauna, 'Nature and functions of the papillary ridges of the digital skin', *Anat Rec.* **119**, 1954).
[2] S. B. Holt, 'Dermatoglyphic patterns', from *Genetical Variation in Human Populations*, ed. by G. A. Harrison, Pergamon Press, 1961.

hand. The 'heart line' marks the position of the metacarpo-phalangeal joints. Laterally, this line shows a human peculiarity in turning downwards towards the interval between the index and middle fingers. In doing so, it separates the index finger from the metacarpo-phalangeal region of the other fingers, and is a superficial expression of the functional individuality which has been acquired by the human forefinger. In apes and monkeys the 'heart line' runs transversely straight across the palm, and this more primitive disposition is still to be observed in a small proportion of human beings. The 'head line' or middle crease of the palm, is a compensatory flexure line developed between the other two main creases. In the region of the face and neck, with the movements of facial expression, flexure lines appear momentarily in the skin at right angles to the pull of the fibres of cutaneous muscles. In other words, their pattern is determined by the arrangement of the underlying musculature. With advancing age, as the result of the progressive loss of the elasticity of the skin, these creases become established as permanent wrinkles. In so far as the pattern of the creases is a manifestation of the actions of different combinations of facial muscles expressing different types of emotion, they evidently provide some basis for the ability to read the character of an individual from his face.

3. PIGMENTATION OF THE SKIN

Even in the white races of mankind most regions of the skin contain a brown pigment. This is a melanin pigment which is mainly found in the form of small intracellular granules in the deepest layers of the stratum of Malpighi. In dark races it is much more abundant, and tends to spread through the greater part of the stratum, becoming more diffused in the superficial layers. The chemical origin of the pigment is still in doubt. There are a number of allied compounds in the body which might theoretically provide the basis for its formation, such as *dioxyphenylalanin* and *epinine* which can be converted into a melanin pigment by an oxidase in the epidermis. The presence of this enzyme can in fact be detected colorimetrically by the application of dioxyphenylalanin—the so-called *Dopa* reaction.[1] It is absent in the corium generally, and in the epidermal cells of albinos. It is also absent in white hairs. Exposure to sunlight stimulates the production of the oxidase, which is therefore responsible for the tanning of the skin, and, as might be expected, it is especially abundant in the skin of dark races.

Although, in pigmented skin, melanin is found in the epithelial cells of the stratum of Malpighi, it is not elaborated by these cells. The pigment is actually synthesized in the dendritic cells which occur in this layer and these are the only cells of the epidermis which give a positive Dopa reaction. They are therefore sometimes called epidermal melanoblasts. The melanin granules which accumulate in the cytoplasm of the dendritic cells stream along their branching processes, and are thus conveyed to the epithelial

[1] The term 'Dopa' is derived from the initial letters of the elements of the compound term *Di-oxy-phenyl-alanin*.

cells (Fig. 101). As already noted, the processes of the dendritic cells terminate in expansions which are closely applied to the surface of the epithelial cells, and it is from these expansions that the cells of the stratum of Malpighi appear to take up the granules by actively ingesting them. Melanogenesis is thus a specific function of the dendritic cells, at least in areas of normally pigmented skin, and also in the skin generally of the dark races of mankind such as the Negro. Dendritic cells of quite similar morphology are also present in the epidermis of the white races, but here they have comparatively little inherent melanogenetic property though this function may be accentuated temporarily (to a mild degree) under the

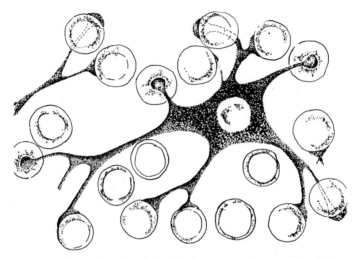

Fig. 101. Diagram illustrating the relationship between a pigmented dendritic cell of the epidermis and the adjacent epithelial cells. The processes of the dendritic cell terminate in expansions which are applied as caps to individual epithelial cells. Pigment granules are carried across from the dendritic cell into the epithelial cells at these sites of contact. (From R. E. Billingham, *Journ. Anat.* **82**, 1948.)

influence of an appropriate stimulus such as exposure to ultra-violet light. If, in parti-coloured animals, an autograft of pigmented skin is transferred to a non-pigmented area, it continues to elaborate pigment and retains its colour. In other words, the specific function of pigment formation is not related to any particular conditions of vascular or nerve supply. Further, it has been observed that, if pigmented epidermis is transplanted in this way, the pigmentation gradually spreads over the surrounding unpigmented skin. On the other hand, unpigmented skin transplanted into a pigmented area appears to be invaded and eventually replaced by the pigmented epidermis. This interesting phenomenon was at one time attributed to the proliferative extension of the pigmented at the expense of the unpigmented skin. What actually happens, however, is that the melanogenetic dendritic cells of the pigmented epidermis at the margin of the graft establish contact with the non-melanogenetic cells of the unpigmented skin (perhaps by forming direct syncytial connexions with them) and 'infect' them permanently with melanogenetic properties. The curious manner in which pigmentation thus spreads has been described in detail by Billingham and Medawar, and has been compared by them with the mechanism of a virus infection.[1]

The specificity of certain cells in relation to melanin production is emphasized by the behaviour of malignant tumours which occasionally develop in a pigmented area of skin. These usually spread rapidly throughout the body, and continue to produce melanin in large quantities in whatever tissues they invade.[2]

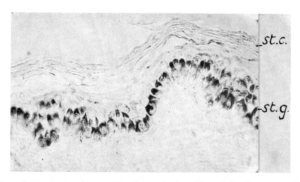

FIG. 102. Microphotograph of a section through the skin of the forearm of a negro. The section is unstained, and shows the melanin pigment in the deeper layers of the stratum germinativum (*st.g.*). Superficially there is a comparatively thin stratum corneum (*st.c.*).
× 350.

Besides the pigment in the cells of the epidermis, branched cells packed with granules of melanin are occasionally seen in the superficial parts of the dermis (where they are quite common in lower mammals). These are called *dermal melanoblasts*, and show some resemblance to the mobile chromatophores of lower vertebrates. Unlike the latter, however, they do not appear to alter their shape or the concentration of their pigment granules on exposure to light or under endocrine influence. Their significance is uncertain. They provide the basis for the bluish colour seen over the back of the sacrum in new-born infants of Oriental races, and for the blue pigmentation characteristic of certain monkeys. The blue tint is due to the fact that the dark melanin pigment is seen through the translucent connective tissue of the dermis. It should be noted that connective-tissue cells containing melanin granules are sometimes found in the dermis which do not give the Dopa reaction. It is probable, therefore, that these have not themselves elaborated the pigment but have taken it up from some other source.

Pigmentation may fluctuate quite rapidly in certain areas of the skin in man, as shown in the dark lines under the eyes which appear or become

[1] R. E. Billingham and P. B. Medawar, 'Pigment spread and cell heredity in guinea pigs' skin', *Heredity*, **2**, 1948; 'Infective transformations of cells', *Brit. Journ. Can.* **2**, 1948.
[2] It may be noted that when the pigmented tissue of the iris is cultivated *in vitro*, the cells continue to elaborate melanin.

markedly accentuated during periods of nervous tension. This pigmentation, from its surface appearance, is probably due to the accumulation of melanoblasts in the dermis. Nothing is known of the physiological basis of the phenomenon.

Racial variations in skin pigmentation.[1] In white races melanin pigment is often fairly abundant in certain localized areas of the skin, e.g. in the areola at the base of the nipple and in the region of the external genitals. In early pregnancy (and in certain pathological conditions) this pigmentation becomes markedly accentuated, and it may in some cases spread to other parts of the skin which normally are unpigmented. At the end of pregnancy it disappears to some extent, but never entirely.

A spectrophotometric analysis of the human skin of different races has shown that its tint is the resultant of at least five pigments—melanin, melanoid (a diffuse substance allied to melanin), carotene, oxyhaemoglobin, and reduced haemoglobin. The amount and distribution of these pigments show some variation in different parts of the body, and it seems probable that the difference in skin colour in dark- and yellow-skinned races is mainly due to the amount of the melanin. In addition, the turbidity of the epidermis introduces a scattering effect, so that the tint of the skin is also determined to some extent by its texture.[2]

It is commonly assumed that a dark skin serves as a protection against the rays of the sun, for, in general, there is a broad correlation between degree of pigmentation and a tropical habitat. This correlation, however, is by no means consistent, though it is possible that in some cases exceptions can be explained on the basis of a recent migration. Of course, the pigmentation of the skin in dark races is an inherited character in individuals, but even among them the degree of pigmentation can be considerably modified by environmental conditions. Melanin is already present in the deepest layers of the epidermis at birth, particularly in those regions which in the adult are most heavily pigmented. In dark races (e.g. Negroes) pigment formation begins in the fifth month of intra-uterine life, but the skin remains relatively very light until birth. Immediately after birth the skin darkens quite rapidly, so that it may attain the colour characteristic of the adult within a few days or weeks.

Skin colour offers an obvious criterion for the differentiation of the races of mankind, and it was used by the older anthropologists in an attempt to draw up a natural classification of races. Three main groups were recognized—*Leucoderms*, *Melanoderms*, and *Xanthoderms* (or, in other words, white-, black-, and yellow-skinned races), and further subdivisions have been suggested by reference to the different shades of brown. Such a classification, however, has proved unsatisfactory, for it leads to the unnatural association of diverse groups which are shown by other, and probably more reliable, physical criteria to be not closely related. For example, the Australian Aboriginal and the Negro, in spite of their similar

[1] G. A. Harrison, 'Pigmentation', in *Genetical Variation in Human Populations*, Pergamon Press, 1961.
[2] E. A. Edwards and S. Q. Duntley, 'The pigments and colour of the living human skin', *Amer. Journ. Anat.* **65**, 1939.

skin colour, are representatives of quite different racial groups, and even among the negroid peoples there are racial elements with a comparatively light-brown skin.

Indirect evidence makes it certain that pigmentation of the skin is a primitive trait in man, and it may be inferred that this character was lost to a greater or lesser degree independently by different races during the course of their evolution. The bleaching of the skin has reached its greatest development in the so-called 'Nordic' type of European, which is today found predominantly in Scandinavia, the British Isles, Holland, Denmark, and to a less extent in north France and along the Baltic coast in Germany. In this physical type relative loss of pigment has also occurred in the hair and the iris.

As already noted, the skin of all white peoples normally retains pigment to some degree, and the range of skin colour in European populations is quite wide—varying from an olive or tawny brown in the Mediterranean area to the creamy-whiteness of the 'Nordic' type. Complete absence of melanin pigment (which affects not only the skin but the hair and eyes) occurs only as an abnormality in albinos. Albinism is a recessive Mendelian character, examples of which have been observed in many mammals. In man it appears sporadically in all races, white or coloured, and its frequent occurrence among certain native peoples is the basis of stories, which in the past have sometimes been brought back by inexperienced travellers, of 'mysterious' white races isolated in remote lands.

It is a curious fact that the adaptive significance of pigmentation of the skin is still a matter for discussion. A dark skin facilitates absorption from the infra-red part of the spectrum and therefore actually tends to raise the skin temperature in a tropical climate. On the other hand, it does provide a protection against the ultra-violet end of the spectrum and so avoids severe skin burns, and also probably accounts for the diminished frequency of skin cancer among dark-coloured races of mankind as compared with members of the white race who are exposed for long periods of time to tropical sunlight.

4. HAIR

One of the distinctive features of *Homo sapiens* is his relative hairlessness. To some extent this appearance is illusory, for a close inspection of the 'naked' surfaces of the body will show that they are usually covered with fine downy hairs in considerable density. Indeed, the only areas of skin completely devoid of hair are the palmar surface of the hand and the plantar surface of the foot, the dorsal surface of the terminal part of the fingers and toes, the red border of the lips, and certain parts of the external genital organs. Human nakedness, therefore, is not so much due to the absence of hair as to its very fine and inconspicuous character.[1] Before birth, at least, there seems to be little difference between man and lower Primates in density of hair growth. It has been shown that in a human foetus of 6 months the number of hairs to the square centimetre on

---

[1] Complete nakedness is not known in any normal mammal, with the exception of one or two cetaceans.

the back is 688, while the corresponding figures for a chimpanzee and gibbon at the same relative stage of development are 420 and 440.[1] In the adult, also, the density of the hair of the scalp in man has been estimated to be 312 per square centimetre, as against 307 in the large anthropoid apes.[2]

In a human foetus at the fifth or sixth month of development, almost the whole of the body is covered by a coat of extremely fine hair called the *lanugo*.[3] This is usually all shed before birth, and during the first few months of life it is replaced by short downy hair termed the *vellus*. In man

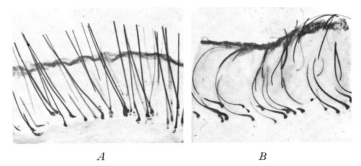

A   *B*

FIG. 103. Microphotographs of thick sections through the scalp of *A*, a Chinese, and *B*, a Bushman, to show the contrasting features of the hair. In the Chinese the hair is quite straight, while in the Bushman it is strongly curved. Note the hair bulbs at the base of the hairs.

the vellus is retained over most of the skin except the scalp and eyebrows, and except where it is superseded at puberty by the coarser hair of the axilla, pubes, and (in the male) over the chest and face.

Among modern races hairlessness is developed to the greatest degree in Mongolian peoples, in whom the relative lack of facial and body hair is a well-recognized characteristic. On the other hand, the most hairy types of man are the Australian aboriginals and the (now extinct) Tasmanians. Hair differs quite markedly in texture and often in colour over different parts of the body, and it provides the most conspicuous of the secondary sexual characters in man. The moustache and beard of the male are known to develop under the influence of sex hormones and, in old women in whom ovarian activity becomes diminished, these hairy characters also tend to develop to some degree. The sexual differences in the disposition of the pubic hair are likewise dependent on endocrine control.

Racial variations in the texture of the scalp hair have formed the basis for a commonly accepted classification of modern human types. Of these, three main groups have been recognized, the *Leiotrichi*, or straight-haired, the *Cymotrichi*, or wavy-haired, and the *Ulotrichi*, or woolly-haired. The Leiotrichi are represented by many Oriental peoples, in whom the hair

[1] F. Meyer-Lierheim, 'Die Dichtigkeit der Behaarung beim Fetus des Menschen und der Affen', *Zeitschr. f. Morph. und Anthr.* **13**, 1911.
[2] A. H. Schultz, 'The density of hair in Primates', *Human Biology*, **3**, 1931.
[3] A similar temporary growth of hair also occurs in other mammals which are relatively hairless in the adult, e.g. the elephant.

of the head is straight or 'lank', and approximately circular in cross-section. The Cymotrichi include Europeans, Mediterranean peoples of North Africa, Indians, and Australian aboriginals. In these the hair tends to be wavy to a variable degree and broadly oval in cross-section. In the Ulotrichi, which embraces all the negroid types, the hair is of a frizzy character and often tightly curled; in section it is considerably flattened.

By measuring the greatest and smallest diameters of a section of hair and expressing one as a percentage of the other, a cross-sectional index of hair is obtained which has been found useful by anthropologists for comparative studies. The index varies from 40–50 in Negroes to 80–90 in Oriental people such as the Japanese. It has been suggested that the shape in cross-section is related to the curling of the hair. Mechanically, the leiotrichous type of hair is the strongest and most rigid; the flat ulotrichous hair is the weakest and most liable to bend. While growing up from its follicle the latter probably bends under the tension of the tonically contracted arrector pili muscle which is indirectly inserted into it (see p. 311). Consequently it continues to grow in a curved direction and appears on the surface with a well-developed kink. There is a close relation between the texture of the hair and the length to which it grows, the straighter type usually reaching a greater length.

The primary purpose of a hairy covering to the body is warmth. In preventing loss of heat by providing a non-conducting felting of fur, and by diminishing surface evaporation, it plays an important part in the temperature regulation of the mammalian body. Compensation for the virtual disappearance of a hairy coat in man is provided by the development of a continuous layer of subcutaneous fat in the superficial fascia.

Another function of hair is sensory. Hair follicles are supplied at their base by a fine nerve plexus which is stimulated by the slightest movement of the free part of the hair. Where hairs are fine and *sparsely* scattered (and so capable of separate stimulation) they serve the purpose of very delicate tactile organs. It has been suggested, indeed, that the relative hairlessness of man is directly associated with an increasing sensitivity of the surface of the skin, and with an enhancement of tactile acuity.

In most mammals special tactile hairs or *vibrissae* are developed in the facial region, and (particularly in arboreal animals) also on the limbs near the carpus and tarsus. Vibrissae are commonly arranged in definite groups, and at the base of each one the skin is elevated into a small tubercle. The nerve and vascular supply to these tubercles is particularly rich, the blood vessels at the root of the hair showing sinusoidal dilatations which provide the basis for an 'erectile' mechanism whereby the hairs can be raised up and held rigid. In man (who in this respect seems to be unique among mammals) vibrissae are absent. They are present in their full complement in lower Primates, but gradually disappear in the ascending zoological scale leading up to man.[1] Supposed remnants have been described in the human embryo,[2] but their interpretation is doubtful. It has generally been

[1] F. Wood-Jones, *Man's Place among the Mammals*, London, 1929.
[2] See A. H. Schultz, who describes a small tubercle in a human foetus occupying the position of a carpal vibrissa, *Amer. Journ. Phys. Anthr.* 7, 1924.

assumed that the stiff hairs of the eyebrows are representative of the supra-orbital tactile vibrissae of lower mammals. They resemble the latter in their early embryological development, for hair appears here during the second month of foetal life, whereas hair over the greater part of the body does not develop till the fourth month. Their mode of development and the structure of their follicles, however, indicate that they are morphologically different.

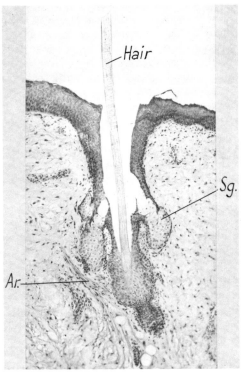

FIG. 104. Microphotograph of a section through a hair follicle. Sebaceous glands (*Sg.*) are seen opening into the follicle, and a portion of an arrector pili muscle (*Ar.*) is also evident. × 60.

In certain mammals the hair is modified to provide weapons for attack or defence. Such are the quills of the porcupine and the spines of the hedgehog. In man some hairs have a mechanically protective function. Within the nostrils and the auditory meatus they are directed outwards and help to prevent the entrance of foreign particles, while the eyelashes guard the eyes in a similar manner.

Hair tracts. With the single exception of the eyelashes, hairs do not grow vertically out of the skin, but always at a slant. Hence they normally lie flat on the surface and their trend follows a definite plan. Primitively all the hairs of the body are directed cranio-caudally over the dorsal parts of the head, neck and trunk, caudo-ventrally over the ventral parts, and post-axially and distally on the limbs. Such a 'stream-line' arrangement is related to simple movements of the body in a forward direction, and this disposition of the hair is found in most small mammals of a generalized type. In other cases it is disturbed and complicated by the formation of divergent streams, reversals of hair direction, and whorls. These variations are distinctive in many different groups of mammals, so that the mapping out of hair tracts on the surface of the body is used by systematists in comparative anatomical studies.

The underlying cause of the deviations of hair tracts is very uncertain. In some cases they seem to have a direct relation to the evolutionary disappearance of certain surface features. For example, in man hair tracts converge to a focal point over the coccyx, suggesting that at one time they were continued down on to a tail. Again, the divergent line or 'parting' of hair extending on the ventral surface of the trunk from the nipple down to the inguinal region seems to mark out the mammary line along which, in lower mammals, a series of nipples is generally found. As is well known, accessory nipples are not uncommonly seen along this line as abnormalities in man.

In many cases the disposition of hair tracts has an obvious relation to habits of posture and movement. It has also been suggested that they are directly related to the toilet habits of an animal, the hair trend coinciding with the direction in which it licks or scratches its fur. However, a careful analysis of hair trends in different species makes it evident that such a relationship is by no means consistent.[1]

The anatomy of hairs. It has been noted that the development of a hairy coat is a distinctive mammalian character. There is reason to believe that in the evolutionary origin of mammals it superseded a scaly covering in their reptilian precursors. Indeed, modified epidermal scales are still developed in certain living mammals—a relic of the reptilian phase of evolution. These are seen on the bare tails of many marsupials, insectivores, and rodents (e.g. the common rat) as polygonal and slightly imbricated thickenings of the stratum corneum arranged in a tessellated pattern. In one mammal only (the scaly ant-eater) do scales reach a high degree of differentiation and specialization, and in this case they cover almost the whole body.

Where well-defined scale formations are found in the skin of mammals, the hairs bear a very regular relation to them. A central hair emerges from under cover of the middle of the free edge of each scale, and one or two smaller hairs on either side. Thus the hairs are arranged usually in groups of three or five. Even in skin where no scales are apparent a similar arrangement of hair follicles may be found, particularly in the foetus, and it seems probable that this grouping is determined by the original development of hairs in relation to epidermal scales. Remnants of these scales are possibly also represented by a slightly thickened area of epidermis which can be often seen in microscopic sections at the base of a hair on the side from which it slopes.

A hair is formed embryologically as a down-growth of the ectoderm into

[1] For a discussion on hair tracts in marsupials see W. Boardman, *Proc. Linn. Soc. N.S.W.* **75**, 1950.

the subjacent mesoderm (Fig. 107). Up to the fifth or sixth week of development in the human embryo, the ectoderm consists of a single layer of cubical cells. This layer then becomes doubled by the formation superficially of a layer of flattened cells. Increasing proliferation of the deeper layer leads to a gradual increase in the thickness of the epidermis, and by the fourth month it has become definitely stratified.

Hairs first appear in the region of the eyebrows, lips, and chin at the end of the second month, while over the rest of the body they do not start their growth till the fourth month.

The first sign of a developing hair is a knot-like thickening formed by a local proliferation of the stratum germinativum (Fig. 107 a). This penetrates more and more deeply into the mesodermal corium as a slender epithelial peg, the lower end of which swells to form the bulb of the hair. The latter comes into contact with a condensation of mesodermal cells which invaginates its free end slightly to form a small vascular papilla. The epithelial cells of the bulb which immediately cap this papilla are columnar, and it is from their active proliferation that the actual hair is developed. The hair bud now sprouts up from these cells and, in doing so, canalizes the originally solid epidermal peg and forms the lumen of the hair follicle. Lastly, the cells of the shaft of the hair, as the latter rises above the level of the bulb, become progressively cornified.

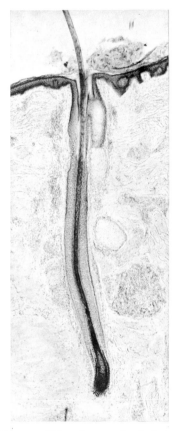

In structure the shaft of a mature hair consists mainly of a fibrillated basis of elongated tapering cells tightly packed together. They have undergone a considerable degree of cornification, but their nuclei may still be visible. Upon the

FIG. 105. Microphotograph of a hair in its follicle. × 40.

amount of melanin pigment which they contain depends the colour of the hair. Typically the hair shaft is really a tubular structure hollowed out by a medullary cavity, the latter being filled with cells of irregular shape among which may be minute air bubbles. In senescence these cells undergo atrophy, and there may be an increase in the bubbles which (with the loss of pigment) makes the hair appear white by reflected light. The presence of air bubbles in pigmented hair is partly responsible for imparting to it a characteristic sheen. Many hairs contain no medulla; this is the case, for example, with the temporary lanugo of the foetus, and also with the downy vellus hairs which succeed it.

The surface of the hair shaft is covered with a delicate cuticle of extremely thin scales which are laid on in imbricated fashion with the free edges directed up towards the free end. In human hair these scales are fairly simple and are disposed circularly round the shaft. Their pattern shows distinctive variations in different mammals, however, and in some species (e.g. bats) they may have elaborate tuft-like fringes.

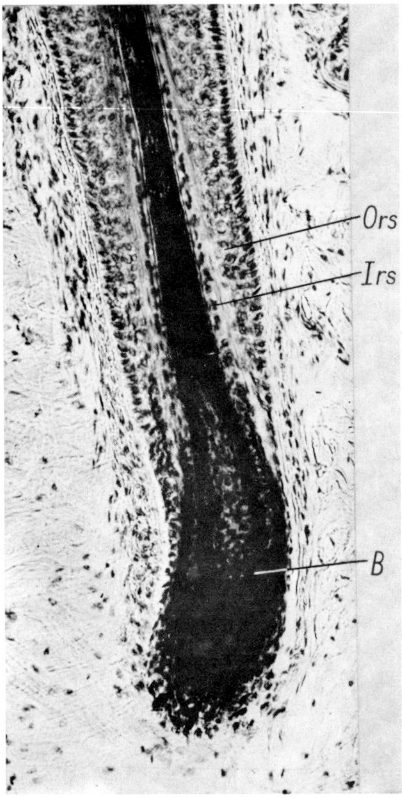

The part of the hair which lies below the surface is termed the *hair root*. This is ensheathed in a pit, the *hair follicle*, the wall of which (as we have just noted) is derived from a down-growth of the epidermis. The base of the hair shaft is slightly swollen to form the *bulb*, and here is found the germinal matrix from which the growth of the hair takes place. It is composed of columnar and polygonal cells which actively proliferate; as they become successively displaced upwards in hair growth they are gradually converted by a process of keratinization into the fibre-like cornified cells of the hair shaft. The cells of the germinal matrix fit as a cap over the soft vascular papilla of the corium, from the blood vessels of which they receive their nutriment.

The wall of the hair follicle is composed of several layers of polygonal cells which are continuous at the surface of the skin with the stratum of Malpighi. Approaching the bulb of the hair, they form the *outer root sheath*, and here they thin out eventually to form a single

FIG. 106. Microphotograph of the base of a hair follicle, showing the hair bulb (*B*), the outer root sheath (*Ors.*) and the inner root sheath (*Irs.*). × 150.

cell layer which becomes continuous with the soft mass of the germinal matrix in the bulb. From the latter, cells are reflected on to the surface of the hair shaft to form the *inner root sheath*. This consists of layers of cells which closely invest the base of the hair but rapidly disappear up the shaft as they become cornified (Figs. 106 and 107).

Hair growth. The duration of life of a hair varies quite considerably. Hairs which grow naturally longer have a longer life, and those of the human scalp may last several years, while shorter hairs over other parts of the body may have a life cycle of a year or less. In many mammals a thick hairy coat is grown annually during the winter to be shed when summer approaches, and in some cases (e.g. the arctic hare and arctic fox) the seasonal coats have different colours. In some animals, on the other hand, certain types of hair, such as the mane and tail of horses, apparently persist throughout the life of the individual.

Normally, each hair which is shed is replaced by a successor. The process is as follows. The root of the old hair undergoes absorption as the

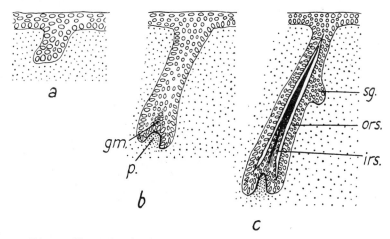

FIG. 107. Diagram illustrating the development of a hair. *a*. The first appearance of a hair is indicated by a local proliferation of the ectoderm. *b*. The ectoderm grows down to form a solid follicle the lower end of which (the bulb) is invaginated by a vascular papilla (*p.*) of mesoderm. Over this papilla the ectodermal cells form the germinal matrix of the hair (*gm.*). *c*. The hair is sprouting up from the germinal matrix, and in doing so it canalizes the solid hair follicle. The wall of the follicle is formed by the outer root sheath (*ors.*), and a layer of ectodermal cells is reflected on to the base of the hair shaft to form the inner root sheath (*irs.*). A developing sebaceous gland is also indicated (*sg.*).

result of the activity of the cells of the root sheath. These cells proliferate and cut off the hair shaft at the bottom of the follicle from the germinal matrix covering the papilla. The detached shaft then becomes gradually pushed up towards the surface until it falls out or is rubbed off, and the cells of the germinal matrix provide a new hair sprout which grows up into the follicle. Sometimes it happens that the hair papilla disappears entirely when the hair itself dies, in which case an epithelial down-growth from the bottom of the follicle leads to the formation of a new papilla and a new hair bulb.

It is known that during life, and even in the foetus before birth, individual hairs are constantly being shed and replaced. Each hair has a life cycle of active growth followed by a period of quiescence and death, and in young growing animals all the hairs of the body may follow the same cyclic phase. In immature white rats, for example, the hairs show a very short cycle of thirty-five days, the phase of active growth and the quiescent phase being about seventeen days each.[1] The old hairs may still remain for some time in the follicles while the new hairs grow up, and a single

[1] E. O. Butcher, 'The hair cycles in the albino rat', *Anat. Rec.* **61**, 1935.

follicle may be occupied by as many as three hairs at one time. It is uncertain what factor determines these cycles—it has been shown that they are not disturbed by removal of the gonads. Experimental studies by Durward and Rudall have shown that the cycles of hair growth in the rat proceed continuously in a series of waves from the ventral to the dorsal surface of the body, and in a zone where active growth is taking place all the hairs grow simultaneously.[1] Denervation of the skin has no effect on this pattern of growth waves. Moreover, if a strip of skin crossing the direction of the growth wave is resected and the edges of the denuded area sutured, the wave is arrested temporarily at the line of the suture but reappears on the other side at its normal rhythm. If an area of skin is lifted up, rotated, and replaced in position, the wave of growth becomes correspondingly reversed. It is concluded from these experiments that the pattern of hair growth represents an inherent property of the skin which is not disturbed by altering the local topography. In man (at least in the adult) replacement is stated to occur in scattered individual hairs, so that some hairs in one group may be growing, while others remain inactive.[2]

The rate of growth of a hair varies with its texture. Coarse hairs which ultimately reach a greater length grow faster than finer hairs which are normally short. Trotter found that the hair on the legs of children grows at the rate of 1·42 mm. per week, and the rate of growth increases steadily with age, reaching 1·85 mm. per week at the age of 40–45. Axillary and pubic hair grows at the rate of 2·2 mm. per week, but in this case the rate of growth is not affected by age. Hair growth is more rapid in the summer than in the winter, and also during the night than during the day. Contrary to popular assumption, it is not affected by repeated cutting (as in daily shaving), or by exposure to the sun.

There is a curious superstition that the vitality of the hair enables it to continue growing even after death. This seems to have arisen from legends relating to certain famous personages such as Charlemagne and Napoleon I, who are reported to have grown beards while resting in their tombs. Such stories are probably to be explained by the fact that the natural shrinkage and retraction of the skin in a dead body exposes to view a part of the hair root which in life remains hidden inside the hair follicle. It is remarkable, however, that hair may continue to grow at its usual rate even though the other tissues of the body are seriously affected by inadequate nutrition. It has been found, for example, that in young rats in which normal growth is suppressed experimentally, the hair continues to develop and grow normally.[3] Similar observations made on chicks have demonstrated that, even if the body weight is held constant by dietetic methods for as long as eighty days, the feathers still continue to grow.[4]

[1] A. Durward and K. M. Rudall, 'Studies on hair growth in the rat', *Journ. Anat.* 83, 1949.
[2] M. Trotter, 'The life cycle of hair in selected regions of the body', *Amer. Journ. Phys. Anthr.* 7, 1924.
[3] C. M. Jackson, 'Structural changes when growth is suppressed by under-nourishment in the albino rat', *Amer. Journ. Anat.* 51, 1932.
[4] J. C. Drummond, 'Observations upon the growth of young chickens', *Biochem. Journ.* 10, 1916.

Hair pigment. The morphogenetic factors underlying the distribution of pigmented hair in lower mammals so as to produce total integrated patterns of a distinctive kind remain unknown. The appearance of stripes and certain other markings has suggested a relation to dermal segments innervated by different segmental nerves, but a close study has shown that this is not so. The production of hair pigment is determined by cells which have a specific distribution in different areas of the skin, and this again must be the result of inherited potentialities dependent on the natural selection of pigmentary factors. The latter, by constucting a pattern which is of direct advantage to the animal (such as a protective or warning coloration), provides for the continuance of the species.

It is interesting to note that pigmentation of the hair may be quite independent of the pigmentation of the skin. In parti-coloured animals coloured patches of fur often overlie pigmented patches of skin, but the correspondence is by no means constant. In the experiments referred to on page 299, in which autotransplants of pigmented skin into unpigmented areas were made, it was found that while pigmentation spreads into the surrounding areas of unpigmented skin, the hair covering the latter remains white.

Hair usually manifests obvious senescent changes before any other tissue of the body in its tendency to whiten and fall out without replacement by new growth. Little is known of the metabolic processes which underlie these changes. Whitening (as stated above) is apparently due to a loss of pigment in the shaft of the hair and also to the accentuation of the minute air bubbles often present in the medulla. It first affects the scalp hair, usually in the region of the temples, while the pubic hair is the last to go grey with old age. The onset of greying is subject to very wide individual as well as racial variation. A statistical study has shown (for example) that in European people the average age at which grey hair begins to appear on the head is 34, while in Negroes the corresponding age is 44.[1] Loss of hair pigment in senility occurs without a corresponding depigmentation of the skin, and its cause must therefore be somehow related to the germinal matrix of the hair itself. The cells of the matrix are either incapable of elaborating the oxidase necessary for the formation of melanin, or else they are unable to take up melanin brought from some other source. The rate at which hair can become completely grey must in any case be ultimately determined by the rate at which it grows or is replaced. It is not possible, therefore, to accept stories which purport to describe how, under exceptional emotional stress, hair has been turned completely white in a night.

Hair muscles. Hairs are capable of a limited movement by the action of small muscles, the *arrectores pili* (Fig. 104). These are slender bundles of unstriped muscle fibres which at one end are inserted into the outer root sheath in the lower part of the follicle, and on the other take origin indirectly from the deep aspect of the epidermis. Each muscle lies on the side towards which the hair slopes, and by pulling the root in this direction it levers the free part of the hair into an upright position.

The arrectores pili are innervated by the sympathetic nervous system.

[1] F. Boas and N. Michelson, 'The greying of hair', *Amer. Journ. Phys. Anthr.* **17**, 1932.

When they contract, the skin immediately round the base of the hair is pushed up, while at the site of their origin from the epidermis the skin is pulled down in local depressions. This produces a surface pimpling of the skin commonly called 'goose skin', an appearance which is seen on exposure to cold or in certain emotional reactions. A sebaceous gland usually occupies the angle between the muscle bundle and the hair follicle, and in many sections the muscle can be seen closely wrapped over the convex fundus of the gland, and may even be inserted into its capsule. It appears certain that the gland will be subjected to pressure when the muscle contracts; the latter, therefore, probably plays a part in promoting its activity by helping to expel its secretion into the lumen of the hair follicle.

5. CUTANEOUS GLANDS

Sebaceous glands (Fig. 104). Opening into each hair follicle towards the surface of the skin are one or more lobulated sebaceous glands, which show a tendency to lie on the side towards which the hair slopes. Each gland is formed developmentally as a diverticulum of the epithelial lining of the follicle, and is thus composed of polygonal cells derived from the stratum of Malpighi. It is enclosed in a thin capsule of connective tissue. The lobules of sebaceous glands are solid masses of cells which from the periphery towards the centre become progressively filled with fat granules. The cytoplasm of the central cells becomes entirely distended with large droplets of fat, and finally the cells as a whole undergo disintegration. An oily mass is thus produced, made up of a mixture of cellular débris and fatty matter. This is the *sebum* which, collecting in the centre of each lobule, is discharged through a short duct into the lumen of the hair follicle, or on to the adjacent surface of the skin.

Sebum provides a lubricant—and possible nutriment—to the hair shaft. More important, probably, is its function in providing a protective medium over the surface of the skin, which renders the latter more impervious to the effects of moisture, as well as to the possibility of desiccation in a dry atmosphere. It keeps the skin supple, and since it has bactericidal properties it perhaps also serves an antiseptic function. The manner of secretion of the sebum entails a continual disintegration of the cells of the sebaceous gland, which is therefore a gland of the holocrine type (see p. 280). This loss is made good by active proliferation of the peripheral cells which lie adjacent to the duct.

Generally speaking, sebaceous glands are developed as appendages to hair follicles, and more than one gland may open into a single follicle. The only areas of skin which contain no sebaceous glands are the palms of the hands and the soles of the feet. In some cases, however (e.g. in the skin of the nose), they attain to a large size in relation to which the associated hair may be so small as to appear quite accessory. Lastly, in skin covering moist surfaces, as over the lips and the glans penis, sebaceous glands are found independently of hairs. The absence of any relation between the size of a hair and the size of its associated sebaceous gland indicates that the two are not very closely related functionally.

There is no direct evidence that the secretory activity of the sebaceous glands is under nervous control. Disturbances of secretion have been noted in certain affections of the nervous system, but it is uncertain how far these are secondary effects. Estimations of sebaceous secretion have shown that it is not affected by lesions of the peripheral nerves or by interruption of the sympathetic nerve supply.[1] The secretion seems rather to be the result of a continuous growth process of the cells of the sebaceous glands (analogous to the growth of hairs).

Sweat glands (Figs. 97, 99). With the exception of moist surfaces such as the lips, the skin almost everywhere contains sudoriferous or sweat glands. For the most part these are simple coiled and unbranched tubes with a fine, even lumen, and a long winding duct which opens on the surface. The secretory part of each gland is coiled up into a compact ball (0·1 to 0·5 mm. in diameter) lying usually in the dermis. In the palms of the hands and the soles of the feet they are situated more deeply in the superficial fascia, and the ducts are here correspondingly longer. The secretory tube is lined by a cubical epithelium, underlying which is a thin layer of elongated epithelioid cells which are believed to be muscular and hence are termed myo-epithelial cells.[2] The duct of the gland, which is also lined by cubical cells, runs up fairly directly to the epidermis. Through the latter the secretion is discharged on to the surface of the skin by a spirally wound channel which is formed as a simple unlined intercellular cleft between the epidermal cells.

In connexion with the papillary ridges of the hands and feet the sweat glands are very numerous, and the coiled tubes are here massed together compactly in the superficial fascia.[3] They are particularly abundant in the palmar and plantar pads of arboreal mammals, serving to moisten the skin in order to provide a good gripping surface. In most sweat glands the secretion is discharged in droplet form from the secretory cells into the lumen of the tube. These are therefore epicrine glands (see p. 281), and they secrete a very watery fluid containing only about one per cent of solids. Of the latter approximately half consists of inorganic salts (mostly sodium chloride) and the other half organic material (mostly urea). Contrary to a long held assumption, the amount of urea excreted by the sweat glands, even when fully active, is too trivial to be of any practical significance for the elimination by this route of nitrogenous matter from the body.

Other glands occur in the skin which in some respects are transitional in structure between sudoriferous and sebaceous glands. They are glands of the apocrine type, and their mode of secretion involves the fragmentation and disintegration of the superficial or projecting part of the secretory cells; this results in the liberation of granules (probably lipoid in nature) which are formed in the cytoplasm. The secretion is much less watery than that

[1] J. Doupe and M. E. Sharp, 'Sebaceous secretion', *Journ. Neurol. Psych.* **6**, 1943.
[2] There is some doubt whether the myo-epithelial cells of sweat glands are true unstriped muscle cells. Unlike the latter, they are derived embryologically from the ectoderm.
[3] The number of sweat glands is recorded as about 110 per sq. cm. over most of the body (but with a variation in different areas of skin from 60 to 200), and about 400 per sq. cm. in the skin of the palms and soles. According to Sappey and Krause the total number of sweat glands in the human body is about 2 million.

of the epicrine glands. The apocrine sweat glands are also larger (1 to 4 mm. in diameter), more elaborate and often branching, and they extend more deeply into the superficial fascia. Moreover, they are developed in close association with hairs, and sometimes open by their ducts actually into hair follicles. They are found in the axilla (where they develop at puberty and in the female have been stated, but on very slender evidence, to show a cyclic activity in relation to menstruation), and in the scrotal and peri-anal skin. Their functional significance is not known, but there seems little doubt that they are related to the scent glands of lower mammals.

In the external auditory meatus are specially modified sweat glands concerned in the production of *cerumen* or wax, and hence called *ceruminous glands*. Although they resemble the apocrine type of sweat gland morphologically, their secretion is of a very different nature. The function of cerumen is evidently to provide a sticky surface covering the skin of the auditory meatus, in which particles of dust can be caught up which might otherwise reach the delicate tympanic membrane.

Sweat glands are developed embryologically as down-growths from the stratum germinativum of the epidermis in the intervals between the dermal papillae (Fig. 100). They appear first over the scalp and forehead, towards the end of the fourth month of foetal life, and then in the palmar and plantar skin. During the fifth month, the blind extremities of the sweat glands become coiled, and not until the seventh month is a communication established with the surface of the skin by the opening up of a spiral channel among the cells of the epidermis.

Although morphologically they appear to be excretory structures, the main function of sweat glands is to provide a mechanism for temperature control by surface evaporation. Direct observation has shown that as soon as the temperature of the body is raised about half a degree above its normal level, they become active and perspiration ensues. In individuals who have become acclimatized to a hot environment, either by residing in the tropics or artificially in a climatic chamber, the sweat glands have been shown to respond more rapidly and also more efficiently.[1] That this response is under nervous control is shown by the fact that perspiration does not occur over a denervated area of skin. It might be expected that the density of sweat glands would show a racial variation related to habitat, being greater in tropical people, but in spite of some statements to the contrary there is no evidence that this is so. Congenital absence of sweat glands has been recorded on several occasions. Individuals who suffer from this anomaly are compelled to adopt special means to keep down their body temperature in hot weather.

Not all the sweat glands of the body are concerned with temperature regulation, that is to say they are not all 'thermal' glands. Those of the palm of the hand and the sole of the foot do not react to any marked extent to thermal stress, but they do react rapidly to emotional stress. The significance of this is not quite clear, but since it seems probable that in arboreal mammals the palmar and plantar glands serve to moisten the foot pads so

---

[1] R. F. Hellon, R. M. Jones, R. K. Macpherson, and J. S. Weiner, 'Natural and artificial acclimatization to hot environments', *Journ. Physiol.* **132**, 1956.

as to ensure a firmer grip in climbing, and since it may be supposed that the need for this becomes particularly insistent in the emotional stress associated with escape and concealment from enemies, the same type of reaction in man is perhaps an atavistic phenomenon.

In mammals generally, with notable exceptions such as the horse, sweating as a temperature-regulating mechanism seems uncommon, typical sudoriferous glands being absent except over the pads of the feet. Epicrine sweat glands predominate over the body surface only in man and the higher primates. Numerous transitional forms between sweat and sebaceous glands occur in lower mammals, however, so that it may be difficult to determine morphologically whether they belong to one or other category. It is certain that in many species the 'sweat' glands, unlike those of man, are apocrine glands which secrete an appreciable amount of albuminous material.[1]

6. NAILS

One of the distinctive anatomical characters of the Primates is the replacement of sharp, recurved claws by flattened nails. As we shall see, this feature is not fully developed in all members of the Order, for some retain claws on certain digits.

Nails are epidermal appendages developed as a modification involving mainly the stratum lucidum of the epidermis. Each nail is closely applied to the dorsal surface of the terminal phalanx, its free edge projecting slightly over a recess of skin called the *hyponychium*, and its root overlapped by a cutaneous fold—the *nail fold*. Under the latter is the *nail groove* which is continued forwards on either side to accommodate the lateral margins of the nail.

The dorsal surface of the nail is marked by fine longitudinal ridges, and at its base is the pale area of the 'half-moon', or *lunula*. This is well exposed in the thumb, but towards the little finger it becomes progressively more obscured by the nail fold, beneath which it is frequently hidden altogether. The lunula marks the area of the nail bed which is actually concerned with the formation and growth of the nail. Its white appearance is due to the fact that here the tissue which is in the process of transformation into the translucent nail substance is rather opaque.

In a longitudinal section, the Malpighian stratum of the epidermis covering the nail fold can be followed into the bottom of the nail groove where it becomes directly continuous with the nail bed underlying the root and the body of the nail. The stratum corneum extends for a short distance on the upper surface of the root of the nail as the *eponychium*, while the stratum lucidum appears to be continuous with the thin margin of the root of the nail itself. The epithelium of the nail bed, in which only the stratum of Malpighi is represented, is quite sharply divided into two zones. Proximally is the fertile or *germinal matrix* which is concerned with the formation of the nail, and which (as already noted) corresponds to the lunula. Here there is no clear boundary between the laminated substance of the

[1] For a detailed account of cutaneous glands, &c., see S. Rothman, *Physiology and Biochemistry of the Skin*, Univ. of Chicago Press, 1954.

nail and the underlying epidermal cells, for the latter are in process of being converted into the former. The distal part of the nail bed, which is called the *sterile matrix*, provides a smooth inert surface over which the body of the nail continuously glides forwards as it is pushed by the proliferative activity of the germinal matrix behind. The nail substance is here separated by a sharp line of demarcation from the underlying epithelium. As a section shows clearly, the nail increases in thickness from its root up to the distal margin of the lunula; therefore it maintains a uniform thickness up to its free edge (Fig. 108).

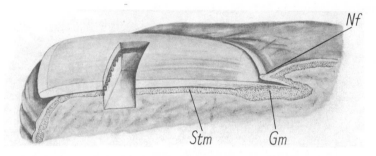

FIG. 108. Diagram of a section of a finger-nail, illustrating its relation to the nail bed. A section has been cut out of the nail to show the fine dermal papillae of the nail bed. The lunula of the nail is seen to correspond to the germinal matrix (*Gm*) from which the nail grows. The rest of the nail bed is made of the sterile matrix (*Stm*). The root of the nail is overlapped by the nail fold (*Nf*).

The epithelium of the sterile matrix is rather thin. Beneath it the dermis is raised up into a regular series of papillary ridges which run longitudinally, and are therefore seen in transverse sections. They are not reproduced to any appreciable extent on the surface of the epidermis.[1] The vascularity of the dermis shows itself in the characteristic pink colour seen through the translucent nail. The germinal epithelium of the fertile matrix is relatively thick, and the deeper cells, which are columnar or cuboidal, display active mitosis. In the more superficial layers the cells become progressively flatter and show a gradual transition into the thin, horny lamellae which form the basis of the nail substance. Granules have been described in the cells, staining brown with carmine, which are believed to represent the material from which the corneous matrix of the nail is formed. This material is called *onychogen*, but little is known of its chemical nature.

At the root of the nail the remains of nuclei can still be seen in the horny lamellae, but as the nail moves forwards from the lunula these disappear. It is possible, however, by suitable treatment with caustic soda, to macerate the mature nail substance into thin scales in which traces of cell structure may still be detected.

[1] The longitudinal grooves seen on the surface of a nail are commonly said to be related to the dermal papillary ridges of the sterile matrix. This can hardly be so, for they can be seen to extend forward from the area of the lunula where these ridges are too fine to cause them.

The lamellae of the nail lie flat in relation to the nail bed. Occasionally they show local distortions or irregularities as the result of uneven growth; minute air bubbles may also collect between them, giving rise to the white flecks which are not uncommonly seen in nails from the surface. Gross disturbances of the proliferative activity of the germinal matrix (such as may occur with local injuries or in acute illnesses) sometimes lead to the formation of an irregular transverse groove on the surface of the nail, which moves forward gradually with its growth.

The growth of nails. The rate of growth of the nails is not the same in all digits, and it shows considerable seasonal variation. Finger-nails grow about four times as fast as toe-nails. Studies on finger-nails,[1] made on numbers sufficiently large to allow of a statistical analysis of results, have shown that there is a relation between the rate of growth and the length of the digits. Thus, the nail on the middle finger grows the most rapidly, and that on the little finger the most slowly, while the growth rate of the thumb-nail is slightly less than that of the second and fourth digits. Contrary to usual statements, there is no significant difference between the right and left hands. The average rate of growth of the thumb-nail in the winter is $95\,\mu$ per day, and in summer $115\,\mu$, or about $1\frac{1}{2}$ inches a year. The nail-biting habit is associated with a remarkable acceleration in growth which is about 20 per cent higher than normal. Lastly, there is some evidence that the rate of nail growth is affected by nutrition.[2]

Head and Sherren[3] have shown that the growth of the finger-nails is retarded to a very considerable degree when the hand and fingers are immobilized in a splint or as the result of muscular paralysis. On the other hand, if an immobilized hand is bandaged and massaged daily, the growth is somewhat accelerated. Sensory nerve injuries alone produce no change in the rate of growth.

The first signs of developing nails in the human foetus are seen towards the end of the third month of intra-uterine life. At this date the nail-groove appears as a shallow furrow in the epidermis covering the dorsal surface of the terminal phalanx. In the area enclosed by the fold the superficial layer of the epidermis at first becomes thickened and cornified, but true nails are not formed till two months later. Then, the nail substance differentiates in the proximal part of the nail fold, apparently from the stratum lucidum just as it does during its growth after birth. When first formed, the whole of the nail is covered on its exposed surface by a continuous extension of the stratum corneum of the epidermis, and therefore at this stage actually lies embedded in the skin. The covering layer of stratum corneum disappears before birth, except proximally where it persists as the eponychium, and the distal edge of the nail finally becomes lifted up and freely projecting during the last month of intra-uterine life. Before birth the nails grow very slowly so that, although they have by then had four months of

[1] W. E. Le G. Clark and L. H. D. Buxton, 'Studies in nail growth', *Brit. Journ. Dermat.* **50**, 1938.
[2] L. H. D. Buxton and M. L. Gilchrist, 'The relation of finger-nail growth to nutritional status', *Journ. Anat.* **73**, 1939.
[3] H. Head and J. Sherren, 'The consequences of injury to the peripheral nerves in Man', *Brain*, **38**, 1905.

life, they do not require cutting in a newly born baby. Sometimes, however, at birth the extremely thin free edge of the first formed nail may project to some extent as a fine membrane which soon becomes rubbed off in
the course of the usual manipulations of the infant.

The relation of nails to claws. From the functional point of view,
a flattened nail must undoubtedly be regarded as a degenerate formation, a
retrogression from the more elaborate structure of a claw. Finger-nails are
certainly used for scratching purposes, and they also provide a mechanical
support for the digital pad of the terminal phalanx. Possibly, by acting as

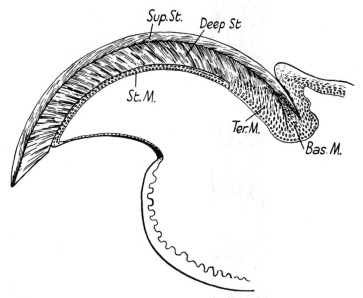

FIG. 109. Diagram showing the structure of a claw as seen in sagittal section. The claw is
composed of two layers of which the deep stratum (*Deep St.*) is the major part, while the
superficial stratum (*Sup. St.*) merely forms a protective sheath. It is the latter stratum
which persists in Primates to form flat nails. The superficial stratum is formed from the
basal part of the germinal matrix (*Bas. M.*) and the deep stratum from the terminal part
(*Ter. M.*). The rest of the claw bed is made up of the sterile matrix (*St. M.*).

a counter-pressure to the digital pad, they form part of a tactile mechanism,
for the dermal papillae of the nail bed are known to be particularly well
supplied with sensory nerve endings. Even these functions, however, can
hardly be ascribed to the toe-nails in man.

The evolutionary origin of flattened nails (or *ungulae*) from sharp claws
(or *falculae*) is strongly attested by morphological evidence.[1] Claws are
found in all primitive and generalized mammals. They are also a characteristic reptilian feature, not only in modern but in extinct forms (as
evidenced by the shape of the terminal phalanx), and especially in the
palaeozoic group of theromorph reptiles from which it is almost certain that

[1] W. E. Le Gros Clark, 'The problem of the claw in Primates', *Proc. Zool. Soc.* **106**,
1936; W. Panzer, 'Das Nagel-Kralle Problem', *Zeitschr. f. Anat. u. Entwicklung*, **98**, 1932.

mammals were originally derived. It may be assumed, therefore, that in the basal mammalian stock which about seventy million years ago became differentiated into the various modern groups of mammals, including primates, the digits were armed with claws.

Apart from its shape, a claw differs from a typical nail in its structure and growth. In many reptiles (e.g. the crocodile and tortoise), the claw is developed from the whole extent of the matrix of the claw bed. In mammals, the germinal matrix is confined to the proximal part of the claw bed, but it is divided into two portions, a basal and a terminal matrix. The latter produces horny lamellae which (as seen in sagittal section) are directed upwards and forwards towards the tip of the claw. These form a deep stratum which is the mechanically important part of the claw; on it depends the maintenance of a sharp and strong point (Fig. 109). The basal matrix produces horny lamellae which lie flat in relation to the surface, and form a thin superficial stratum which does little more than provide a protective cover for the main underlying deep stratum.

In the early arboreal Primates, the digital pads became increasingly specialized for tactile functions by broadening out, acquiring a rich sensory nerve supply, and developing a complicated pattern of papillary ridges. With the assumption of these functions, the sharp, laterally compressed claws became opened out and flattened. The terminal part of the germinal matrix and the deep stratum of the claw shrank until they finally disappeared, and with this change the double-layered claw became replaced by a single-layered nail. In other words, a flattened nail represents only the thin protective layer which ensheaths the main part of a claw.

Some of the stages in the transformation of a claw into a nail may be seen in certain of the more primitive Primates in existence today. In the small marmoset, for example, all the digits with the exception of the big toe are armed with sharp claws. These are slightly less compressed than the typical claws of lower mammals, and the deep stratum of the claw substance has dwindled to a thin and indistinct layer which, it seems, can be of no functional significance. In the larger monkeys, the deep stratum has finally disappeared altogether and true nails are found, but they still retain something of a claw-like character in their longitudinal curvature and their lateral compression. In typical lemurs, it is interesting to note, all the digits are provided with flat nails with the exception of the second toe. This retains a pointed claw which is used for scratching and cleaning the fur—hence it is called the 'toilet digit'. Lastly, in the small Primate, *Tarsius*, in which the digital pads have become elaborated into specialized tactile disks, the nails have become reduced to minute horny plaques, with the exception of the second and third toes which are used as toilet digits.

7. THE INNERVATION OF SKIN

The rich nerve supply of the skin obtrudes itself on the notice of the student of anatomy in the first stages of his dissection. Searching for and tracing out cutaneous nerves in the human body is an exacting task which requires considerable skill and patience, for many of them are extremely fine. These nerves are branches of the main mixed trunks which lie in the deeper tissues, and they enter the superficial fascia by piercing the deep fascia at relatively fixed points. Ramifying freely beneath the skin, they are each distributed to cutaneous areas of fairly constant extent, though the latter may show individual variations sufficiently marked to require attention in the clinical examination of peripheral nerve injuries.

The overlapping of the territories of adjacent cutaneous nerves is often quite considerable, so that the area of complete anaesthesia following the section of one nerve is never as extensive as the area of its supply delineated by anatomical dissection. Surrounding such an area of complete anaesthesia is a zone of variable width in which, while the finer elements of tactile sensation are lost, cruder forms of sensation such as pain still persist. It thus appears that the overlap is mainly confined to the fibres of the cutaneous nerves which conduct the latter type of impulse. It is interesting to note that down the midline of the body (except in the region of the lips) there is usually no overlap between nerves of either side.

Anatomical areas of cutaneous nerve distribution may be mapped out over the surface of the body by several methods. The most obvious of these are gross dissection and the study of sensory loss following nerve injuries. Dissection has certain disadvantages, for the terminal branches of a nerve are so fine as to be traced with difficulty, and, since adjacent peripheral nerves and their branches often effect direct anastomoses, the course of the constituent fibres derived from different main branches cannot always be followed. Experimental methods on normal living individuals are available which provide accurate results, and incidentally allow of a comprehensive study of individual variations. Cutaneous nerves may be stimulated by direct application to the skin immediately overlying them of a finely pointed unipolar electrode and the passage of a weak faradic current (using an induction coil and a 4-volt accumulator). This gives rise to a characteristic tingling sensation extending over the area of skin supplied by each cutaneous nerve.[1] It has further been found that, if a nerve is stimulated with an alternating current of a certain strength, the skin area which it innervates is rendered insensitive to touch. By this method, it is a simple matter to map out the area precisely by testing the skin with light tactile stimuli.[2]

With the exception of the branches of the spinal nerves that mainly preserve their simple segmental arrangement (e.g. the intercostal nerves), most cutaneous nerves are composed of fibres which are derived from several segments of the spinal cord. By reference to the cutaneous anaesthesia which follows lesions of nerve roots or of the spinal cord, it is possible to map out the skin into areas corresponding to each segment. In the case of the thoracic segments represented by the intercostal nerves, the overlap of adjacent segmental territories is quite considerable, and it is probable that the overlap is equally marked in other segmental areas. The segmental

[1] W. Hughson, 'Electrical stimulation of cutaneous nerves', *Anat. Rec.* **23**, 1922.

[2] I. M. Thompson, V. T. Inman, and B. Brownfield, 'On the cutaneous nerve areas of the forearm and hand', *Univ. of California Publ. in Anatomy*, **1**, no. 7, 1934; V. T. Inman and R. C. Combs, 'A simple apparatus for stimulating human nerves', *Journ. Anat.* **72**, 1937–8.

areas are regularly disposed in a series of annular bands on the trunk, but in the limbs their primitive and orderly arrangement has been disturbed to a certain extent.

It will be apparent that, while a knowledge of the cutaneous areas of peripheral nerve trunks is essential for the diagnosis of peripheral nerve lesions, a knowledge of the segmental areas is equally important for the diagnosis of lesions involving the spinal cord or nerve roots. It is also of considerable use in the interpretation of what is termed *referred pain*. This is the pain or tenderness felt subjectively over an area of skin as the result of inflammation or some other pathological stimulation involving some deep-seated structure, such as a visceral organ. In the latter case, the pain is not felt in the viscus itself, but is referred or projected by the patient to the area of skin which is innervated from the same spinal segment.[1] For example, in the early stages of acute appendicitis, pain and tenderness may be felt in the skin round the umbilicus. This area is supplied by cutaneous fibres from the tenth thoracic nerve, and the appendix (or the peritoneum in relation to it) is also innervated from the tenth spinal segment. The delineation of an area of tenderness in the skin may thus provide guidance in the localization of a visceral affection whose site is otherwise obscure.

Nerve endings of skin and subcutaneous tissues. Cutaneous nerves (despite the name) supply not only the skin proper, but also the subcutaneous tissues and even (in some cases) adjacent joints and tendons. It is convenient here to consider some of the nerve endings in these contiguous regions as well as those in the skin proper, partly because of the difficulty of defining any precise boundary between the dermis and the subcutaneous tissue. Accompanying the cutaneous nerves there are also sympathetic fibres which innervate the cutaneous blood vessels, the arrector pili muscles and the sweat glands.

In the skin and subcutaneous tissues, sensory nerves terminate in a variety of nerve endings. Many of these are free, naked endings, ramifying in the dermis and in the deepest layers of the epidermis (Fig. 110 A, F). Others are more highly organized in the sense that they are insulated in connective-tissue capsules to form sensory 'corpuscles', or end in relation to what appear to be special sensory cells. Each nerve fibre usually bears on its terminal branches only one type of ending, and as many as thirty-seven end organs of the same kind have been counted on the collaterals of a single fibre.

Of the sensory corpuscles in man we may briefly refer to some of the commoner types. The largest are the *Pacinian corpuscles* (or lamellated corpuscles), distinguished for the fact that they are the only sensory nerve endings which are readily visible to the unaided eye (Fig. 110 P). They

[1] The mechanism underlying referred pain is still debated. It has commonly been explained on the basis of a convergence of afferent impulses from structures of a common segmental origin on the same group of sensory nerve cells in the corresponding segment of the spinal cord. However, the problem is complicated by the fact that referred pain has sometimes been observed to involve the territory of quite a distant spinal segment. See D. C. Sinclair, G. Weddell, and W. H. Feindel, 'Referred pain and associated phenomena', *Brain*, **71**, 1948, and D. C. Sinclair, 'The remote reference of pain aroused in the skin', ibid. **72**, 1949.

are oval in shape, and vary in size from 1 to 4 mm. in length and up to 2 mm. in breadth. In structure they consist essentially of numerous thin lamellae (up to fifty or more) of connective tissue arranged in a closely concentric formation and separated by intervals containing tissue fluid. Each corpuscle is pierced at one pole by a coarse medullated nerve fibre which, losing its myelin sheath, ends in terminal expansions in its centre. Pacinian corpuscles are found in the superficial fascia at some little distance from

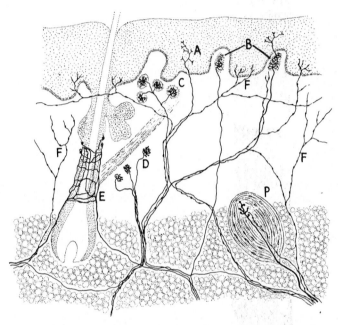

FIG. 110. Diagram of a section through the skin, showing different types of sensory nerve ending. Above is seen the epidermis and below the subcutaneous layer of fatty tissue. Between is the corium or dermis of the skin. *A*. Tactile nerve ending in the epidermis. *B*. Meissner's corpuscles situated in dermal papillae. From the corpuscle on the right a nerve fibre is continued into the epidermis to form a tactile ending. *C*. Krause's end organs, which are believed to respond to cold stimuli. *D*. Nerve endings resembling Ruffini's end organs. *E*. Nerve endings forming a characteristic palisade formation around the base of a hair follicle. *F*. Fine varicose fibres which form a rich plexus everywhere permeating the dermis, also penetrating into the deeper layers of the epidermis. These fibres subserve the sensation of pain. *P*. Pacinian corpuscle innervated by a main nerve fibre concerned with pressure sensation, and also by an 'accessory fibre' of the pain type.

the surface of the skin, and they are particularly numerous under the digital pads of the fingers. They are also found quite frequently in the fibrous capsules of joints, in the walls of large vessels, in the periosteum, and in the peritoneal mesenteries of the abdominal viscera. There is little doubt that they are specific end organs which respond to the mechanical stimuli of pressure and tension. This has been particularly well shown by electro-physiological studies of the passage of impulses along the nerve fibres of Pacinian corpuscles which have been exposed to different kinds of stimulation.[1] In the subcutaneous tissues the corpuscles respond to tactile pressure, while in joint capsules and neighbouring structures they probably provide a proprioceptive mechanism whereby movements may be registered.

Situated quite superficially in the dermis, and often within the dermal papillae, are *Meissner's corpuscles* (Fig. 110 B). Since they are mainly found in cutaneous areas which are particularly sensitive to touch, they have been regarded as specifically related to tactile sensation and are sometimes called 'tactile corpuscles'. They are cylindrical structures reaching a length of not more than 0·1 mm., and within them are transverse lamellae of connective tissue between which are a few flattened cells.[2] Each corpuscle receives a medullated nerve fibre which winds spirally round its lower part and, after losing its myelin sheath, ends by ramifying inside it.

Pacinian corpuscles and the corpuscles of Meissner are relatively numerous in certain areas of the skin in man. Less frequent types of nerve ending which may be noted are the *end organs of Krause*, in which nerve a fibre ends as a globular network in a fine capsule of connective tissue (Fig. 110 c), and the *corpuscles of Golgi-Mazzoni*, which are somewhat similar to Pacinian bodies but much smaller and much less elaborately organized. The corpuscles of Krause are said to be particularly numerous in the skin of the external genitals and the area surrounding the nipple. Finally, mention should be made here of the *end organs of Ruffini* which are also found in the dermis. They somewhat resemble Meissner's corpuscles, but they are really free nerve endings of non-medullated fibres arranged in a terminal brush formation, and they are not encapsulated. Recent studies have made it evident that these different types of organized nerve ending are not all such clear-cut entities as some histologists have supposed, for there are occasionally to be found intermediate types which it may be difficult to assign to one category or the other.

Relation of nerve endings to sensation. Of great anatomical interest is the question of the relation of different types of sensory nerve ending to different forms of sensation. That such a relation exists has appeared probable from the observation that in the skin there are minute areas sensitive to different sensory stimuli, which have a discontinuous punctate distribution. This suggests that each sensory spot may be related to a specific type of nerve ending.

The punctate nature of cutaneous sensation is most easily demonstrated in the case of temperature. If a hot or cold metal point is placed in contact with the skin, sensations of heat or cold are only felt at certain spots, which are therefore called heat and cold spots. Between these spots the skin seems to be relatively insensitive to temperature stimuli. A similar punctate distribution has been claimed for painful and tactile sensations, so that the whole skin has been spoken of as 'a mosaic of tiny sensorial areas' each of which responds specifically to one type of stimulus. The accuracy of this conception has been called into question. 'Pain spots' are so numerous that

[1] J. A. B. Gray and P. B. C. Matthews, 'Response of Pacinian corpuscles in the cat's toe', *Journ. Physiol.* **113**, 1951.
[2] While Meissner's corpuscles are characteristically encapsulated, quite similar nerve endings may sometimes be found in a dermal papilla with no apparent capsule.

it is difficult to plot them all out individually, while 'touch spots'—in so far as they have any punctate arrangement—seem to be related to hair follicles. In regard to heat and cold spots, also, it has been stated that they vary in position from one time to another, suggesting that, while the entire skin surface is in fact sensitive to thermal stimuli, at any particular moment only certain scattered points are actively responsive. However, this is not in accordance with the experience of some observers who, though admitting that all the heat and cold spots in a given area of skin may not be detected at one examination, maintain that they remain fixed and recognizable for long periods of time. In any case, the number of these spots seems to show considerable individual variability. It should be noted that the testing or mapping of sensory spots in the skin is complicated by the fact that the nerve endings relating to them are distributed in depth as well as in surface extent. Thus it is possible for a 'cold spot' to overlie a 'heat spot', and this may perhaps explain the variable response sometimes found to follow a local stimulus applied under different conditions.

Experimental studies on the anatomical basis of cutaneous sensation have been advanced in support of the conception that the different morphological types of nerve ending are each responsive mainly to one specific kind of stimulus. It has thus been supposed that there are specific touch endings, and other endings related to the sensations of pressure, heat, cold, or pain. In other words, it is suggested that the skin is sprinkled with peripheral analysers which, so to speak, 'sort out' the various stimuli which impinge upon it so that finally, by the activity of the higher centres of the brain, the individual can recognize them and distinguish them from each other. In recent years, however, more critical studies have raised some doubts whether the relationship of different morphological types of receptor to different sensory modalities is really as simple and precise as some authorities had supposed. For example, it has been found that in certain areas of skin in which sensory discrimination is quite well developed there may be no histological evidence of a marked diversification of sensory nerve endings. But while this indicates that *morphologically differentiated* nerve endings are not essential for sensory discrimination, it does not dispose of the conception of a functional specificity of some sort for cutaneous nerve fibres in general. It should perhaps be emphasized that this conception must not be taken to imply an absolute specificity in the sense that some nerve endings respond selectively only to one kind of stimulus and not at all to others. It envisages no more than a preferential sensitivity, each functional type of ending responding to one kind of stimulus at a somewhat lower threshold than other types of ending, and, of course, the degree of specificity of different endings may vary quite considerably. It may also be presumed that the specific endings with their fibres are scattered in a matrix of non-specific endings, the latter providing for a background of general activity on which, so to speak, the effects of stimulation of the specific receptors are high-lighted.

It has been generally assumed that painful sensations in the skin are served by the free endings of very fine medullated and non-medullated nerve fibres in the epidermis and the sub-epidermal tissues. Several reasons have been advanced for this assumption. They are the characteristic type of nerve ending in the surface of the cornea of the eye or in the ear drum, where normally only painful sensations can be elicited.[1] They have been found to be the only type of ending present at the margin of ulcers where sensation is also limited to pain. Direct stimulation of the fine fibres in the rabbit's ear (in which they can be seen after staining with methylene blue *in vivo*) gives rise to what is evidently a painful response. Lastly, there is experimental evidence showing that, in the nerve roots and the spinal cord, pain impulses are conducted (at least in part) by the finest medullated or non-medullated fibres.

There is good evidence that the free nerve endings in the epidermis are almost entirely concerned with touch; in other words, so far as sensation is concerned, the epidermis is essentially a tactile mechanism. This can be shown by some very simple experiments.[2] In the first place, a sensation of touch results from the lightest stimulation of the epidermis with a wisp of cotton-wool. Again, shaving off thin slices of epidermis is quite a painless operation until the vascular dermis is reached and bleeding occurs. If the skin is blistered, tactile sensation remains so long as a thin film of epidermis is left intact, but as soon as the whole thickness of the epidermis is removed, it can no longer be appreciated, and a local stimulus applied to the denuded area only elicits painful sensations. Experimental embryological evidence supports the implications of these observations, for it has been found that in the developing limbs of tadpoles the first appearance of a definite reaction to light tactile stimuli is associated with the invasion of the epidermis by free nerve endings.[3] In most parts of the skin, free nerve endings are limited to the deepest layers of the stratum of Malpighi, but in the hand, where tactile sensation is very acute, they reach up as far as the stratum granulosum. A similar plexus of free nerve endings is found among the epidermal cells of each hair follicle (Fig. 110 E), and the relation of hairs to tactile sensation is well established. When the exposed part of a hair is touched and slightly displaced, the movement is transmitted to its root and stimulates the nerve endings. If it is recognized that the exposed part of a hair is the long arm of a lever with the fulcrum at the skin surface, it will be appreciated that the force resulting from a slight lateral pressure applied to the exposed part will be considerably magnified at the hair root.

While it may be accepted that the free nerve endings in the epidermis are mainly responsive to touch, there is histological evidence that 'pain fibres' (which are characterized by their fine varicose appearance in stained material) do reach the deepest layers of the stratum of Malpighi. But painful cutaneous sensation, it is agreed, is mainly served by a rich subepidermal

[1] It is to be noted, however, that if painful impulses from the cornea are eliminated by interruption of the pathways for those impulses in the central nervous system, a sensation of touch may be evoked by applying a tactile stimulus to the cornea. This observation evidently lessens the strength of the argument so far as the cornea is concerned, though histological studies make it probable that the sensation of touch is mediated by more deeply lying fibres.

[2] D. Waterston, 'The sensory activities of the skin for touch and temperature', *Brain*, **46**, 1923; 'Observations on sensation', *Journ. Physiol.* **77**, 1933.

[3] K. A. Youngstrom, 'Studies in the developing behaviour of Anura', *Journ. Comp. Neur.* **68**, 1938.

plexus of extremely fine varicose nerve fibres which everywhere permeate the dermis immediately below the epidermis (Fig. 110 F). Similar
plexuses are found in other tissues which are responsive to painful stimuli,
e.g. the tunica adventitia of blood vessels, periosteum, and the sheaths of
tendons. In general it may be said that impulses are initiated in 'pain fibres'
by any stimulus which exceeds normal physiological limits, and which
therefore leads to actual injury of the tissues, or tends to do so. Thus pain
sensation serves predominantly a protective function. It is therefore interesting to note that the cutaneous end organs which have been related to
other kinds of sensation are not only supplied by their own specific nerve
fibres, but are in many cases also innervated by a fine 'accessory fibre' of
the pain type. This perhaps provides the anatomical basis for the fact that
if a stimulus such as heat, for example, becomes so excessive as to endanger
the tissues, the sensation of warmth is replaced by a sensation of pain, due
to the stimulation of the accessory fibre. The actual termination of a 'pain
nerve fibre' is a rich arborization of free endings which may extend over
an area as large as 0·75 cm. in diameter, each overlapping and interlocking
with adjacent endings. The large size of the ending explains why painful
stimuli cannot be localized with the same precision as tactile stimuli, and
the overlap explains why, after interruption of a cutaneous nerve, the area
of analgesia is smaller than the area of anaesthesia.[1]

It has been noted above that Meissner's corpuscles are commonly regarded as tactile organs, and this seems to be established by the fact that
they are very numerous in areas particularly sensitive to touch, such as the
skin of the finger-tips. They are not the only mechanisms involved in this
sensation for, as we have seen, free endings in the epidermis and the nerve
endings related to hair follicles also respond to touch. In some areas of
the skin, indeed, the corpuscles are very sparse. Woollard has demonstrated that many of the fibres which terminate freely in the epidermis
are really continuations of nerve fibres which have entered Meissner's corpuscles (Fig. 110 B). Hence there can be no doubt that they are concerned
with the same type of sensation.[2] In certain regions of the epidermis in
lower mammals (e.g. in the pig's snout) the nerve endings terminate in
disk-like expansions closely applied to specialized epidermal cells. These
end formations are called *Merkel's corpuscles*. It is doubtful whether
corpuscles of this specialized type occur in man, but somewhat similar
disk-like expansions have been described—particularly in relation to the
epidermal cells of hair follicles.

As the result of experiments combined with histological studies of human
skin, it has been reported that in some cases cold spots coincide with the
distribution of complex nerve endings resembling Krause's end organs
(Fig. 110 C). It has already been noted that the latter are particularly
numerous in the skin of the external genitals and the area surrounding the

[1] G. Weddell, 'The pattern of cutaneous innervation in relation to cutaneous sensibility',
Journ. Anat. **75**, 1941. For a short general review of the peripheral mechanism of pain
sensation, see S. S. Tower, 'Pain: Definition and properties of the unit for sensory reception', *Res. Publ. Assoc. for Res. in Nervous and Mental Disease*, **23**, 1943.

[2] H. H. Woollard, 'Observations on the terminations of cutaneous nerves', *Brain*, **58**,
1935; 'Intra-epidermal nerve endings', *Journ. Anat.* **71**, 1936.

nipple. This perhaps explains why a cold stimulus applied to these areas gives rise to an unusually intense and vivid sensation. The identification of the end organs for the appreciation of warmth has been a matter of considerable difficulty. Experimental evidence has suggested that they are situated fairly deeply in the skin, for there is a rather long latent period in the response to a thermal stimulus. Their depth has actually been measured by stimulating them with an extremely fine needle pushed directly into the skin; by this method a sensation of warmth is aroused at a depth of 1·75 to 2·5 mm. (as compared with 0·25 mm. for the sensation of pain). In some cases heat spots have been identified with closely clustered nerve endings which appear to be similar to the end organs of Ruffini (Fig. 110 D), but the evidence for this relationship is by no means conclusive.

Although the evidence is not inconsiderable for the general thesis that the main modalities of cutaneous sensation may be served by nerve endings of different morphological or functional types, there is considerable uncertainty regarding the anatomical basis of the less well-defined sensations which may be experienced (such as itching or tickling), the different qualities of pain produced by different kinds of stimulus, or the abnormal sensations (such as pins and needles) which occur as the result of pressure and interference with the blood supply to the skin. However, there is reason to suppose that the pattern of stimulation of the mosaics of nerve endings in the skin (and thus the pattern of impulses to the central nervous system to which this gives rise) is the important factor underlying these sensory phenomena. For example, the ordinary sensation of pain which is experienced as the result of a needle prick is based on the stimulation of a complex pattern of interweaving fibres and nerve endings present in the skin. But in tender scars it may be found that a prick arouses a painful sensation of a particularly unpleasant quality, and in these cases it has been observed histologically that the nerve endings at the site of the scar are isolated instead of forming interlocking networks. In this abnormal condition, therefore, it is clear that the stimulus of the prick must give rise to a very different pattern of impulses reaching the central nervous system. It should be emphasized, however, that the solution of these problems of cutaneous sensibility is still far from clear, and much work remains to be done in attempting to correlate histological studies with experimental observations.[1]

The general pattern of sensory innervation of the skin. Two important anatomical features of cutaneous innervation have been made clear by the study of whole preparations of the skin in which the nerve fibres and their endings have been stained intravitally with methylene blue. In the first place, there is a most intricate subcutaneous plexus from which fibres enter any localized area of skin from all directions. It is because of this that, during sensory recovery after interruption of a sensory nerve, a circumscribed patch of anaesthesia in the arm or leg does not shrink from above downwards but from all sides, the perimeter of the area contracting

---

[1] For a discussion emphasizing the difficulties of the problem of the peripheral apparatus of cutaneous sensation, see D. C. Sinclair, 'Cutaneous sensation and the doctrine of specific energy', *Brain*, **78**, 1955.

more or less evenly as sensory recovery spreads towards its centre. The second point which has been established is the principle of the multiple innervation of sensory spots. Thus, Meissner's corpuscles commonly occur in groups and are innervated by fibres approaching them from different directions. Some of the fibres reach their termination more or less directly; others may pursue a tortuous course through the subcutaneous plexus before they reach the group. Similarly, histological studies have suggested that each cold spot is represented anatomically by a cluster of several Krause's end organs, while each hair follicle is supplied by two or more sensory nerve fibres. This multiple innervation provides a mechanism for the summation of stimuli and thus facilitates the discrimination of stimuli of different intensity. It may also provide the mechanism whereby a stimulus can be accurately localized, and it is of considerable importance for understanding the process of sensory recovery after section of a cutaneous nerve.

If a sensory nerve, such as the superficial part of the radial nerve in the forearm, is cut, loss of sensation occurs in the cutaneous area of its distribution. As regeneration of the cut nerve approaches completion and new nerve fibres grow down to reach the anaesthetic area, sensation once more returns. Complete sensory recovery, however, is not abrupt in the re-innervated portions of skin; it is a gradual process. In a classical study of this process, first published in 1908, Head described two consecutive stages.[1] On the completion of the first stage the skin only reponds to sensory stimuli of a gross character. That is to say, the individual can appreciate pain and the wider ranges of temperature. On the other hand, the sensations are somewhat crudely felt, they cannot be accurately localized, and they have no quality of gradation. This type of sensation Head termed *protopathic*. Later on in the regeneration of the nerve the crude sensibility is superseded by the development of a more refined type of sensation to which the term *epicritic* was applied. This allows the recognition of light tactile stimuli, precise localization, the appreciation of temperature differences close to that of the body, and the perception of gradation in the intensity of the stimuli. Head supposed that the two forms of sensation are served by different sets of fibres in the peripheral nerves, which regenerate at different rates. An evolutionary significance was also imputed to the two categories, the protopathic representing a primitive sensory system showing itself on the effector side as crude mass reactions in response to painful and pleasurable stimuli, and the epicritic a later evolved discriminative system which allows of more delicately graded responses.

Head's conception of a dual mechanism for cutaneous sensation has now been discarded. The comprehensive studies of Trotter and Davies, Boring and others, failed to confirm some of his original observations and gave no support for the thesis that sensory recovery occurs in two definite stages.[2] Repeated attempts have also failed to show any anatomical evidence

[1] H. Head, *Studies in Neurology*, 1920.
[2] W. B. Trotter and H. M. Davies, 'Experimental studies on the innervation of the skin', *Journ. Physiol.* 38, 1909; E. G. Boring, 'Experimental studies on the innervation of the skin', *Quart. Journ. Exp. Physiol.* 10, 1916.

of two sets of cutaneous receptors and two sets of sensory fibres. In 1924 Boeke and Heringa studied the anatomy of regenerating nerve endings in correlation with the return of sensation after a peripheral nerve lesion.[1] Their observations led them to believe that 'protopathic sensation' is merely the expression of incompletely regenerated nerve fibres, and that, when the same fibres have completely recovered their normal connexions, epicritic sensation becomes established. The principle of the multiple innervation of sensory spots now seems to provide a more satisfactory explanation for the gradual character of sensory recovery. Some fibres (which run a circuitous course in the subcutaneous nerve plexus) will have farther to go than others before they can reach and re-innervate the end organs in a sensory spot. Consequently there will always be a preliminary stage in sensory recovery when multiple innervation has not yet been established. At this stage, therefore, recognition of small differences in the intensity of stimuli and accurate localization will be limited. These aspects of cutaneous sensation will not return completely until the sensory spots have been re-innervated by their full quota of nerve fibres.

It is interesting to note that in certain areas of the body, e.g. the glans penis and the nipple region, Head's protopathic form of sensibility is alone represented in the normal individual. This is presumably related to some local difference in the pattern of cutaneous innervation.

8. THE BLOOD SUPPLY AND LYMPHATIC DRAINAGE OF SKIN

The cutaneous blood vessels which vascularize the skin are relatively quite numerous. A good blood supply is required not only for the nourishment of the epidermal cells which are undergoing continual proliferation, but also for the special requirements of temperature regulation and protection. In regard to the former, it has already been noted that heat loss at the surface of the body is controlled to a very considerable extent by the dilatation and constriction of cutaneous arterioles, which lead to a flushing or blanching of the skin. The protective function of the cutaneous vessels is shown in the intense vasodilatation which occurs locally in response to injurious stimuli, and which clearly provides for a rapid mobilization of defence mechanisms in order to withstand and repair the injury. Compared with a tissue such as muscle, however, the capillary density in the skin is low, in relation to its lower and less variable metabolic level.[2]

Cutaneous vessels of supply are generally more numerous in those areas where the epidermis is thicker, e.g. in the palms of the hand and the soles of the feet. They anastomose very freely to form a plexus in the superficial fascia. From the plexus branches take origin to form a fine sub-papillary plexus in the superficial part of the dermis, and from this, again, terminal arterioles arise to reach the dermal papillae. In the latter the capillaries are

---

[1] J. Boeke and G. C. Heringa, 'Tactile corpuscles and protopathic sensibility of the skin in a case of nerve regeneration', *Konin. Akad. v. Wetensch., Amsterdam*, **27**, 1924.
[2] A. Krogh, *The Anatomy and Physiology of Capillaries*, London, 1929.

arranged in characteristic loops. These loops can be seen and studied during life, and, by observing changes in their appearance, important information has been obtained with regard to the reactions of capillary blood vessels in general to various types of stimulus and in inflammatory conditions. The method of observation consists of dehydrating the surface of a small patch of skin with alcohol, applying a drop of cedar-wood oil, illuminating the skin with a concentrated beam of light, and examining it with a binocular microscope. By this means it is possible to see not only the capillary loops, but also the sub-papillary plexus.[1]

A characteristic feature of the vascular plexuses in the skin is the large number of arteriovenous anastomoses found here (see p. 204). It is believed that these provide a mechanism for maintaining the temperature of exposed parts since, when the muscular wall of the anastomotic connexions is relaxed, they allow a much more rapid circulation than could take place through a capillary bed only. It is interesting to note, therefore, that they are absent in the skin of cold-blooded vertebrates such as reptiles.

The lymphatic drainage of the skin is effected by closely meshed superficial plexuses of lymphatic capillaries which lie immediately deep to the sub-papillary plexus of blood vessels. Three lymphatic plexuses can be recognized, (1) superficial, which begins as minute channels draining the dermal papillae and forming a network at the base of the papillae, (2) intermediate, occuping the middle third of the dermis, and (3) deep, occupying the deeper layer of the dermis and the superficial zone of the subcutaneous tissue.[2] The vessels of the deep plexus are supplied with valves, but the latter are absent in the superficial and intermediate plexuses. The density of the lymphatic plexuses varies in different areas, being greatest in the sole of the foot and the palm of the hand. From the deep plexus collecting vessels arise and run proximally in the superficial fascia in close topographical relation to the superficial veins. At certain well-defined points the cutaneous lymphatic vessels pass through lacunae in the deep fascia and so join the deeply situated vessels, but, before doing so, they may be interrupted in their course by superficial lymph nodes. It is of some practical importance to note that peripherally there is no anastomotic communication of any significance between the cutaneous and deep lymphatic plexuses— the deep fascia forming an effective barrier between them (see p. 248).

9. THE REGENERATION OF SKIN

Skin shows a high capacity for regeneration and repair; skin grafting is therefore a common procedure in plastic surgery. If a patch of epidermis is shaved off, provided fragments of the stratum germinativum are left here and there, repair takes place very rapidly by the proliferative activity of these remnants. Otherwise, a new covering of skin is formed over the denuded area by a gradual extension from the margins of the wound. Hairs will not grow from the healed surface unless the deep portions of the original hair follicles have been left intact. On the other hand, sweat glands

---

[1] See T. Lewis, *The Blood-vessels of the Human Skin and Their Responses*, London, 1927.
[2] G. Forbes, 'Lymphatics of the skin', *Journ. Anat.* **72**, 1937–8.

can be regenerated from new epidermis after the whole thickness of the skin has been removed. The grafting of skin on areas denuded by severe injuries such as burns is a therapeutic technique of great importance. Under ordinary circumstances only autografts (taken from some other part of the same individual) are successful. Homografts (taken from another individual) almost invariably break down as the result of immunity reaction.[1] Attempts have been made to overcome this difficulty by inhibiting the formation of antibodies or neutralizing their effect.

It has already been mentioned that the papillary ridge pattern of the fingers is accurately reproduced if the epidermis is scraped or burnt off. If, however, the underlying dermis is destroyed or severely damaged, although papillary ridges are regenerated they do not conform to the original pattern. The morphological specificity of different types of epidermis has been clearly demonstrated by a series of grafting experiments in animals. Thus, skin from the sole of the foot transplanted to the chest still maintains its characteristically high rate of mitotic activity and continues to produce a thick stratum corneum where this is no longer needed. On the other hand, the corneal epithelium transplanted to the same region preserves its thin transparent character.[2] In a series of transplantation experiments Boeke[3] has shown that nerve endings (sensory corpuscles) are differentiated afresh in grafted skin, and that the type of corpuscle which is found under these conditions is determined by the grafted epidermis and not by the regenerating nerve fibres in the region where the graft is made. If, for example, a portion of skin is transplanted from a duck's bill on to its leg, the sensory corpuscles which become differentiated in the graft are those which are normally characteristic of the bill (Herbst's corpuscles).

    [1] P. B. Medawar, 'The experimental study of skin grafts', *Brit. Med. Bull.* **3**, 1945.
    [2] R. E. Billingham and P. B. Medawar, 'A note on the specificity of the corneal epithelium', *Journ. Anat.* **84**, 1950.
    [3] J. Boeke, *Problems of Nervous Anatomy*, Oxford Univ. Press, 1940.

XIII

THE TISSUES OF THE NERVOUS SYSTEM

ONE of the most outstanding characteristics of a living organism is its power of integration, that is to say, its ability to combine various functional processes in complex and co-operative reactions so as to lead to a unified response adapted to the total environment at any one moment. On the sensory side integration implies the correlation of incoming impulses dependent on stimuli arising outside or inside the organism, while on the motor side it involves the co-ordination of efferent impulses so that the activities of individual muscles or glandular mechanisms are combined harmoniously.

Integration is a property of unicellular organisms, and also of simple multicellular organisms which have no nervous system. In these forms it depends on the transmission of the effects of mechanical and chemical stimuli from one part of the cell to another, or from one cell to its immediate neighbour. Such a method of transmission must inevitably be slow, and the integrated responses dependent on it are of a very simple nature. With the development of a circulatory system a means is provided for co-ordinating the activities of distant parts of the body by chemical activators distributed in the blood stream. Endocrine glands exert their correlating activities in this manner. A far more efficient integrating mechanism is introduced by the development of a nervous system, for this permits of a very rapid transmission of the effects of a stimulus over wide areas, and also allows the building up of extremely intricate and elaborate mechanisms for the correlation of incoming stimuli and the co-ordination of motor reactions.

All living protoplasm shows the properties of *irritability* and *conductivity*. A simple unicellular organism is sensitive to local mechanical or chemical stimuli, and the results of the stimuli are transmitted from the point of their application to other parts of the cell. These properties are developed to a high degree by the cells of nervous tissue. To provide long conducting paths the cytoplasm of nerve cells is drawn out into thread-like processes (nerve fibres) which may extend over a considerable distance, linking up peripheral structures (such as sensory surfaces, sense organs, muscles, and glands) with the centralized organization of the brain and spinal cord. Along these fibres impulses can be propagated with great rapidity, in some cases reaching a rate of about 125 metres a second in man.

For the reception of sensory stimuli nervous elements are modified to form specialized receptor mechanisms which are scattered over sensory surfaces, or concentrated in special sense organs. Many of these receptors are highly sensitive to one particular kind of stimulus, and are therefore selective in function. By this specialization the organism as a whole becomes responsive to a wide range of stimuli (mechanical, chemical, sound waves and light waves, &c.) which have their origin in changes in the external or internal environment of the organism.

Fundamentally, then, the nervous system is made up of cells highly specialized in the direction of sensitivity and conductivity, and it is contrived primarily for the reception of sensory stimuli and their transmission to effector organs, whether muscular or glandular structures. In all vertebrates this transmission takes place almost entirely through a central organization, termed the central nervous system, which is linked up with receptors and effectors by afferent and efferent conductors. The neural composition of the central nervous system is exceedingly complex and provides the anatomical basis for modification of behavioural reactions in close adaptation to the requirements of any given situation. The more elaborate the central nervous system, the greater is the range of adjustments which the organism can effect in relation to its total environment, and the more readily is it able to determine its behaviour in accordance with the results of past experience.

1. THE ANATOMY OF THE NEURON

The ultimate anatomical unit of the nervous system is the neuron, that is to say, the nerve cell with its processes.

In a typical multipolar cell (such as the motor neurons of the spinal cord) the cell body is polygonal in shape, with a centrally placed round nucleus containing a well-defined nucleolus (Figs. 112, 113, 114). The cell processes are of two types—a number of short, richly branched, and often varicose processes called *dendrites*, and a single elongated process—the *axon*. The latter is relatively uniform in diameter and extends for a variable distance before it breaks up into arborizations (*telodendria*). In its course it may give off collateral branches, though these are usually very few.

The neuron shows a dynamic polarity with respect to its processes— nerve impulses being directed towards the cell body by the dendrites and away from the cell body by the axon. Within the central nervous system impulses are conducted from one part to another along a chain of neurons, the terminal arborizations of the axon of one neuron effecting a contact with the dendrites or the cell body of another. Such a connexion is termed a *synapse*. The structural and functional complexity of the synapse is often very great. It is estimated, for example, that the dendrites and cell body of a single motor cell in the spinal cord may be in synaptic relation with as many as a thousand axonal terminals bringing impulses from other nerve cells.

The relative lengths of the axon and dendrites, and the number of the latter, vary in different types of cell. In the type of motor neuron to which we have just referred the dendrites are numerous and relatively short, and the axon is long (Fig. 111 *e*). The latter may even reach a length of 2 feet or more, as is the case with the large nerve cells in the upper part of the motor area of the cerebral cortex whose axons extend in continuity down to the lower end of the spinal cord, or the motor cells of the spinal cord which innervate muscles at the extremities of the limbs. In a second type

of neuron the axon is quite short, so that it is difficult to distinguish morphologically from the dendrites (Fig. 111 *b*). Such nerve cells (which are

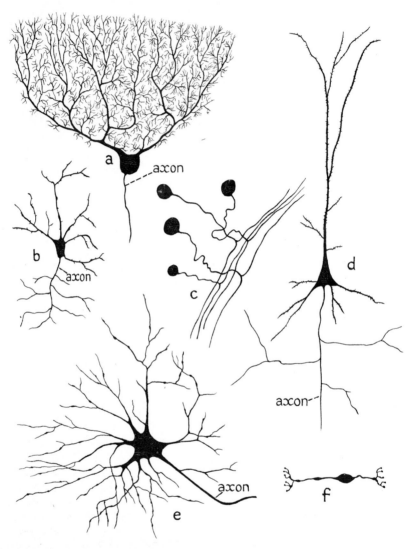

FIG. 111. Outline drawings of different types of nerve cells and their processes as they
appear in Golgi preparations. (*a*) Purkinje cell from the cerebellum; note the elaborate
ramifications of the dendrites. (*b*) A short-axon nerve cell from the central nervous system.
(*c*) Spinal ganglion cells, showing the single processes which divide into central and peripheral processes. (*d*) Pyramidal cell from the cerebral cortex. (*e*) Motor nerve cell from the
ventral horn of grey matter in the spinal cord. (*f*) Bipolar nerve cell from the retina.

usually small) are abundant in the grey matter of the central nervous system,
where, as *intercalated neurons*, they play the part of linking up afferent and
efferent neurons, and in some cases form highly complicated organizations.

A third type of neuron is represented by the cells of the posterior root ganglia of spinal nerves and some of the equivalent sensory ganglia of cranial nerves. Here, the cytoplasmic process which conducts impulses to-wards the cell body (and which must therefore be regarded functionally as a dendrite) is greatly extended, while the efferent process (or axon) is much shorter. When they are first developed in the embryo, these cells are spindle-shaped and bipolar, with a distal process (dendrite) which runs from the periphery to the cell, and a proximal process (the axon) which runs from the cell into the central nervous system. At their attachment to the cell body the two processes become during development gradually approxi-mated and eventually fuse, so that the cell becomes secondarily unipolar, with a single process which bifurcates (Fig. 111 c). The dendrite of such a cell may extend a considerable distance from the sensory surface which it innervates to the spinal ganglion, while the axon often runs a relatively short course from the ganglion into the spinal cord.[1]

The embryonic bipolar character of the sensory ganglion cells is retained in the ganglion associated with the auditory nerve and also by the olfactory sensory nerve cells, while in some lower vertebrates (e.g. certain teleostean fishes) it persists even in the spinal ganglia. Although most neurons in the central nervous system have a primary unipolarity when they are first differentiated embryologically (i.e. a developing nerve cell usually sends out a single process to begin with), it is doubtful whether there are in the adult nervous system of mammals any unipolar cells of the embryonic type, though certain cells in the mid-brain (the cells of the mesencephalic nucleus of the fifth cranial nerve) have been described as such.

The size of the cell body of a neuron varies within fairly wide limits—the smallest cells in man being less than 5 μ in diameter, and the largest reaching as much as 120 μ.[2] The factors determining the size are not fully known, but it is certainly related in part to the length and thickness of the axonal process. In general, the longer the axon the larger the cell body, while somatic motor nerve cells which give off coarse axons are larger than visceral cells which give off fine axons of an equivalent length.

The internal structure of nerve cells and their processes now requires consideration. In the cytoplasm a number of characteristic constituent ele-ments and inclusions can be detected by appropriate methods of staining. These are the Nissl bodies, neurofibrillae, pigment, mitochondria, and the Golgi apparatus.

Nissl bodies. Most nerve cells, when stained with basic dyes such as methylene blue or neutral red, show in their cytoplasm rather irregular oval or sub-angular particles which stand out conspicuously because of their affinity for these stains. They are termed *Nissl bodies* (after the neurologist who first described them in detail in 1894), or *chromophil bodies*. Their size and distribution show some variation in different types of nerve cell, and they are not apparent at all in the smallest neurons.

[1] It should be noted that, although the peripheral process of a sensory ganglion cell is *functionally* a dendrite, in its anatomical features it is apparently identical with the axonal process of a motor nerve cell.

[2] In certain invertebrates (e.g. the squid) some nerve cells may reach a diameter of half a millimetre, so that they can be detected with the naked eye.

In typical large nerve cells (e.g. the motor neurons of the spinal cord) the Nissl bodies fill almost the whole cell (Fig. 113 A). They tend to be arranged concentrically around the nucleus, and at the periphery of the cell they streak out into the dendritic processes. On the other hand, they do not usually extend into the axonal process; indeed, at the point where the latter leaves the cell there is commonly a small zone of clear cytoplasm which is termed the *axon hillock*. This contrast makes it possible sometimes to distinguish axonal from dendritic processes in stained preparations. With different basic dyes the Nissl bodies in normal resting cells of any

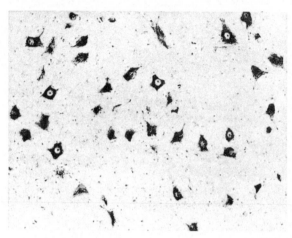

FIG. 112. Microphotograph showing a group of motor nerve cells in the spinal cord, stained with methylene blue. Among the nerve cells are seen the small nuclei of neuroglia cells. × 90.

one particular type show a fairly constant picture, though their appearance varies to a slight extent with the kind of fixative used.[1] They exhibit marked changes, however, with variations in the functional activity of the cell, and also as the result of injury. This reaction is of very great importance since it provides the basis for a valuable anatomical technique in the study of functional localization in the central nervous system, and also in tracing the cell origin of groups of nerve fibres (*vide infra*, p. 386).

If the axon of a nerve cell is cut, within a few days the Nissl bodies appear to break up into a diffuse fine granular deposit scattered throughout the cytoplasm. For a little time they persist at the periphery of the cell, but eventually they disappear altogether. This reaction is termed *chromatolysis* (Fig. 113 B). It is usually accompanied by a general swelling of the cell body and the displacement of the nucleus to an excentric position. The whole process is reversible for, if regeneration occurs, Nissl bodies appear once more—first in the immediate vicinity of the nucleus and subsequently spreading throughout the cytoplasm. Chromatolytic changes

[1] L. Einarson, 'Histological analysis of the Nissl-pattern and substance of nerve cells', *Journ. Comp. Neur.* **61**, 1935.

also occur in extreme fatigue of the nerve cell and as a result of toxic influences.[1]

The nature of Nissl bodies is uncertain. Microchemical tests and studies of the absorption spectrum have shown that they consist essentially of nucleoprotein with organically combined iron.[2] Together with their reactions to basic dyes this suggests the possibility that their composition is similar to that of the nuclear chromatin. That they are a product of nuclear activity is also suggested by the observation that in the regeneration which

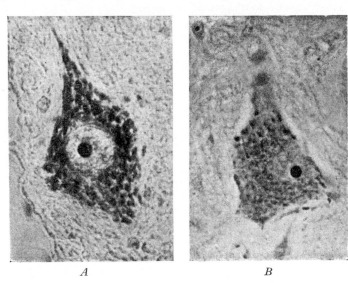

A   B

FIG. 113. Microphotograph showing the appearance of (*A*) a normal motor nerve cell of the spinal cord, and (*B*) a nerve cell which has undergone chromatolysis two weeks after its axonal process has been cut. Note that in the degenerate cell the Nissl granules have disappeared or become dispersed in fine particles, and that the nucleus has become displaced to an excentric position. × 700.

follows chromatolysis Nissl substance first appears in the neighbourhood of the nucleus. According to Einarson,[3] indeed, the material from which it is elaborated is formed primarily round the nucleolus inside the nucleus, migrates towards the periphery of the nucleus, and then diffuses gradually through the nuclear membrane to form Nissl bodies in the cytoplasm. Later studies by Hydén served to emphasize further the part which the nucleus plays in the elaboration of Nissl substance.[4] According to Gersh and Bodian the immediate cause of chromatolysis is the breaking up of the nucleoprotein content of the Nissl bodies by an intracellular

[1] In certain types of cell the Nissl bodies normally form a peripheral ring at the margin of the cell, giving a picture which may simulate chromatolysis.

[2] F. M. Nicholson, 'The changes in amount and distribution of the iron-containing proteins of nerve cells following injury to their axons', *Journ. Comp. Neur.* **35**, 1923.

[3] L. Einarson, 'Notes on the morphology of the chromophil material of nerve cells and its relation to nuclear substance', *Amer. Journ. Anat.* **53**, 1933.

[4] H. Hydén, 'Protein metabolism in the nerve cell during growth and function', *Acta Phys. Scand.* **6**, 1943.

enzyme.[1] The resulting increase of osmotic pressure in the cytoplasm leads to imbibition of water and accounts for the swelling of the cell body and the displacement to the periphery of the nucleus. Recovery is accompanied by a re-synthesis of the nucleoprotein.

The question now arises whether the appearance of Nissl bodies in fixed and stained preparations represents the appearance of actual structures in the living cell, or whether they are precipitation artefacts. The latter view has been generally held for many years, but there has now accumulated some evidence that the ordinary histological picture gives a fairly accurate reproduction of real structures in the cell.

If a fresh nerve cell is examined microscopically with dark-ground illumination, the cytoplasm is seen to be filled with refractile granules. These granules are composed of a colloidal fluid suspended in the semi-solid gelatinous matrix of the cytoplasm, and it has been suggested that, on the death of the cell, the colloidal fluid is coagulated to form the Nissl bodies. Wiemann has demonstrated what appear to be Nissl bodies in fresh nerve cells which have been photographed in ultra-violet light.[2] Even in this case, however, it is open to argument whether the appearance is not the result of coagulation in a cell which is necessarily injured in the process of its microscopical examination and its exposure to ultra-violet light. Obviously it is hardly possible to avoid an objection of this kind, since a living nerve cell must be removed from its normal environment before it can be subjected to direct anatomical study. However, further indirect evidence for the reality of Nissl bodies has been obtained by studying the effects on living nerve cells of ultra-centrifugation. Beams and King[3] found that, when spinal ganglion cells are centrifuged at 400,000 times the force of gravity, the Nissl bodies are displaced to the centrifugal pole. The fact that, as seen in stained sections, they still retain a discrete form after displacement certainly leads to the inference that in the living cell they are definite masses at least comparable in appearance to those seen in usual histological preparations. Under the electron microscope Nissl bodies appear to consist of a dense formation of closely packed and convoluted double membranes (the so-called 'endoplasmic reticulum') among which are clusters of minute granules or ribosomes (see p. 17). The resemblance of this formation to similar structures in gland cells supports the conception that the Nissl substance in a nerve cell is intimately concerned with synthetic functions.[4]

Whatever interpretation of the Nissl bodies may be accepted, the fact remains that in many types of nerve cell which are normal they are always to be found with appropriate histological treatment. Alterations in their usual pattern can therefore be taken to indicate extreme functional disturbance or a definite pathological condition. It may be assumed, from their

[1] I. Gersh and D. Bodian, 'Some chemical mechanisms in chromatolysis', *Journ. Cell. Comp. Phys.* **21**, 1943.

[2] E. Wiemann, 'Studien am Zentralnervensystem des Menschen mit der Mikrophotographie im ultravioletten Licht', *Zeitschr. f. d. ges. Neurol. u. Psych.* **98**, 1925.

[3] H. W. Beams and R. L. King, 'The effects of ultra-centrifuging the spinal ganglion cells of the rat, with special reference to Nissl bodies', *Journ. Comp. Neur.* **61**, 1935.

[4] S. L. Palay and G. E. Palade, 'The fine structure of neurons', *Journ. Biophys. Biochem. Cytol.* **1**, 1955.

chemical constitution and from the changes they show in relation to functional activity, that they are in some way concerned with the nutrition of the cell and its processes. Possibly they serve in part for the storage of oxygen, but the fact that their main constituent is ribose nucleic acid (as well as the reactions which they show in relation to the degeneration and regeneration of the axonal process of the cell) suggests important functions in connexion with protein synthesis.

Neurofibrillae. In sections of nervous tissue treated by methods of silver impregnation the cytoplasm of nerve cells and their processes shows a very definite fibrillary structure. Extremely fine protoplasmic fibrils can be seen crossing through the cell body and streaming out into its axon and

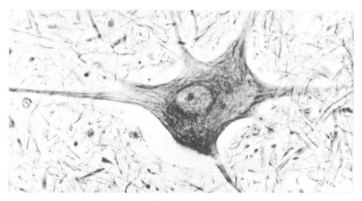

FIG. 114. Microphotograph of a motor nerve cell from the spinal cord, showing the neurofibrillar structure of the cytoplasm. Stained with protargol (Bodian's method). × 500.

dendrites, and the axon itself appears to consist of a bundle of such fibrils embedded in a structureless matrix. They are distinctive characters of nerve cells, and there is some evidence that they become differentiated embryologically in relation to the onset of functional activity (as suggested by Cowdry[1] in the developing chick).

As with the Nissl bodies, however, it has been a matter of much discussion whether the neurofibrillae seen in histological preparations have any counterpart in living nerve cells, or whether they are entirely products of coagulation to be seen only in damaged and fixed tissues. In this case, also, it seems almost impossible to arrive at a final proof of their existence during life, since any manipulation which is necessarily required for the examination of living nerve cells may injure the latter and so vitiate a positive conclusion.

Several histologists claim to have observed neurofibrillae in fresh neurons. De Rényi,[2] for example, has described them in the large nerve fibres of Crustacea. By methods of microdissection he demonstrated that,

[1] E. V. Cowdry, 'The development of the cytoplasmic constituents of the nerve cells of the chick', *Amer. Journ. Anat.* **15**, 1914. See also a series of papers by W. F. Windle and his collaborators in the *Journal of Comparative Neurology*, **63**, 1935–6.

[2] G. S. de Rényi, 'Observations on neurofibrils in the living nervous tissue of the lobster', *Journ. Comp. Neur.* **48**, 1929.

in the lobster, a neurofibrillar structure is clearly visible in practically all nerve fibres with a diameter of more than 20 μ. The fibrils run a wavy course in closely packed bundles, but, if subjected to pressure, they straighten out and float apart. They appear therefore, to be separate filaments with no anastomosing interconnexions. Their behaviour when pulled apart with a needle further suggests that they are not merely linear rows of closely approximated granules, but 'true thread-like structures of a highly viscous nature' embedded in a semi-fluid matrix (axoplasm). These conclusions, however, have been criticized on the grounds that the nerve fibres which were examined had necessarily been exposed to conditions which resulted in a coagulation of the axoplasm.

In his studies of the fresh uninjured nerve fibres of cephalopods Young[1] was unable to find definite neurofibrillae. On the other hand, he states that there are faint visible striations in the axon, which he suggests are due to the presence of longitudinally orientated micellae, and which, by coagulation, probably produce the more definite fibrils which can be seen quite easily in damaged neurons. Studies of nerve fibres in the squid by the electron microscope also indicate that the fibrils are made up of oblong particles of axoplasm arranged in a linear aggregation.[2] Of very considerable interest in this connexion is the observation of Weiss and Wang[3] that neurofibrillae exist as discrete continuous filaments in living nerve cells which have been cultivated *in vitro*. Here the criticism that they are coagulation products can hardly be raised, though it is of course open to argument whether living cells in tissue cultures reproduce accurately the appearance of living cells in the body, or whether the fibrillae seen in living cells are really identical with those which are displayed by silver impregnation. It is to be noted that other observers have been unable to demonstrate visible fibrillae in unfixed nerve cells; for example, Hughes found no evidence of them in fresh cells examined with the phase-contrast microscope, though they became evident in the same cells after fixation.[4]

The suggestion was at one time made that neurofibrillae are the ultimate conducting units of the neuron, and, in view of this theory, it was of some importance to establish their reality as structurally continuous filaments. It is now well known, however, that the passage of a nerve impulse along a nerve fibre is a surface phenomenon, involving the propagation of a wave of depolarization along the surface of the axon. Hence the controversy on neurofibrillae within the axon assumes less significance from the functional point of view. If, as seems probable, neurofibrillae are nothing more than an expression of the longitudinal orientation of micellae in an otherwise homogeneous substance, their definition may be expected to vary in relation to any factor which alters temporarily or permanently the physical consistency of the axoplasm. This no doubt accounts for the varied interpretations

[1] J. Z. Young, 'The structure of nerve fibres in cephalopods and crustacea', *Proc. Roy. Soc.* **121**, 1936.
[2] A. G. Richards, H. B. Steinbach, and T. F. Anderson, 'Electron microscope studies of squid giant nerve axoplasm', *Journ. Cell. Comp. Phys.* **21**, 1943.
[3] P. Weiss and H. Wang, 'Neurofibrils in living ganglion cells of the chick, cultivated *in vitro*', *Anat. Rec.* **67**, 1936.
[4] A. F. W. Hughes, 'The effect of fixation on neurons of the chick', *Journ. Anat.* **88**, 1954.

which have been placed on their appearance by different anatomists using different techniques.

Even if the existence of neurofibrillae as morphological entities in the living cell is open to question, it remains true that their appearance in mature nerve cells which have been treated by silver impregnation is a characteristic and normal feature. Subsequent references in this chapter to neurofibrillae or neurofibrillar differentiation are to be accepted with this implication in mind.

Pigment. Granules of yellow pigment composed of a lipochrome are commonly found in the larger nerve cells, often forming a circumscribed clump in close relation with the nucleus. The fact that this pigment increases with advancing age is taken to indicate that it is a by-product of cellular activity rather than an element of nutritional importance.

In certain cells in the mid-brain (particularly a layer of nerve-cells called the substantia nigra) a melanin pigment is present. Melanin is also found scattered in some cells of spinal and sympathetic ganglia. The significance of this pigment is not known. It first appears a few years after birth and reaches its maximal development at puberty.

Other constituents of nerve cells. Among the other contents of nerve cells may be mentioned mitochondria, lipid inclusions, the Golgi apparatus, and the occasional occurrence of colloid globules. The mitochondria are similar to those of the cells of other tissues. Their disposition in the cytoplasm corresponds closely with that of the Nissl bodies, and they maintain this arrangement even when the Nissl bodies disappear in chromatolysis. In inflammatory and toxic conditions they undergo rapid changes and may even disappear altogether. It may be assumed that they play a part in synthetic processes in the nerve cell such as those involved in the formation of pigment and the final elaboration of Nissl substance. The Golgi apparatus of nerve cells forms an intricate network in the cytoplasm, usually in close relation to the nucleus (see Fig. 4). Like the Nissl bodies it becomes fragmented and dispersed when the axon is cut, and its appearance may therefore be taken to indicate whether the nerve cell is in a normal and healthy condition.[1]

The presence of inclusions of a colloid material in certain nerve cells has assumed some importance recently, since it forms the basis of the conception that these cells, besides their purely neural functions, are capable of an endocrine secretory activity. This process has been termed *neurocrinie*, and has been described as occurring in certain groups of cells in the hypothalamus (a region at the base of the brain where the pituitary gland is attached). Indeed, the colloid content of these cells is so conspicuous in certain vertebrates that the term 'Zwischenhirndrüse' (diencephalic gland) has been used to describe them. Beyond the observation that this accumulation of colloid is formed independently of the pituitary gland, and shows seasonal variations in lower vertebrates, there still remains considerable uncertainty regarding its real significance.[2] However, there is a good deal of

---

[1] W. G. Penfield, 'Alterations of the Golgi apparatus in nerve cells', *Brain*, **43**, 1920.
[2] R. Gaupp and E. Scharrer, 'Die Zwischenhirnsekretion bei Mensch und Tier', *Zeitschr. f. d. ges. Neur. u. Psych.* **153**, 1936; E. Scharrer, 'Vergleichende Untersuchungen

evidence that the colloid secretion is conveyed down the pituitary stalk along connecting tracts of nerve fibres to the posterior lobe of the pituitary gland and that it provides at least some of the hormonal products formerly presumed to be elaborated entirely by the posterior lobe itself (see p. 276).

As we shall see, the integrity of the cell body of a neuron is essential for the maintenance of the vitality of its processes. It does not, however, play an essential part in the transmission of a nervous impulse from one neuron to another. This fact, which was first demonstrated by Bethe in 1897, has been confirmed by Young by an ingenious experiment on a nerve ganglion in the squid.[1] He was able to sever the cell bodies from their processes without interrupting the synaptic junction between these processes and other neurons. Stimulation of the latter showed that synaptic transmission could still take place, and it was inferred that 'synaptic excitation of one nerve fibre by another does not depend on the presence of the nucleus, Golgi bodies, Nissl substance, or any other substances present in the nerve cell body'.

2. NERVE FIBRES

The essential element of a nerve fibre is the axonal process of a nerve cell. As we have noted, it consists of a substance of semi-fluid consistency, the axoplasm, with a fibrillary structure which may be considerably enhanced under the influence of fixatives. There is some reason to suppose that the axoplasm is normally maintained under a slight positive pressure which is controlled from the cell body of the neuron. If a fresh nerve fibre is cut, the axoplasm actually flows out from the stump, and the diameter of the proximal part of the axon rapidly shrinks following the release of pressure. It has also been observed that if a living nerve is constricted locally, the axons above the level of the constriction become slightly swollen as though the axoplasm is dammed back. In a peripheral nerve the fibres are each enclosed in an extremely delicate sheath, the *neurilemma* or *sheath of Schwann*, beneath which is a chain of elongated cells, the *Schwann cells*, having a close relationship with the axon. Many of the axons are also immediately surrounded by a sheath of fatty tissue, the *medullary sheath* (Fig. 115). Finally, outside these coverings each nerve fibre is encased by a delicate sheath of fine connective tissue, the *endoneurial tube*.[2]

Neurilemma. This fine, membranous sheath is probably derived from the connective tissue in which the individual nerve fibres are embedded. It seems to serve the purpose of holding in place the underlying Schwann cells and the medullary sheath.

über die zentralen Anteile des vegetativen Systems', *Zeitschr. f. Anat. u. Entwickl.* **106**, 1936; R. Roussy and M. Mosinger, 'La neurocrinie hypophysaire et les processes neurocrinies en général', *Ann. d'Anat. path. et d'Anat. normale*, March 1937.
    [1] J. Z. Young, 'Synaptic transmission in the absence of nerve cell bodies', *Journ. Physiol.* **93**, 1938.
    [2] There has been considerable controversy regarding the definition and nomenclature of the sheaths of peripheral nerves. For a discussion of this matter, and for a review of the changes which occur during the degeneration and regeneration of nerve fibres, see J. Z. Young, 'The functional repair of nervous tissue', *Physiol. Rev.* **22**, 1942.

Medullary or myelin sheath. This is a complex substance composed of a lipoid material called *myelin*, with a protein admixture. It is relatively thick in somatic motor fibres and in most somatic sensory fibres (Fig. 118 A). In certain sensory and autonomic fibres it is much finer, while some peripheral nerve fibres (particularly the post-ganglionic fibres of the sympathetic system) are apparently unmyelinated. Even in 'unmyelinated fibres' the axons are also covered by a surface membrane of lipoid substance, but it is too fine and its actual lipoid content too meagre to be detected by ordinary histological methods. While it thus appears that the contrast usually made between myelinated and unmyelinated fibres is to some extent

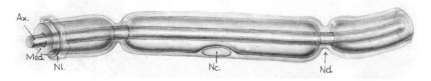

FIG. 115. Diagram illustrating schematically the essential structure of a medullated nerve fibre. The axis cylinder or axon (*Ax.*) is enclosed in a medullary sheath (*Med.*), and this again is surrounded by a fine sheath of neurilemma (*Nl.*). The nucleus (*Nc.*) of a Schwann cell is seen beneath the neurilemma; its cytoplasm is not depicted. The medullary sheath is interrupted at intervals by the nodes of Ranvier (*Nd.*). It should be emphasized that the distance between nodes is relatively greater than this schematic diagram indicates. For a somewhat more realistic representation of the appearance and proportions of medullated fibres, see Fig. 117.

arbitrary, it is nevertheless a matter of convenience to retain these terms for ordinary histological description. The thickness of the myelin sheath bears a relation to the diameter of the axon, for the coarser the latter the thicker the sheath. But this relationship is not constant, and in different vertebrates the thickness of the axon and the medullary sheath may vary independently. The structure and composition of the medullary sheath also show variations, particularly in the proportions of the lipoid and protein components, and the latter have been demonstrated by the polarizing microscope to be arranged in concentric layers of oriented lipid molecules alternating with layers of oriented protein molecules. The origin of this curiously patterned formation depends on the relation of the axon to the Schwann cells (see p. 345). The medullary sheath is not continuous, for it is commonly interrupted at fairly regular intervals of about half a millimetre, the distance between the intervals varying directly as the thickness of the sheath. These interruptions are called the *nodes of Ranvier*, and here the nerve fibre shows constrictions, the neurilemma dipping in to come into contact with the axon (Fig. 115).

Within the central nervous system a large proportion of nerve fibres are also myelinated, and though nodes of Ranvier are also present here, they do not show the typical character of nodes in peripheral nerves. Fibres of the central nervous system usually run in well-defined tracts, and they make up the 'white matter' of the brain and spinal cord. The 'grey matter', on the other hand, is predominantly composed of nerve cells and non-myelinated fibres. At their termination where they are about to break up into motor or sensory endings or into synaptic arborizations, all medullated fibres lose their visible myelin sheath.

The significance of the myelin sheath is related to its effect on the transmission of impulses along the axon. The essential basis of this transmission is the propagation of a wave of increased permeability along the plasma membrane at the surface of the axon. This, again, is accompanied by a transfer of ions and a transitory change in the local electrical potential. In medullated nerves the axon membrane is freely exposed only at the nodes of Ranvier; between the nodes it is ensheathed by the relatively thick insulating layer of myelin substance. Because of the low electrical capacity of the myelin sheath, the speed of the impulse is accelerated over the internodal segments, and the transmission of the impulse thus becomes a matter of a rapid 'saltatory conduction', sweeping along from node to node.[1] It will be understood, then, that the speed of transmission of a nervous impulse is directly related to the internodal distance, and since the latter is related to the thickness of the myelin sheath it follows that the rate at which the impulse travels is greater in the more coarsely medullated fibres (and, of course, much slower in non-medullated fibres). By restricting the electrical activity to the nodal regions of the axon, the myelin sheath also provides for an economy of energy expenditure. Lastly, by its insulating properties it serves to reduce the loss of electrical activity through dispersion into surrounding tissues.

Schwann cells. All peripheral nerve fibres are invested with these cells whose thinned-out cytoplasm closely envelopes the axon, or its medullary sheath if present. The flattened, elongated nuclei of the cells can be seen at fairly regular intervals along the course of each fibre, one being found in each internodal segment of medullated fibres. In the past there has been much argumentation on the precise relationship between the Schwann cells and the axon, and whether the cytoplasmic investment is complete or whether it forms only an interrupted or fenestrated sheath, and so forth. The examination of transverse sections of peripheral nerves with the electron microscope has resolved some of these arguments in a somewhat unexpected way. From the studies of Geren[2] and Robertson,[3] it now appears that when during development an axon comes in contact with a Schwann cell, it indents the surface membrane of the latter, and subsequently becomes infolded into the cell (Fig. 116 A and B). This inclusion leads to the formation of an invaginated reduplication of the cell membrane to form a sort of 'mesentery' called the *mesaxon*. Such a disposition is characteristic of non-medullated nerve fibres, and several fibres may become infolded into one Schwann cell.[4] In the case of medullated fibres, the

---

[1] R. Stämpfli, 'Saltatory conduction in nerve', *Physiol. Rev.* **34**, 1954; B. Katz, 'The properties of the nerve membrane and its relation to the conduction of impulses', *Structural Aspects of Cell Physiology*, Cambridge, 1952.

[2] B. B. Geren, 'The formation from the Schwann cell surface of myelin in the peripheral nerves of chick embryos', *Journ. Exp. Cell Res.* **7**, 1954.

[3] J. D. Robertson, 'The ultrastructure of adult vertebrate peripheral myelinated fibres in relation to myelinogenesis', *Journ. Biophys. Biochem. Cytol.* **1**, 1955.

[4] In the case of the exceedingly fine olfactory nerve fibres, several hundreds may be invested by a single Schwann cell.

mesaxon becomes much elongated and spirally wound about the axon. In this manner a closely packed helicoid formation of double membranes is produced, and it is this formation which provides the essential basis of the

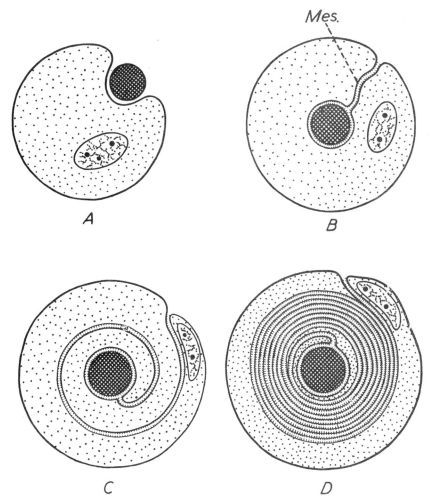

FIG. 116. Schematic diagram (based on the electron microscope studies of B. B. Geren and J. D. Robertson) illustrating the relation of the axon of a nerve fibre to a Schwann cell, and the construction of the myelin sheath. The axon sinks into the Schwann cell (*A* and *B*), and is connected with the surface cell membrane by a double membrane called the mesaxon (*Mes.*). In the formation of a myelin sheath, the mesaxon becomes tightly wound in a helicoid formation (*C* and *D*).

myelin sheath with its alternating layers of lipid and protein material (Fig. 116, c and d). It is thus apparent that the Schwann cell cytoplasm not only forms a surface investment of the myelin sheath, it coils its way through the latter finally to come into immediate relationship with the plasma membrane of the axon. At the nodes of Ranvier the myelin spiral becomes unwound in progressive layers, leaving a gap of exposed axon at each node of about 0·5 μ. The mechanism whereby the spiral formation develops in growing and maturing nerve fibres remains entirely obscure.

While Schwann cells are constantly found along peripheral nerve fibres,[1] they are entirely absent within the central nervous system. This has been supposed to be of considerable practical significance because of the important role which they appear to assume in the regeneration of nerve fibres in peripheral somatic nerves. Since this role was assumed to be essential for the regenerative process, the absence of Schwann cells was taken to explain why regeneration of nerve fibres does not normally occur in the brain and spinal cord (see p. 352).

Endoneurial tubes. The fibres of a peripheral nerve are embedded in a matrix of connective tissue which provides for them a supporting framework. Each fibre is thus enclosed in its own endoneurial tube of connective tissue. As will be seen, these tubes play an important (though passive) part in the regeneration of a cut nerve, for they maintain the original pattern of the constituent nerve fibres after the actual nerve fibres have degenerated and disappeared, and provide pathways along which newly growing fibres may be guided to their appropriate destinations.

If a nerve fibre is traced to a peripheral ganglion such as a spinal ganglion, it is interesting to note that its sheaths are continued round the cell body of the neuron. Thus each spinal ganglion cell (in common with other types of cell) is covered with a lipoid surface film of molecular dimensions which evidently corresponds to the medullary sheath. Outside this is an incomplete layer of cells (sometimes called 'satellite cells') identical with the Schwann cells of a peripheral nerve fibre. Lastly, each ganglion cell is enclosed in a delicate capsule of connective tissue which is continuous with the endoneurial tube of the peripheral fibre.

Degeneration and regeneration. If a nerve fibre is interrupted by injury or disease, the distal part which is thereby cut off from the cell body of the neuron rapidly degenerates. Within a few hours the axon becomes irregular and varicose in appearance. It then rapidly undergoes fragmentation and is ultimately absorbed altogether. The myelin sheath also disintegrates; this process begins during the first few days following the lesion, and in two or three weeks the sheath has everywhere broken down to form fatty droplets scattered along the course of the degenerating fibre. The removal of the débris of the axon and myelin sheath is effected mainly by the activity of macrophage cells which are evidently derived from the surrounding connective tissue (or, according to some authorities,[2] by a transformation of some of the Schwann cells). At the height of the degenerative process the endoneurial tubes of the distal part of a cut peripheral nerve are seen to be filled with great numbers of these macrophages performing their scavenging functions. Meanwhile, the Schwann cells show

---

[1] Apparent exceptions are found in the optic nerve and in the proximal few millimetres of the auditory nerve. But the optic nerve is really a brain tract which has become 'drawn out' from the cerebrum, while the proximal part of the auditory nerve is likewise an extruded part of the brain.

[2] P. Weiss and H. Wang, 'Transformation of adult Schwann cells into macrophages', *Proc. Soc. Exp. Biol. Med.* **58**, 1945.

a remarkable proliferative activity, forming masses of slender protoplasmic strands which extend in parallel formation along the course of the degenerated portion of the nerve. When the absorption of the remains of the axon and myelin is approaching completion and the macrophages begin to disappear, the closely packed, longitudinal columns of Schwann cells make a very conspicuous histological picture (Fig. 119). They are sometimes termed Schwann bands, or the 'bands of Büngner', and if regeneration is delayed they may persist unchanged for a long time occupying the otherwise empty endoneurial tubes. In the proximal stump of a cut nerve a similar process of degeneration may extend up for a short distance (half an inch or so), presumably as the local result of the damage to the nerve.

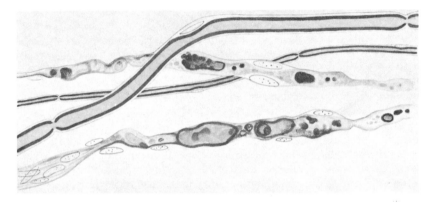

FIG. 117. Drawing illustrating the appearance in osmic acid stained material of two normal medullated fibres (coarse and fine), and two fibres undergoing degeneration ten days after separation from their cells of origin. The myelin in the degenerating fibres has become disintegrated into irregular droplets, and the nuclei of the Schwann cells show commencing proliferation.

In the central nervous system, degeneration of nerve fibres cut off from their cell of origin follows a much slower course, and the débris of the disintegrated myelin sheaths may persist in considerable quantity for a year or more. In the absence of Schwann cells, also, the formation of 'bands of Büngner' does not occur; instead, there is a proliferation of neuroglial cells (gliosis) along the track of the degenerating fibres.

In the first stages of regeneration of a cut peripheral nerve, the axons in the proximal stump send out irregular swollen processes which may branch in quite a complicated manner. These sprouts give a superficial impression in histological preparations of 'exploring' their way. Some go astray into the surrounding tissues where they may eventually become absorbed; others penetrate the scar tissue at the site of the injury and so find their way into the degenerated peripheral stump of the cut nerve. The successful fibres rapidly extend down in parallel bundles along the course of the degenerated nerve, and many of them eventually complete the formation of motor and sensory endings, with a corresponding restoration of function. Under favourable conditions, the regenerating axons appear in the peripheral stump after a latent period of about a week, and then proceed to grow down within the original endoneurial tubes at the rate (in man) of about 3–4 mm. a day, but the rate tends to fall off progressively as the process of regeneration approaches completion.[1] Under special circumstances, regenerating fibres may grow more rapidly. Wislocki and Singer[2] have shown that the antlers in the Virginia deer are richly innervated by fibres of the trigeminal nerve, which supply sensory endings of touch and pain. But the antlers are shed and replaced annually. This means that every year the nerves which supply them are severed at their base and must undergo a process of regeneration to keep pace with the newly growing antler. The antlers grow very rapidly, and it is estimated that in the larger deer, such as the elk and caribou, the rate of growth of the nerve fibres may reach 20 mm. a day—a rate far exceeding the average rate which has been recorded in the regenerating nerves of other mammals.

In the regeneration of fibres of peripheral somatic nerves, the strands of proliferated Schwann cells serve more than one function. Not only do they fill the old endoneurial tubes of the degenerated nerve and thus preserve their patency until the regenerating nerve fibres reach them again, they also provide smooth surfaces to which the growing axons cling and along which they glide as they push their way towards the periphery. In other words, the Schwann bands appear to provide the basis for a kind of stereotropism by forming a mechanically easy pathway which guides the axons during the process of their regeneration. The Schwann cells facilitate rapid regeneration in yet another way, for they migrate out from both stumps of a cut nerve, forming outgrowths which may unite to form a connecting bridge and thus serve to conduct the growing axons from the central stump to the peripheral part of the nerve with the minimum delay. The suggestion has been made (and was indeed generally accepted for many years) that the Schwann cells elaborate a specific neurotropic substance which, by a chemotactic process, attracts growing nerve fibres from a distance. This hypothesis was strongly supported by experiments on rabbits carried out in the Biological Institute of Madrid in which portions of degenerating nerve, or pieces of pith soaked in extract of Schwann tissue, were grafted in the brain. In these experiments newly growing nerve fibres were reported to sprout out in an astonishing way towards the grafts, although under normal conditions axonal regeneration has not been found to occur to any significant extent in the mammalian brain. However, similar experiments carried out with careful controls suggested that the earlier conclusions were based on a misinterpretation of the histological picture.[3] It seems probable, in fact, that the intrinsic neurons of the central nervous system in mature mammals are only capable (at the most) of an extremely limited degree of regeneration, even when they are presented with the facilities available to regenerating peripheral nerve fibres (see p. 353). One of the most convincing

[1] H. J. Seddon, P. B. Medawar, and H. Smith, 'Rate of regeneration of peripheral nerves in man', *Journ. Physiol.* **102**, 1943.

[2] G. B. Wislocki and M. Singer, 'The occurrence and function of nerves in the growing antlers of deer', *Journ. Comp. Neur.* **85**, 1946.

[3] W. E. Le Gros Clark, 'The problem of neuronal regeneration in the central nervous system', *Journ. Anat.* **77**, 1942–3.

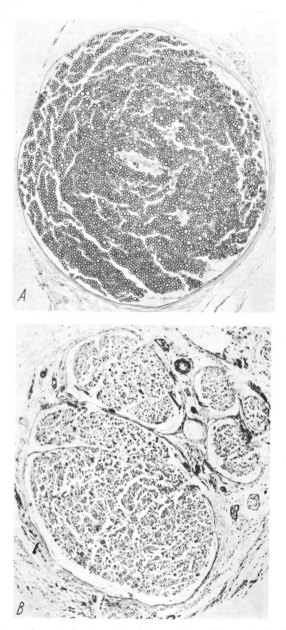

FIG. 118. Sections through the hypoglossal nerve of a rabbit, stained with osmic acid. *A*. Normal. Note the difference in the calibre of individual nerve fibres. *B*. Two weeks after section of the nerve, showing that the medullated fibres have undergone total degeneration. Note that the whole nerve is enclosed in a sheath of connective tissue, the epineurium. $A = \times 100$; $B = \times 80$.

demonstrations that the Schwann cells do not exert a chemical attraction on growing nerve fibres has been given by Weiss and Taylor. In an ingenious experiment, these investigators introduced the central stump of a cut nerve into an artery which had been excised at the point where it divides into two equal branches. The regenerating nerve fibres thus had the choice, when they reached the bifurcation, of travelling down one or other of the branches. Into one of the branches was placed a portion of proliferated Schwann tissue obtained from another degenerating nerve. In spite of this 'bait', however, the number of growing axons which penetrated the two branches was approximately equal.[1]

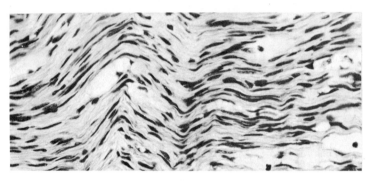

FIG. 119. Longitudinal section of the peripheral portion of a nerve which had been cut three weeks previously. The nerve fibres have undergone complete degeneration and have disappeared. The endoneurial sheaths which they occupied are now filled with columns of proliferated Schwann cells. The latter form long, attenuated protoplasmic strands arranged in closely packed bundles, with long, narrow nuclei. They provide the pathway along which new fibres grow during the course of regeneration. × 330.

Although the proliferation of Schwann cells appears to facilitate the regeneration of somatic nerves, it is interesting to note that in the case of non-myelinated fibres of autonomic nerves degeneration may not be accompanied by such a significant increase in these cellular elements.[2] It is evident, therefore, that their proliferation is not an *essential* preliminary to the outgrowth of new axons, even though they appear to facilitate this process when it does occur in somatic nerves. The Schwann cell proliferation in the latter may well have some relation to the disintegration and final absorption of the myelin sheaths.

Regenerating nerve fibres following interruption of a peripheral nerve have many hazards to overcome before they can establish their original terminal connexions. Indeed, considering these hazards, it is perhaps remarkable that the restoration of function in nerve injuries is as good as it often is. If dense scar tissue is formed at the site of the injury (as commonly occurs in an infected and septic wound), the outgrowing axons from the central stump may be unable to penetrate it in order to reach the peripheral

[1] P. Weiss and A. C. Taylor, 'Further experimental evidence against "neurotropism" in nerve regeneration', *Journ. Exp. Zool.* **95**, 1944.
[2] J. Joseph, 'Absence of cell multiplication during degeneration of non-myelinated nerves', *Journ. Anat.* **81**, 1947.

stump. In such a case the surgeon will find it necessary to excise the
scar, freshen the cut ends of the nerve, and approximate them by suture.
In injuries which involve the destruction of a considerable part of a nerve,
the resulting gap between the central and peripheral stumps may be too
great to be crossed successfully by the regenerating nerve fibres. This process can be facilitated, however, by interposing a graft of some kind to
bridge the gap and thus provide a guide to the growing fibres. Experimental evidence has shown beyond doubt that the most favourable type
of graft is a strip of nerve taken from some other part of the body (e.g. a
cutaneous nerve whose loss would be of little importance). Such an autograft, when placed in position, undergoes the usual degeneration seen
in a peripheral nerve, and its proliferated Schwann tissue is then able to
conduct the regenerating axons into the peripheral stump. It will be
understood that, even under the most favourable conditions, many of the
outgrowing axons from the central stump of a cut nerve are likely to 'miss
their way' and fail to reach the peripheral stump. Compensation for this
kind of wastage is provided by the fact that each axon in the central stump
sends out large numbers of separate sprouts. This prodigality usually ensures the re-innervation of the peripheral stump and, indeed, the latter
may receive many more fibres than it previously contained. Thus, a single
endoneurial tube (which normally contains a single axon) may be found
to contain a bunch of as many as fifty growing fibres. It has commonly been
stated that when one of these fibres reaches its destination and effects a
functional contact with an end organ, the remainder undergo absorption
and disappear. However, while this occurs to some extent, it appears that
many redundant fibres which effect no peripheral connexions may survive
almost indefinitely.[1] Even if the peripheral stump is adequately re-
innervated, the functional end result of regeneration is still governed by a
number of other factors. In the first place, fibres may fail to reach their
appropriate destinations. The endoneurial tubes appear to play a most
important part in guiding the growing nerve fibres to their ultimate objective,
but in a mixed nerve motor fibres may find themselves in endoneurial tubes
which originally contained sensory fibres and so are conducted to skin
instead of to muscle; as a result they never establish functional end connexions. Sensory fibres may also be led astray in the same way and be
conveyed to destinations not properly their own. Such a 'mix-up' is
inevitable to a greater or lesser degree in the process of regeneration which
follows complete interruption of a nerve, and functional recovery is correspondingly imperfect. If a nerve is simply crushed, however, so that while
the axons are effectively destroyed at the site of injury the pattern of the
endoneurial tubes is maintained in continuity, regeneration occurs much
more rapidly and functional recovery is likely to be much more perfect. If
regeneration is unduly delayed, fibres may not be successful in establishing
their ultimate functional connexions even if they do reach their appropriate
structures, owing to the excessive deposition of fibrous tissue (fibrosis)
which is apt to occur in long denervated structures. In a paralysed muscle,

[1] D. H. L. Evans and J. G. Murray, 'A study of regeneration in a motor nerve with a
unimodal fibre diameter distribution', *Anat. Rec.* **126**, 1956.

for example, the fibrous connective tissue in which the muscle fibres are embedded progressively increases in quantity and density. The regenerating motor nerve fibres then tend to become lost in a dense jungle of collagenous fibres, and many of them (after wandering hither and thither in an attempt, so to speak, to find their objective) will fail to reach the motor end plates of the muscle fibres.[1]

One of the most remarkable features in nerve regeneration is the vigour and 'drive' manifested by the regenerating nerve fibres. The sprouting axons not only stream out in great numbers, they give the appearance of pushing their way through adjacent tissues with considerable force. If they meet with obstructions, such as dense fibrous tissue, the streams may become dammed up in great pools of axoplasm, or they may form elaborate coils and become distorted by beaded swellings. The experimental evidence makes it clear that the formation of new axoplasm in regenerating fibres is the result of synthetic processes in the cell body of the neuron, and that the actual growth takes place at the base of the axon where it arises from the cell. Moreover, there is evidence that a centrifugal convection of axoplasm is a normal phenomenon in the intact neuron, serving to replace protein systems of the axon which are in a continual state of flux in relation to their normal metabolic activities.[2] The persistent attempts of regenerating nerve fibres to penetrate dense fibrous tissue (e.g. in amputation stumps) frequently lead to the formation of tender and painful scars, or to localized swellings called neuromata, and, in order to prevent these complications of nerve injuries, various attempts have been made to suppress the regenerative 'urge' of nerve fibres by the application of chemical inhibitors such as formaldehyde or gentian violet.[3] But it is by no means easy to prevent altogether the outgrowth of fibres from nerve stumps. It may be noted, further, that there seems to be no limit to the number of times a nerve can be interrupted and still continue to show vigorous regenerative powers. A peripheral nerve may be cut or crushed again and again in an experimental animal, and each interruption is followed by the outgrowth of new fibres. In other words, it appears that the fibres are capable of a more or less continuous cycle of degeneration and regeneration if this is demanded of them by artificial methods.[4]

The vigorous and almost immediate regenerative response of fibres of peripheral nerves following their interruption is in marked contrast with the apparent lack of regenerative capacity in fibres of the mammalian central nervous system. This contrast has been well demonstrated in rabbits by cutting the facial nerve and inserting the proximal stump into the substance

[1] For general reviews of work on nerve regeneration, see *Peripheral Nerve Injuries*, Med. Res. Council Special Report Series, No. 282, London, 1954; L. Guth, 'Regeneration in the mammalian peripheral nervous system', *Physiol. Rev.* **36**, 1956.

[2] J. Z. Young, 'The history and shape of a nerve fibre', *Essays on Growth and Form*, Clarendon Press, 1945; P. Weiss and H. B. Hiscoe, 'Experiments on the mechanism of nerve growth', *Journ. Exp. Zool.* **107**, 1948; P. Weiss, 'The concept of perpetual neuronal growth and proximo-distal substance convection', from *4th Internat. Neurochem. Symp.*, Pergamon Press, 1961.

[3] L. Guttmann and P. B. Medawar, 'The chemical inhibition of fibre regeneration', *Journ. Neur. Psych.* **5**, 1942.

[4] D. Duncan and W. H. Jarvis, 'Observations on repeated regeneration of the facial nerve in cats', *Journ. Comp. Neur.* **79**, 1943.

of the brain through a small trephine hole in the skull. Regenerating fibres from the stump actively push their way through the brain in many directions; they may even find their way into the ventricular cavities and stream out over the surface of the choroid plexuses. On the other hand, the immediately adjacent intrinsic fibres of the brain which have been interrupted by the operation remain passive and inert, showing no obvious attempt at regeneration. These experiments suggested strongly that (as already noted) the intrinsic neurons of the mammalian central nervous system are incapable of repair—a matter of very considerable clinical importance. However, more recent experimental studies have led to the conclusion that regeneration of fibres in the spinal cord and brain may occur to a limited extent if a local cellular reaction is induced at the site of the lesion.[1] It is not easy to assess the results of these experiments in view of the negative observations recorded by numerous investigators on clinical and experimental material. But if regeneration of nerve fibres does occur in the adult mammalian central nervous system, it has not so far been demonstrated to be adequate for a full and effective restoration of normal functions.

3. NEURAL MECHANISMS AND NEURAL PATTERNS

While the anatomical unit of the nervous system is the nerve cell and its processes, the functional unit is the reflex circuit. The elementary basis of a reflex action may be most conveniently indicated by reference to a section through a segment of the spinal cord, as shown in the accompanying diagram (Fig. 120).

A reflex arc consists of the following elements: (1) a *receptor* which responds to a stimulus of some kind, and represented, for example, by a sensory nerve ending in the skin, (2) an *afferent conductor* or sensory nerve fibre which enters the spinal cord by way of the dorsal root, and comes into synaptic relation with (3) an *efferent conductor* or motor neuron, whose cell body is situated in the ventral horn of grey matter in the spinal cord, and (4) an *effector*, represented, for instance, by a muscle fibre. In a simplified system of this kind, a relevant stimulus applied to the receptor gives rise to a nerve impulse which eventually leads to a contraction of the muscle fibre. This is spoken of as a reflex response. The response is automatic and relatively invariable, being predetermined by the nature of the stimulus and conditioned by a set pattern of neurons in the nervous system. However, simple reflex arcs do show a certain degree of functional plasticity, for it has been demonstrated by direct experimental evidence that use and disuse can lead to an increase or decrease in the efficiency of synapses, even in the simplest (monosynaptic) type of reflex.[2] Clearly, this phenomenon may have important implications in regard to habit formation and learning processes.

[1] W. F. Windle and W. W. Chambers, 'Regeneration in the spinal cord of the cat and dog', ibid. **93**, 1950; W. F. Windle, 'Regeneration of axons in the vertebrate central nervous system', *Physiol. Rev.* **36**, 1956.
[2] J. C. Eccles, *The Neurophysiological Basis of Mind*, Clarendon Press, 1953.

It will be readily understood that such a simple system may be elaborated to provide the anatomical basis of more complicated activities. Two or more afferent neurons may terminate in relation to a single motor neuron. In this case it is possible for the application of stimuli of different modalities to lead to an impulse which is conveyed to the effector by a final common path, and subliminal stimuli, which by themselves are too weak to be effective, can by reinforcing each other produce a definite response. Again, the contraction of a muscle (the effector) may initiate a second reflex action

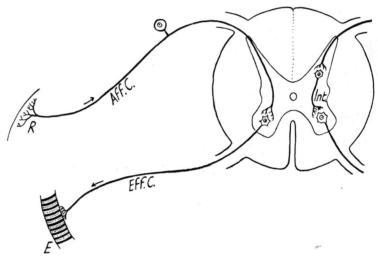

FIG. 120. Diagram showing the essential elements of a reflex arc. *R*. Receptor, represented by a free nerve ending in the skin. *Aff.C.* Afferent conductor. *Eff.C.* Efferent conductor. *E*. Effector, represented by a striped muscle fibre. *Int*. Intercalated neuron.

by the stimulation of sensory (proprioceptive) nerve endings in the muscle itself, and thus form the basis of a *chain reflex* in which one reaction provides the stimulus for another reaction in a whole sequence of motor activities. The afferent neuron may not be confined to one segment of the spinal cord, for it commonly divides into ascending and descending processes which connect directly or indirectly with motor neurons in adjacent segments, providing an anatomical basis for intersegmental responses.

A further complication of great theoretical importance is the interposition of an additional neuron in the reflex circuit between the afferent and efferent conductors. Such an element is called an *intercalated neuron*. The significance of the intercalated neuron lies in the fact that it introduces an additional element of uncertainty into the type of response evoked by a given stimulus, so that it is no longer so predictable. The nature of the response is rendered susceptible to modification since, through this additional element, the effects of other stimuli to which the organism may be simultaneously exposed are more readily brought to bear on the reflex circuit. Further, a single intercalated neuron may make synaptic contacts with a number of different motor neurons, thus leading to a wide dispersal of the impulses which it receives; or it may have the effect of focusing or concentrating a number of afferent impulses on a single motor neuron, leading to an intensification of the response. A reference to the accompanying diagram (Fig. 121) will serve to illustrate some of these points.

In the diagram, a^1, a^2, and a^3 represent afferent conductors which terminate in relation to two intercalated neurons, i^1 and i^2. The effects of a stimulus conveyed by a^2 can be transmitted to either i^1 or i^2, resulting in contractions of either muscles m^1, or m^2 and m^3. Suppose, now, that coincidentally with a subliminal stimulus arriving by way of a^2 another subliminal stimulus (derived perhaps from an impulse of quite a different

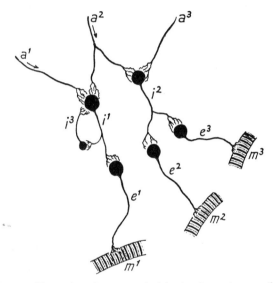

FIG. 121. Diagram illustrating the anatomical basis of certain neural mechanisms. (For explanation, see text.)

category) arrives by a^1. The latter may lead to a summation effect by raising the excitatory state of i^1 sufficiently to allow it to be 'fired off' by the stimulus from a^2. In a case such as this, the initial stimulus now leads to a contraction of m^1. This phenomenon whereby a neuron is, so to speak, sensitized by a stimulus from one source so that it can now be activated by a subliminal stimulus from another source is termed *facilitation*. Another example of this is illustrated in the diagram by the intercalated neuron, i^1. The axon of this neuron gives off a collateral which makes a synaptic contact with the cell body of i^3, and the axonal process of the latter returns to the cell body of i^1. Thus a closed circuit is formed which, once started into activity, may lead to a continuous re-excitation of i^1 by subliminal stimuli, so that it will be immediately activated by other impulses arriving by a^2. In this way it will be seen that the effects of past stimuli will partly determine the response to succeeding stimuli, and the suggestion has been made that it is possible in such a mechanism (combined with the effects of repeated activity on the functional efficiency of synapses) to envisage the rudimentary basis of habit and memory. These closed circuits, it may also be noted, by converting the effects of a single impulse into a prolonged bombardment, provide the anatomical basis for the sustained nervous capacity (as recorded electrically) which is sometimes found to follow the application of a single stimulus.

It seems probable, from physiological studies, that the phenomenon of facilitation plays an essential part in all normal nervous activities. It may be a spatial facilitation, depending on the simultaneous arrival of impulses from different sources, or temporal facilitation, depending on the arrival of successive impulses along the same nerve fibre. But immediately after excitation a neuron goes into a refractory state, that is to say, a state of subnormal excitability, and while in this condition it ceases to respond to a stimulus. This was at one time presumed to provide the essential basis of the *inhibition* of nervous activity which is a common phenomenon in the functions of the central nervous system. Inhibition, it was supposed, is the opposite of excitation, and probably results from the temporary absence of the facilitation which normally is a prerequisite for eliciting a response to a single stimulus. If this were the case, the phenomenon of inhibition, like that of facilitation, must ultimately depend on the correct timing of the impulses which arrive at a synapse. In recent years, however, evidence has accrued in favour of the conception that inhibition is a positive rather than a negative phenomenon, depending on the activity of a special category of inhibitory neurons. The latter are believed to be short-axon cells of the intercalated type, impulses from which tend to produce a hyper-polarization of the surface membrane of the nerve cell at the synapse and thus to depress its excitability.[1]

In neural mechanisms of the sort indicated (but, of course, vastly more complicated) lies the ultimate morphological basis of sensory *correlation centres* of the central nervous system, where incoming impulses are correlated and integrated in such a manner as to call forth an adaptive response. This response will be the resultant of all the afferent stimuli to which the organism may be exposed at any one moment, and will also be influenced by the effects of past experience which leaves its traces in synaptic connexions, and (perhaps) in the activity of short-circuiting intercalated neurons. In other words, correlation centres permit the interaction of numerous afferent impulses of diverse types, and so enable the organism to respond to *patterns* of excitation rather than to the effects of individual stimuli.

On the effector side it will be noted that in the diagram the intercalated neuron, i^2, establishes synaptic connexions with two efferent neurons which control the action of separate muscles, m^2 and m^3. If these muscles are antagonistic, or concerned with incompatible types of motor reaction, the impulse can only be effective along one path or the other, and the route which it follows will be determined by the effects of past reactions as well as by other impulses to the influence of which neurons e^2 and e^3 may be exposed at the time of excitation. If muscles m^2 and m^3 represent synergic elements which play a part in effecting one composite movement, the

[1] J. C. Eccles, P. Fatt, and S. Landgren, 'The inhibitory pathway to motoneurones' *Progress in Neurobiology*, Cleaver-Hume Press, 1956.

intercalated neuron i^2 may serve to bring them into simultaneous and co-ordinated contraction. Mechanisms of this type form the basis of *co-ordination centres* in the central nervous system, through which all the individual muscles which are involved in any movement are controlled and co-ordinated so as to effect an integrated and harmonious combination.

Correlation and co-ordination centres in the brain reach an extreme complexity of structure. Histologically they consist of multitudes of nerve cells, embedded in a rich matrix of interlacing fibres which form what is termed *neuropil* (Fig. 126). The latter is mainly composed of the dendritic and axonal processes of the intercalated neurons. It has been surmised that, when afferent impulses are first conveyed to centres of this kind, they will tend to be dispersed throughout the neuropil in all directions, leading to a diffuse and generalized motor reaction. That this is indeed the case seems to follow from Coghill's studies on the development of behavioural reactions in *Amblystoma*.[1] The first responses to external stimuli shown by this larval amphibian are total reactions, in which the body and limbs act as a whole. Only later do separate limb reflex movements become individualized as simple and isolated reactions independent of total movements of the body.

It may be supposed, therefore, that in the maturation of the higher functional levels of the nervous system, definite paths become 'crystallized out' from a diffuse matrix of nerve fibres so as to form the anatomical basis of individuated motor responses. This is no doubt the physical basis of education, expressed in the simplest terms, for the nature of the functional pattern which is finally established in a correlation centre is in part determined by the effects of individual experience. The passage of nerve impulses in certain directions presumably leads to a process of continued facilitation at certain synaptic junctions by short-axon intercalated neurons, and this tends to encourage the passage of impulses along the same path on subsequent occasions when a similar stimulus is applied. This being so, it will be easily understood how the frequent repetition of a reaction will establish a set response to a given situation in the form of a motor habit.

The brief and schematic account of the development of functional patterns in the nervous system which has been given above suggests that elaborate behavioural reactions are not so much the result of a co-ordinated combination of activities originally dependent on simple and separate reflex mechanisms, as an expression of the primary complexity of the nervous centres involved. In other words, although it is convenient to abstract the segmental reflex arc as the elementary functional unit of the nervous system, this cannot be taken to imply that the segmental reflex arc really provides the basis for the evolutionary or individual development of more complicated types of motor response. Coghill's classical studies on *Amblystoma* are more welcome to the 'Gestalt' school of psychological thought, which regards a behavioural response as a response of the animal as a whole to a situation as a whole, than to the 'reflexologist' who would analyse a response into an assemblage of primarily independent reflex

[1] G. E. Coghill, *Anatomy and the Problem of Behaviour*, Oxford Univ. Press, 1929.

units secondarily harmonized by some integrating mechanism at a higher functional level of the nervous system.

Reference should be made to the fact that some local reactions of the peripheral nervous system can occur as the result of an impulse whose passage is limited to the terminal branching of a sensory axon. Such a reaction is termed an *axon reflex*, and it is illustrated in a peripheral dilator mechanism of blood vessels. Sensory fibres in subcutaneous tissues may dichotomize so that one terminal is situated in the skin, and another ends in relation to an arteriole. If the former is stimulated (e.g. by some chemical irritant) an impulse can pass up to the point of branching, and then pass down in the other terminal as an *antidromic* impulse to the vessel wall, causing vasodilatation. An axon reflex can still occur after section of the main sensory fibre proximal to the point at which it dichotomizes, but it disappears after sufficient time has elapsed to allow degeneration of the peripheral branches.[1]

4. HISTOGENESIS OF THE NERVOUS SYSTEM

The whole of the central and peripheral nervous system is derived from the embryonic skin or ectoderm. Such a derivation might perhaps be anticipated on general grounds, for if the skin is the anatomical boundary between an organism and its external environment, the nervous system simply extends this function in so far as it provides a mechanism whereby the organism receives external impressions from the outside world and in turn expresses its own individuality on the environment.

As we have noted in a previous chapter, in the earliest stages of the formation of the embryo a flat band of thickened ectoderm appears along the mid-dorsal line—the *neural plate*. This rapidly sinks into a longitudinal groove the margins of which (the *neural folds*) rise up and fuse dorsally to complete a closed *neural tube* (Fig. 122). Mesodermal tissue insinuates itself between the tube and the surface ectoderm from which it was originally derived, so that the neural tube becomes finally submerged into the body of the embryo. The marginal cells of the neural plate do not become actually incorporated in the wall of the tube, but remain at its dorsal surface to form a ridge projecting on either side. This is the *neural crest*, which stretches almost the whole length of the neural tube.

The only elements of the nervous system which are not derived directly from the neural tube and crest are the olfactory sensory cells, and the sensory ganglion cells of some of the cranial nerves. They arise from separate ectodermal thickenings or *placodes* which sink in from the surface in the region of the olfactory pits and the branchial grooves. These placodes may be regarded morphologically as outlying detached parts of the original neural plate.

It has been noted in an earlier chapter that the primary formation of the neural tube in the embryo is initiated by the inductive action of the 'organizer', and that the latter is localized in the primitive streak tissue which

[1] For a statement on the problem of axon reflexes, see J. Doupe, 'The mechanism of axonal vaso-dilatation', *Journ. Neurol. and Psych.* **6**, 1943.

ultimately forms the notochord and the axial mesoderm (p. 32). It may be inferred that gross developmental anomalies of the central nervous system such as spina bifida (in which a portion of the spinal cord remains exposed on the surface of the back) or anencephalus (in which most of the cerebral hemispheres are undeveloped) are the result of some inherent defect in the organizing capacity of the underlying mesoderm.

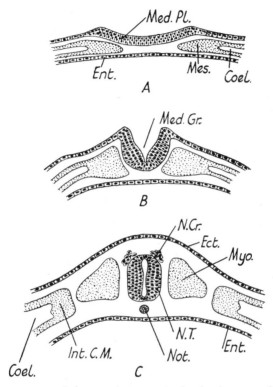

FIG. 122. Diagram representing successive stages in the development of the neural tube (*N.T.*) from the medullary plate (*Med.Pl.*) of the embryo. *Ent.* Entoderm. *Mes.* Mesoderm. *Coel.* Coelomic cavity. *Med.Gr.* Medullary groove. *N.Cr.* Neural crest. *Ect.* Ectoderm. *Myo.* Myotome. *Not.* Notochord. *Int.C.M.* Intermediate cell mass.

Differentiation of the neural tube. Even before the neural plate becomes a closed tube, its anterior end has expanded considerably in preparation for the development of the brain. We will confine our attention, however, to the spinal cord portion of the neural tube where the histogenetic differentiation of the central nervous system has its simplest expression.

The wall of the neural tube consists at first of a few layers of columnar cells. Their rapid proliferation soon gives rise to many layers, and the wall becomes differentiated into three fairly distinct zones (Fig. 123). The innermost of these is the *ependymal zone*, which surrounds the lumen of the tube, and the cells composing it are considerably elongated in a peripheral direction. Some of them remain to form the ependymal epithelium which, in the adult, everywhere lines the central canal of the spinal cord and the ventricular cavities of the brain. Others form *germinal cells* which undergo mitotic proliferation and migrate outwards into the *mantle zone*. This zone is richly cellular. Its constituent elements are all derived ultimately from the germinal cells of the ependymal zone and become differentiated in two directions—to form *neuroblasts* or embryonic nerve cells and *spongioblasts* or embryonic neuroglial cells.

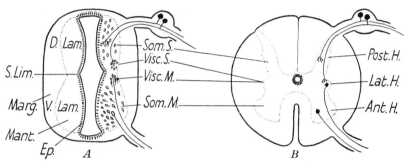

Fig. 123. Diagram showing the essential organization of the neural tube and spinal cord as seen in transverse section. In the neural tube of the embryo (*A*), the lateral wall is divided into a dorsal lamina (*D.Lam.*) and a ventral lamina (*V.Lam.*) by the sulcus limitans (*S.Lim.*). It also shows three zones, the ependymal zone (*Ep.*), the mantle zone (*Mant.*), and the marginal zone (*Marg.*). In the spinal cord (*B*), the mantle zone forms the grey matter which consists of a posterior horn (*Post.H.*), anterior horn (*Ant.H.*), and a lateral horn (*Lat.H.*). The relative position of motor and sensory nerve cells is indicated. *Som.S.* Somatic sensory neurons. *Visc.S.* Visceral sensory neurons. *Visc.M.* Visceral motor neurons. *Som.M.* Somatic motor neurons.

The *neuroblasts* are characterized by their early assumption of a piriform shape, due to the incipient outgrowth of an axonal process. The spongioblasts, on the other hand, develop a series of branching protoplasmic processes which interlace freely to form a closely meshed network. The outermost or *marginal zone* of the neural tube is at first mainly composed of such a network. It provides a scaffolding for the support of tracts of fibres which later grow up and down the spinal cord linking up one part with another, and connecting its whole length with the brain. These tracts do not begin to appear until the second month of foetal life. At this stage, short intersegmental fibres which form the basis of primitive spinal cord reflexes are laid down. In the third month longer tracts connecting the cord with the lower functional levels of the brain are completed. Finally, in the fifth month there appear the important motor fibres of the pyramidal tract through which spinal cord activities are eventually brought under the control of the cerebral cortex. It is interesting to note that, in many instances, the order in which these neural paths become established in the embryo repeats the sequence of their evolutionary history.

The mantle and marginal zones of the developing neural tube correspond to the grey and white matter of the adult cord—the former containing motor and sensory nerve cells, and the latter the fibre tracts. The differentiation of the grey matter requires brief attention since it shows the fundamental plan on which the motor and sensory divisions of the central nervous system are arranged.

In its early developmental stages the neural tube (as seen in transverse section) has relatively thick lateral walls which are connected together dorsally and ventrally by thin *roof* and *floor plates*. Each lateral wall then becomes topographically marked off into dorsal and ventral laminae by the development of a groove, the *sulcus limitans*, on its inner surface.

In the dorsal (or alar) lamina are differentiated the sensory neurons of the spinal cord, and into it grow the fibres of the sensory roots of the spinal nerves, which arise from the posterior root ganglion cells. The alar lamina becomes the posterior horn of the grey matter of the adult spinal cord. The ventral (or basal) lamina contains the motor neurons whose axons grow out and emerge from the neural tube to form the motor or ventral roots of the spinal nerves. Along the line of the sulcus limitans, that is to say at the junctional region of the dorsal and ventral laminae, are developed the neurons of the autonomic nervous system (which innervate visceral mechanisms).

Primarily, therefore, the arrangement of nerve cells in the lateral wall of the neural tube from dorsal to ventral aspect is as follows: somatic sensory, visceral sensory, visceral motor, and somatic motor (Fig. 123). This simple arrangement persists in the spinal cord, and to some extent also in the brain stem. In the brain generally, however, it becomes considerably modified and obscured by a series of developmental complications associated with distortions of the neural tube in this region, and with subsequent migrations of some groups of nerve cells from one lamina to another.

The fate of the neural crest. We have noted the formation of the neural crest from those marginal ectodermal cells of the neural plate which fail to become actually incorporated in the wall of the neural tube. In early stages of development, the cells migrate ventrally from their position on the dorsolateral aspect of the neural tube, but in normal preparations their subsequent history is not easy to follow, for they become mingled with mesenchyme cells from which they may be difficult to distinguish. Most of them, however, provide the basis for the development of the posterior root spinal ganglia and the corresponding ganglia of the cranial nerves. Taking up a position alongside the neural tube, they send out a process to the periphery which forms a sensory receptor ending, and a process which enters the dorsal lamina of the neural tube. In this way they form bipolar neurons, which, as already noted (p. 335), subsequently become secondarily unipolar by the approximation and fusion of the two processes close to the cell body.

Neural crest cells are also believed to form the cells of the ganglia of the autonomic nervous system and the Schwann cells of all peripheral nerve fibres. Experimental evidence in support of this interpretation was originally brought forward by Harrison.[1] He excised the neural crest in amphibian larvae, and observed that the posterior spinal roots and the sympathetic ganglia failed to develop, while the motor fibres of the anterior

[1] R. G. Harrison, 'Neuroblast versus sheath cell in the development of peripheral nerves', *Journ. Comp. Neur.* **37**, 1924.

spinal roots grew out with no covering of Schwann cells. Similar experiments were carried out by Müller and Ingvar,[1] who found that removal of the neural crest and the dorsal half of the neural tube was followed by complete absence of the sympathetic nervous system, while, after removal of the ventral half of the neural tube, both the spinal ganglia and the sympathetic nervous system developed normally.

There has remained some doubt whether the dorsal half of the neural tube itself also participates in the formation of sympathetic ganglia and Schwann cells. This question was studied again by Raven,[2] who grafted portions of the neural plate from the embryo of one amphibian genus (*Amblystoma*) to that of another (*Triton*). Since the cells in these two animals are of a different size, their subsequent history could be followed in sections of the developing embryos more easily than in normal material. Raven concluded from his experiments that the sympathetic nervous system is derived both from the neural crest and also from the neural tube, while the Schwann cells are entirely derived from the wall of the neural tube.

Obviously it is a matter of great difficulty to reach a final decision on these questions by methods of experimental embryology, since it is hardly feasible to effect a complete operative removal of the neural crest without the possibility of damaging the dorsal half of the neural tube, and vice versa. In any case, the derivation of sympathetic ganglia from the dorsal lamina of the neural tube seems inexplicable on purely morphological grounds, for the cells of these ganglia are motor in function and might be expected, therefore, to arise rather from the ventral or motor lamina of the neural tube.[3]

Besides the part which the cells of the neural crest play in the development of cranial and spinal ganglia, the sympathetic nervous system, and Schwann cells, they are also believed to be the source of chromaffin tissue cells (which are mainly found in the medulla of the adrenal gland). These cells are really an essential element of the sympathetic nervous system, being equivalent to post-ganglionic neurons. Other derivatives of the neural crest include pigment cells and cartilaginous elements of the branchial arch system. Indeed, in recent years experimental embryological studies have made it apparent that the cells of the neural crest have far wider potentialities for development than was at one time supposed.[4]

Neuronal differentiation. The differentiation of individual nerve cells may be briefly described by reference to the motor neurons of the neural

---

[1] E. Müller and S. Ingvar, 'Über den Ursprung des Sympatheticus beim Hünchen', *Arch. f. mikrosk. Anat. u. Entwickl.* **99**, 1923.

[2] C. P. Raven, 'Experiments on the origin of the sheath cells and sympathetic neuroblasts in amphibia', *Journ. Comp. Neur.* **67**, 1937.

[3] In a paper by D. S. Jones ('The origin of the sympathetic trunks in the chick embryo', *Anat. Rec.* **70**, 1937), evidence is presented in favour of the thesis (previously held by such authorities as Froriep and Cajal) that the cells of the sympathetic ganglia migrate out from the neural tube along the ventral roots of the spinal nerves. While this conclusion might seem to be in accordance with purely morphological expectations, it has been contradicted by the more recent experimental studies of G. Nawar ('Experimental analysis of the origin of the autonomic ganglia in the chick embryo', *Amer. Journ. Anat.* **99**, 1956).

[4] S. Hörstadius, *The Neural Crest*, Oxford Univ. Press, 1950.

tube. As already mentioned, these are formed from neuroblasts which migrate peripherally from the ependymal zone and which soon assume a characteristic piriform appearance by sending out a short pointed process. The latter is the growing axon. This pushes its way through the marginal zone of the neural tube and finally emerges at the surface. Subsequently the axonal processes continue to extend out into the mesodermal tissues of the embryo until they eventually establish contact with the muscle cells which they are destined to innervate.

By what means each motor nerve fibre finds its way to its appropriate muscle fibre is not clear. The growing tip of the axon shows a protoplasmic expansion which is irregular in outline (the *growth cone*), and the study of living tissues has shown that this sends out processes capable of a kind of amoeboid movement. It has been supposed that the growing tip is 'attracted' towards its objective by chemotactic influences of some kind, but there is no conclusive evidence to support this conception.

Soon after the axonal process begins to sprout from the neuroblast, dendritic processes make their appearance at the opposite pole of the cell. Their subsequent differentiation varies in different types of neuron.

The distinctive cytological characters of nerve cells begin to appear very early. In the human embryo, neurofibrillar differentiation first occurs before the end of the fifth week of development, and neurons can then be stained selectively by the technique of silver impregnation. According to Windle and Fitzgerald,[1] in the spinal cord this differentiation first occurs in the segmental reflex systems. Peripheral motor and sensory endings are formed during the seventh week, and the anatomical basis of the first spinal reflex arcs is completed during the eighth week. It may be noted, also, that the first reflex movements of the human embryo have been recorded at or just before the end of the eighth week. It therefore appears that neurofibrillar differentiation in the embryonic nerve cells and their processes is related chronologically to the development of their ability to conduct impulses in response to peripheral stimuli.

This relationship has been demonstrated by the detailed correlation of neurofibrillar differentiation and the development of reflex patterns in lower animals, particularly in the studies of Windle and his collaborators.[2] They showed that differentiation first occurs in the neuronal mechanisms of simple reflex activities, and that its completion is associated with the first appearance of elicited motor responses. They found also that in the brain (of the rat) the motor nuclei of the cranial visceral nerves are the first structures to show the development of demonstrable neurofibrillae, and these are followed by the cells of the somatic efferent nuclei. It is interesting to note, further, that within any one functional system (at least in the spinal cord and brain stem) the process of neurofibrillar differentiation begins

---

[1] W. F. Windle and J. F. Fitzgerald, 'Development of the spinal reflex mechanism in human embryos', *Journ. Comp. Neur.* **67**, 1937.

[2] W. F. Windle, 'Neurofibrillar development of cat embryos', ibid. **63**, 1935–6; W. F. Windle and R. E. Baxter, 'The first neurofibrillar development in albino rat embryos. Development of reflex mechanisms in the spinal cord of albino rat embryos', ibid. **63**, 1935–6; W. F. Windle and M. F. Austin, 'Neurofibrillar development in the central nervous system of chick embryos', ibid. **63**, 1935–6.

on the efferent side and subsequently extends into the afferent side. This suggests that the initial differentiation of neural mechanisms depends on inherent tendencies of the developing nervous tissues rather than on the result of induction by exogenous afferent stimuli.

Myelinization. When neural paths are first developed in the early embryo, the nerve fibres are all unmyelinated. It was observed many years ago by Flechsig that myelinization of fibres in the brain follows quite a definite chronological sequence which is constant for individuals of the same species, though it may vary somewhat in different species. On the basis of this observation, he formulated the 'myelino-genetic law', which also states that fibres forming any particular functional system all become myelinated at the same time. We have now to consider what factors determine the onset of this process.

It is known that, before any fibres at all have acquired a medullary sheath, myelin substance is already present in nervous tissues—in the brain it can be detected in the cytoplasm of neuroglial elements (particularly along the course of blood vessels), while along peripheral nerves it is found in the Schwann cells and in the connective tissue cells of the perineurium.[1] It is probable, therefore, that the substance is carried to the fibres from some extraneous source, and is not elaborated *in situ* by the nerve fibres themselves. Nevertheless, its actual deposition around a peripheral nerve fibre is probably a function of the Schwann cells. By their transparent chamber technique (see p. 6), Clark and Clark[2] have observed that visible medullation in a growing nerve fibre begins at the site of a Schwann cell and proceeds in both directions from it.

Flechsig's observations suggested a relationship between the onset of myelinization and function, and although other factors have been invoked to explain the chronological order in which fibres become myelinated (such as the calibre of the fibres, and the relative development of the vascular supply), this conception has in the past been broadly accepted. That is not to say, however, that nerve fibres (which are myelinated when they become mature) cannot conduct impulses until they have acquired their medullary sheath. Indeed, there is plenty of evidence that they can do so. In the new-born rat, for example, the central nervous system is almost entirely devoid of medullated fibres, and yet the baby animal is capable of a diversity of motor reactions.[3] But the process of myelinization of nerve fibres was assumed to be associated with the completion of their functional development as fully efficient conductors. This, indeed, might seem a reasonable inference from the fact that the velocity with which an impulse travels along a nerve fibre is related to the thickness of the medullary sheath, for it is certain that nervous activities normally depend on a very precise timing of the arrival and departure of impulses in centres of the brain and spinal cord.

[1] F. Tilney and L. Casamajor, 'Myelinogeny as applied to the study of behaviourism', *Arch. Neurol. & Psych.* **12**, 1924.

[2] E. R. Clark and E. L. Clark, 'Microscopic observations on new growth and medullation of peripheral nerves in the living mammal', *Anat. Rec.* **70**, 1938 (abstract).

[3] A. W. Angulo y Gonzalez, 'Is myelinogeny an absolute index of behavioural capability?' *Journ. Comp. Neur.* **48**, 1929.

Tilney and Casamajor[1] studied this question in kittens, and arrived at the conclusion that the deposition of myelin is coincidental with the establishment of normal function in different fibre systems. Some years later, Langworthy[1] examined the same problem by comparing the development of behaviour patterns and myelinization in kittens and opossums, and also in the human foetus and infant. His results substantiated in general those of Tilney and Casamajor. In opossums, for example, he found that the appearance of the righting reflex is approximately synchronous with the medullation of vestibulo-spinal connexions, and when the cerebellar connexions become myelinated the movements of the animal become less ataxic and better co-ordinated. When the young opossum begins to leave the pouch, new behaviour patterns develop rapidly and at the same time the deposition of myelin in the appropriate tracts is greatly accelerated.

The myelinization of the human central nervous system has been studied in some detail by Lucas Keene and Hewer.[1] They note that tracts begin to acquire myelin sheaths in the fourteenth week of intra-uterine life. The motor roots of the spinal nerves become medullated before the sensory roots, and the earliest fibres in the brain stem and spinal cord to mature are those of association tracts such as the posterior longitudinal bundle and the fasciculi proprii (ground bundles). In the fifth month, when 'quickening' movements of the foetus commence, myelinization is markedly accelerated. The spino-cerebellar tracts become well medullated at this time. At birth, the fibres of the pyramidal tract (which is concerned with the initiation and control of voluntary movements) begin to acquire their myelin sheaths. These examples are undoubtedly suggestive of a relationship between myelinization and the onset of functional activity, and, indeed, Langworthy has concluded that the 'initiation of activity in a group of neurons appears to stimulate the laying down of myelin'. It is interesting to note that, in the functional systems of lower levels of the nervous system, myelinization is completed first on the efferent side. This repeats the sequence already noted in neurofibrillar differentiation, and suggests an inherent potentiality on the part of the organism to express itself actively and spontaneously before the afferent paths by which it receives information from the external environment are fully matured. For example, the oculomotor nerves by which movements of the eyes can be effected are myelinated long before birth, while the fibres of the optic nerves do not become mature till about the time of birth. On the other hand, in the higher functional levels of the brain (e.g. the cerebral cortex) afferent fibres become myelinated before the efferent fibres, perhaps an expression of the fact that these 'analytical' mechanisms of the nervous system (which are

[1] F. Tilney and L. Casamajor, op. cit.
[2] O. Langworthy, 'The behaviour of pouch-young opossums correlated with the myelinization of tracts in the nervous system', *Journ. Comp. Neur.* **46**, 1928; 'A correlated study of the development of reflex activity in foetal and young kittens, and the myelinization of tracts in the nervous system', *Contrib. to Embryology*, Carnegie Inst. **394**, 1929; 'Development of behaviour patterns and myelinization of the nervous system in the human foetus and infant', ibid. **443**, 1933.
[3] M. F. Lucas Keene and E. E. Hewer, 'Some observations on myelination in the human central nervous system', *Journ. Anat.* **66**, 1931–2.

comparatively recent in their phylogenetic development) depend entirely for their functioning on an adequate supply of sensory material.

Although the evidence presented in the preceding paragraphs appears to indicate a broad chronological relationship between myelinization and the assumption of normal function, it has proved premature to conclude (as some have been inclined to do) that the relationship is causal in the sense that the onset of function actually initiates the deposition of myelin. Indeed, the progress of myelinization seems to show considerable variation in different mammalian species, and in some cases it by no means accords with expectations based on the conception of a direct functional relationship. Mention has already been made of the relative absence of myelinated fibres in the new-born rat. In the sheep, on the other hand, the degree of myelinization is much more advanced at birth than in many other mammals. This contrast is presumably associated with the great difference in functional maturity of the new-born rat and the new-born lamb. But, according to Romanes,[1] in the sheep foetus the onset of myelinization in the brain and spinal cord is not associated with any marked change in the physiological responses which can be elicited experimentally, and he also notes that in this animal the fibres of the optic nerve acquire myelin sheaths long before birth. However, the *order* of myelinization of the various elements of the nervous system in the sheep foetus agrees in general with that of other mammals. This developmental sequence, it is interesting to note, coincides approximately with the order in which the tracts have become developed phylogenetically. In man, fibres subserving primitive segmental and intersegmental reflexes are the first to mature. This is followed by long ascending and descending tracts connecting the lower functional levels of the spinal cord with the primitive co-ordinating mechanisms of the hind-brain and mid-brain. Later, the dominance of cortical control is reflected in the myelinization of the pyramidal tracts. After birth, the process of maturation still proceeds, and only at puberty are the fibres and connexions of the so-called 'association' areas of the cerebral cortex completely medullated.[2]

5. THE NEURON THEORY

Sections of embryological material showing the early differentiation of the nervous system give the appearance of the neuroblast as a separate cell sending out processes which ultimately establish a secondary contact with other cellular elements—whether these are muscle cells, epithelial cells, or other nerve cells. It was this appearance which primarily led to the enunciation of the *neuron theory* by Waldeyer in 1891. This theory states that

---

[1] G. J. Romanes, 'The prenatal medullation of the sheep's nervous system', *Journ. Anat.* **81**, 1947.

[2] Among other factors which may also play a subsidiary part in the order of myelinization should be mentioned the influence of axial gradients. On the basis of the conception of physiological gradients (which was developed by C. M. Child and summarized in his book *The Origin and Development of the Nervous System*) it may be supposed that there are dominating centres of growth activity in the central nervous system from which processes of growth and differentiation spread out in waves to other parts. In this connexion, it may be noted that in the spinal cord myelinization in general first becomes evident in the cervical region and gradually spreads downwards to the lower end.

each nerve cell with its processes (axon and dendrites) is a separate and independent anatomical unit, establishing relations with other cells by contiguity but having no physical continuity with them.

It is a matter of historical interest that at one time the neuron theory excited considerable controversy, and even latterly an occasional histologist has questioned its validity. Many years ago, Hensen put forward the proposition that the constituent cells of the body are commonly in direct protoplasmic continuity with each other—in other words, the tissues of the body really comprise a syncytium rather than an aggregation of anatomically separate cellular units. He supposed, for example, that in the earliest stages of embryonic development the motor neuroblasts are connected with developing muscle cells or myoblasts by fine protoplasmic bridges and that, when these elements become separated as the result of differential growth, the connecting strands are drawn out into long threads which form the actual nerve fibres. A similar point of view was later maintained by Held,[1] who described protoplasmic networks (which he termed *plasmodesmata*) connecting neuroblasts with each other and providing bridges along which the neurofibrillae of the cell processes were presumed to extend in their outward growth. The existence of such plasmodesmata, however, was strenuously denied by other histologists who maintained that the fibrillated processes of nerve cells are apparently never preceded by such indifferent protoplasmic connexions.

The neuron theory seemed to have been finally established by the tissue culture experiments of Harrison.[2] The work of this investigator is of particular historical interest since it provided the foundation for the development of the tissue-culture technique which is so widely used today in the study of problems of growth and differentiation. He cultivated living neuroblasts in sealed chambers containing lymph, and was able to observe the outgrowth of axonal processes to form what appear to be true nerve fibres, in the absence of other tissues. Nevertheless, the possibility in the living body of direct protoplasmic continuity (which may be established secondarily) between nerve processes at synaptic junctions, and between nerve terminals and muscle or epithelial cells, still remained a matter for consideration. Some histologists, for example, maintained that motor nerve fibres end in a terminal reticulum which extends into continuity with the sarcoplasm of the muscle fibre, and that sensory fibres may have a similar relationship to the protoplasm of sensory epithelial cells. Here a question of technique was involved, for others expressed the opinion that such conclusions were based on the misinterpretation of artefacts, or on a staining method which is not really selective for nerve fibres. Today the evidence is overwhelmingly against the existence of these protoplasmic continuities.

The suggestion that adjacent neurons are in syncytial connexion with each other at synaptic junctions likewise has no serious histological evidence in its favour. Specialized techniques which render the nerve cell and all its processes visible by metallic impregnation (e.g. the Golgi method) failed

---

[1] H. Held, *Die Entwicklung des Nervengewebes bei den Wirbelthieren*, Leipzig, 1909.
[2] R. G. Harrison, 'Observations on the living developing nerve fibre', *Proc. Soc. Exp. Biol. & Med.* **4**, 1907.

to demonstrate that the inter-neuronal relationship at a synapse is any more intimate than mere contact. The Golgi technique, indeed, provided a very forcible argument for the neuron theory, since by this method only an occasional nerve cell here and there is picked out, and the whole neuron can be completely impregnated down to the finest ramifications of its processes so far as the latter can be made visible with the light micro-scope, while adjacent neurons with which it is in synaptic relationship remain completely unstained.

The precise anatomical details of synaptic connexions have now been finally established by studies with the electron microscope. These have clearly demonstrated that the axon terminals are everywhere separated from the cell body and dendrites of the neuron with which they are in synaptic connexion by definite interface plasma membranes. These limiting mem-branes, however, are in virtual contact, being separated by an apparent interval of no more than about 100 Å thick.[1]

Short of actual continuity, the relationship between neurons at their synaptic junction is usually of a most intimate nature. The telodendria of one neuron may form a richly branching basket-work of fibres wrapped round and closely applied to the cell body of another neuron, or they may be closely intertwined with its dendritic processes. In many cases—for example, in the motor neurons of the spinal cord—the telodendria end in minute expansions of an ovoid or annular shape called *boutons* or *end bulbs*, which lie in immediate contact with the cell body or its dendritic processes (Fig. 124). These boutons undergo degeneration when the axon is inter-rupted.

For many years it was accepted that at least in the peripheral plexuses of the autonomic nervous system (e.g. the myenteric plexus in the wall of the intestine) there exists a true nerve net, with direct protoplasmic con-tinuity between the constituent nerve cells. However, by the use of critical histological methods, even this has now been demonstrated to be not the case, and it may even be doubted whether true nerve nets exist anywhere in the animal kingdom (in spite of the fact that their existence had been widely accepted in invertebrates).[2]

Apart from direct histological observations, it has for long been recog-nized that there is a good deal of indirect evidence in favour of the neuron theory, which may be listed as follows: (1) The functional polarity of the neuron allows an impulse to pass in one direction only across the synapse, i.e. from axon to dendrites, whereas within the limits of the nerve fibre itself an impulse can be made to pass in either direction. This suggests an inter-ruption in the actual conducting elements at the synapse. (2) The passage of a nerve impulse is delayed at a synaptic junction, and the latter is particu-larly susceptible to the effects of fatigue. (3) The *immediate* degenerative changes following injury to the cell body or interruption of the axon do not

---

[1] E. De Robertis, 'Submicroscopic changes of the synapse after nerve section in the acoustic ganglion of the guineapig', *Journ. Biophys. Biochem. Cytol.* **2**, 1956.

[2] In some invertebrates, however, giant nerve fibres are found which result from the syncytial fusion of the *axons* of nerve cells. In this case, the nerve cells act as a single functioning unit (see J. Z. Young, 'Structure of nerve fibres and synapses in some inver-tebrates', *Cold Spring Harbor Symposia on Quantitative Biology*, **4**, 1936).

extend beyond the limits of the neuron—in other words, the synapse forms a definite barrier beyond which they cannot pass. (4) The synapse is susceptible to certain pharmacological agents (e.g. nicotine) which have no effect on the nerve fibre itself.[1]

Since there is no protoplasmic continuity between one neuron and another at a synapse, or between peripheral nerve fibres and sensory or muscle cells, the question arises how a nervous impulse is propagated across the structural gaps separating the apposed limiting membranes. Up till comparatively recent years the passage of a nervous impulse had generally been regarded as an excitatory wave of electrical activity which is in some manner continued without interruption across from one cell to another. The

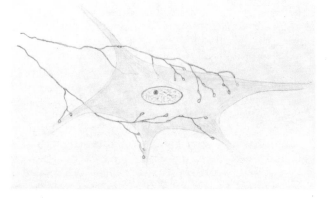

FIG. 124. Diagram showing the appearance of terminal boutons at the synaptic junction between the telodendria of one neuron and the cell body and dendrites of another. Note that some of these boutons are found on the course of the telodendria before they actually terminate. These are sometimes called 'boutons de passage'.

conception of chemical (or humoral) transmitters provided an alternative explanation.[2] It has been shown that when an impulse reaches the end of a nerve fibre, it leads to the liberation there of a chemical substance which is believed to be directly responsible for stimulating the cell which the fibre innervates. In the case of most post-ganglionic fibres of the sympathetic nervous system this substance is noradrenalin with a slight admixture of adrenalin, while in the case of all preganglionic fibres of the autonomic system, post-ganglionic parasympathetic fibres, and also somatic motor fibres, it is acetylcholine. Peripheral nerve fibres have therefore been divided into two categories, *cholinergic* and *adrenergic*, in accordance with the type of substance which they produce when stimulated.

When the hypothesis of humoral transmission was first advanced it led to considerable controversy. For example, the objection was raised that, to

[1] For a general statement on the neuron theory, see J. F. Nonidez, 'The present status of the neuron theory', *Biol. Rev.* **19**, 1944.

[2] For early statements on chemical transmission, see O. Loewi, 'Problems connected with the principle of humoral transmission of nervous impulses', *Proc. Roy. Soc.* **118**, 1935; H. Dale, 'Chemical agents transmitting nervous excitation', *Irish Journ. Med. Soc.*, No. 150, June 1938.

account for the rapid and fleeting changes of activity which follow a succession of impulses delivered at a synapse, it would be necessary to postulate an excessively high rate of diffusion and destruction of the activating chemical substances liberated at the nerve ending. It was further suggested that these substances are by-products rather than the essential factor in the propagation of a nervous impulse. These objections are no longer accepted as valid; indeed, some of those who most strongly opposed the hypothesis are now prepared to apply it to synaptic activity in the central as well as the peripheral nervous system.

6. THE DIFFERENTIATION OF NEURAL MECHANISMS

We have noted that the initial differentiation of nerve cells and their processes in the embryo is apparently coincidental with the assumption of their function as conductors. The correlation between the completion of definite neural patterns and the initiation of specific types of motor response has been worked out in some detail by Coghill on the larval amphibian, *Amblystoma*.[1]

The question arises whether the differentiation of neurons and the outgrowth of their processes are the result of developmental tendencies inherent in the neuroblasts, or whether they are actually determined by exogenous stimuli arising from the surrounding tissues. Chemotropic, galvanotropic, and stereotropic influences have been invoked as causative factors, and a great deal of experimental work has been carried out in order to test such possibilities. One of the results of this work has been to establish the fact that the *primary* differentiation of a nerve cell and its processes does in fact depend on its own morphogenetic propensities, for this can take place in tissue cultures of neuroblasts where the growing cells are completely removed from their normal environment. By means of the tissue-culture technique, Harrison[2] was able to watch day by day the axonal processes sprouting from the embryonic nerve cells. He estimated their rate of growth at between 40 μ and 50 μ an hour, and he was able to see that the protoplasmic expansion or *growth cone* at the tip of the axon has active amoeboid properties. His experiments led him to conclude that the *initial* direction in which the axon grows out is also predetermined—depending on a dynamic polarity of the neuroblast which is probably conditioned normally by its position in the neural tube.

Even in abnormal situations, axonal processes may persist in growing in a direction which under normal circumstances would lead them to their destination. If, for example, a section of the spinal cord of a larval amphibian is removed and replaced by blood-clot so that the developing nerve fibres which ordinarily extend down the cord are deprived of their normal

[1] See G. E. Coghill, *Anatomy and the Problem of Behaviour*, Camb. Univ. Press, 1929.
[2] R. G. Harrison, 'Observations on the living developing nerve fibre', *Anat. Rec.* **1**, 1907; 'The outgrowth of the nerve fibre as a mode of protoplasmic movement', *Journ. Exp. Zool.* **9**, 1910. See also Harrison's general review 'On the origin and development of the nervous system studied by methods of experimental embryology', *Proc. Roy. Soc.* B, **118**, 1935.

environment, they continue to grow straight through the clot. It has also been observed that, if embryonic sensory ganglia are transplanted to abnormal positions in the body, the axonal processes of their cells extend straight out into neighbouring tissues for a considerable distance, even though they may never be able to establish functional connexions with the nervous system.

The outgrowth of embryonic nerve fibres has been studied in the transparent tail of larval amphibians by Speidel,[1] who also noted the tendency of these fibres to grow in a straight line so long as the substratum is fairly homogeneous. When they meet with some obstacle, however, they may be deflected from their course. One gets the impression from Speidel's observations that in such a case the actively amoeboid tip of the axon 'explores' the neighbourhood of the obstacle until it finds a way round. Where the obstacle is too great, however, the axon may be retracted, or it may be absorbed and disappear altogether.

After the initial outgrowth of nerve fibres, the influence of the surrounding tissues in determining the subsequent paths by which they reach their destination is undoubtedly responsible for the establishment of the characteristic patterns of peripheral nerves. Thus, it has been found by several observers that if a limb bud is grafted into some abnormal position so that it receives its nerve supply from an unusual source, the paths followed by the nerves in the grafted limb are the same as those in a normal limb. Like blood vessels, nerves follow paths of least resistance in their growth, and hence make use of fascial planes where there is loose connective tissue.

The fact that growing nerve fibres will often, in circumventing obstacles, pursue a circuitous course to reach their objective, suggests that they are exposed to an attractive influence of some kind. The nature of this influence, however, is uncertain. As already mentioned (p. 348), the conception of chemical tropisms (first suggested in connexion with developing nerves by Cajal) seemed at one time an acceptable explanation to many anatomists. Experiments designed to test this theory, however, have had negative results.

Weiss[2] tried to control the orientation of growing nerve fibres in tissue cultures by means of chemical agents, but he was unable to find that these had any direct effect. On the other hand, the direction of growth could be controlled by modifying the ultra-structure of the substratum. Some of Speidel's observations are apposite to this question, for he noted that, in a fine fasciculus of growing fibres, one fibre may suddenly double back and grow alongside its fellows, but in an opposite direction. It is difficult to explain a phenomenon of this kind on the basis of chemotropism.

The influence of the bio-electric field is another possible factor, and many years ago it was reported by Ingvar[3] that the outgrowth of axonal processes in nerve cells developing in tissue cultures could be determined and controlled by very weak galvanic currents, the processes becoming orientated along the lines of force in the electric field. According to Weiss (who

[1] C. C. Speidel, 'Studies of living nerves', *Amer. Journ. Anat.* **52**, 1923.
[2] P. Weiss, '*In vitro* experiments on the factors determining the course of the outgrowing nerve fibre', *Journ. Exp. Zool.* **68**, 1934.
[3] S. Ingvar, 'Reactions of cells to the galvanic current in tissue cultures', *Proc. Soc. Exper. Biol. & Med.* **17**, 1920.

repeated similar experiments with negative results) the reaction of the growing fibres was due, not directly to electrical stimuli, but to alterations in the ultra-structure of the nutrient medium resulting from the tensional forces which had been set up by the galvanic current.[1] More recently, however, Marsh and Beams[2] have demonstrated that, in explants of portions of embryo chick brain exposed to direct currents, the outgrowth of nerve fibres is directed mainly towards the cathode, even though they do not obviously follow the 'lines of force' in the electric field. It thus appears that electrical potentials may play at least some part in the organization of the developing nervous system.[3]

While the nature of the influence which directs growing nerve fibres to their destination remains a matter of controversy, it is quite clear that the proliferation and differentiation of neuroblasts in the central nervous system can be modified by extraneous stimuli of some sort—whether these arise in the central nervous system itself or in peripheral structures. Some of the evidence for this may be briefly reviewed.

In a study of neurofibrillar differentiation in the embryonic brain, Bok[4] found that, as a certain tract of fibres (the posterior longitudinal bundle) extends down the length of the brain stem, the groups of neuroblasts with which it establishes contact in a caudal direction become successively differentiated in a corresponding sequence. In other words, the neuroblasts appear to be activated by the tract as soon as the latter reaches their level, giving off dendrites which approach the fibres of the tract, and an axon which grows out in the opposite direction. Bok supposed that impulses passing down the tract were directly responsible for inducing this structural differentiation, and he applied the term *stimulogenous fibrillation* to the process. These observations were hardly conclusive in themselves, however strongly they might suggest that such a process does occur. It remained possible, for instance, that the progressive cephalocaudal differentiation of neurons in the brain stem may be related to a metabolic gradient of growth activity, for it is known that in the development of the central nervous system there is a dominant centre of growth at the anterior end of the neural tube whose influence spreads gradually backwards.

The experimental work of Detwiler has thrown considerable light on these problems.[5] He found, for example, that if in *Amblystoma* the brachial region of the spinal cord (which contains the neurons innervating the forelimb) is excised, and is replaced by more posterior segments of the cord (which normally only supply the body wall), there is an increased cellular proliferation in the grafted segments which equals in extent that normally

[1] For an interesting discussion on the directional forces involved in nerve growth, see P. Weiss, 'Nerve patterns: the mechanics of nerve growth', *Growth, Third Growth Symposium*, **5**, 1941.

[2] G. Marsh and H. W. Beams, '*In vitro* control of growing chick nerve fibres by applied electric currents', *Journ. Cell. Comp. Physiol.* **27**, 1946.

[3] For a recent review of the factors initiating and controlling the development of nerve fibres, see A. Hughes, 'The development of the peripheral nerve fibre', *Biol. Rev.* **35**, 1960.

[4] S. T. Bok, 'Die Entwicklung der Hirnnerven und ihrer zentralen Bahnen', *Folia Neurobiol.* **9**, 1915.

[5] It is only possible to refer here to a few of Detwiler's brilliant experimental studies. See his general review of this work, 'Experimental studies of morphogenesis in the nervous system', *Quart. Rev. of Biology*, **1**, 1926, and his book *Neuroembryology*, New York, 1936.

occurring in the excised brachial region. That this proliferation is not in-duced by influences arising in the fore-limb was shown by the fact that it occurred when the limb had been previously removed. It appeared, there-fore, that the proliferation was due to the influence of tracts of nerve fibres descending from the medulla into the upper end of the spinal cord, and this seemed to accord with Bok's conception of stimulogenous fibrilla-tion. The question of axial metabolic gradients still required consideration, however, and Detwiler pursued the inquiry by removing the first few seg-ments of the spinal cord and replacing them by a grafted medulla. In this case, he found that a greater cellular proliferation was induced in the more caudal spinal segments, apparently due to the augmentation of descending tracts arising in the transplanted medulla.

The influence of peripheral structures on the developing central nervous system has also been demonstrated experimentally. Burr[1] found that, if one of the olfactory sensory placodes in *Amblystoma* is removed, the corre-sponding cerebral hemisphere fails to develop normally, and that if an ad-ditional placode is grafted into a position alongside the normal placode, cellular hyperplasia is induced in the olfactory regions of the hemisphere. May and Detwiler[2] transplanted developing eyes into the auditory region of the head, with the result that the optic nerve entered the adjacent cranial nerve ganglion or even the medulla, giving rise to increased growth and proliferation of neuroblasts in these centres. Harrison[3] performed the ingenious experiment of replacing the eye of one species of *Amblystoma* (*A. punctatum*) by the larger eye of another species (*A. tigrinum*), and found that the corresponding optic centres in the mid-brain of the host in-creased in size. Similar results have emerged from Detwiler's experiments in the grafting of limbs in amphibian larvae. In these experiments, he aimed at overloading the peripheral area of nerve supply in order to see whether it led to corresponding changes in the centres of the central ner-vous system. If a limb rudiment is transplanted caudally a sufficient dis-tance, it will derive its nerve supply from new segments of the spinal cord which normally are only concerned with innervating the body wall. This adaptation is accompanied by a marked hypertrophy involving the sensory neurons in the corresponding region of the cord.

It will be observed that the experiments which have been quoted show a response of *sensory* neurons to an artificial increase in peripheral struc-tures. Motor neurons seem to be affected either not at all by these pro-cedures, or only to a very limited extent. Detwiler's experiments have shown, however, that they do respond to stimuli arising within the central nervous system, e.g. as the result of an augmentation of descending bulbar tracts. He has also demonstrated that the segmental arrangement of the spinal nerves is dependent on the myotomes, for if additional myotomes

[1] H. S. Burr, 'The effect of removal of the nasal placode in Amblystoma embryos', *Journ. Exp. Zool.* **20**, 1916; 'Experimental hyperplasia of the cerebral hemispheres in Amblystoma', *Anac. Rec.* **25**, 1923.

[2] R. M. May and S. R. Detwiler, 'The relation of transplanted eyes to developing nerve centres', *Journ. Exp. Zool.* **43**, 1925.

[3] R. G. Harrison, 'Correlation in the development and growth of the eye studied by means of heteroplastic transplantation', *Arch. f. Entwickl.-Mech.* **120**, 1929.

are transplanted into one particular region alongside a section of the spinal cord, the number of spinal nerves which grow out from this section tends to increase to a corresponding degree, while the experimental ablation of myotomes leads to a disappearance of the segmental arrangement.

The lack of response of motor neurons in comparison with sensory neurons in many of these experiments suggests that they are the more stable and conservative elements in the central nervous system, and this is perhaps correlated with the fact that the motor neurons are the first to become differentiated in ontogenetic development. In this connexion it is of interest from the point of view of comparative anatomy to note that sensory mechanisms of the nervous system may become developed to an unusual degree without a corresponding development of motor mechanisms. For example, in some primitive Primates, the eyes have increased to an enormous size with a corresponding elaboration of the visuo-receptive apparatus of the brain, but the motor tracts of the brain remain apparently unaffected, and it may be a matter of speculation how the animal can make use of the additional amount of sensory material with which it is provided. In the later phases of primate evolution it is essentially the sensory side of the brain which is mainly affected. The entire cerebral hemispheres, which reach such an astonishing development in man, are derived from the sensory or alar lamina of the neural tube, and here are built up immensely complicated integrating mechanisms whereby incoming sensory impulses can be analysed and correlated, with the object of securing a motor response which is in every way suitable to the environmental conditions at the time. It has been emphasized by several anatomists that a monkey is unable to perform the delicate and complicated movements of which the human hand is capable, not because it lacks the appropriate muscles and the appropriate motor paths of the peripheral nervous system, but because the higher functional levels of the brain are not developed to an extent sufficient to allow it to make full use of this apparatus.

Much of the experimental work on neural differentiation to which reference has been made in this section has demonstrated that the ingrowth of nerve fibres into the central nervous system provides a direct stimulus for the proliferative activity of the nervous tissues which they reach. There is also evidence that primary centres of proliferation may attract the ingrowth of nerve fibres, since it has been observed embryologically that active cell division in the central nervous system often precedes in time the development of the corresponding afferent fibres. Hence the relation between the two processes is reciprocal.

Herrick[1] found that, in *Amblystoma*, histological differentiation of the cerebrum begins apparently spontaneously in a number of independent and isolated foci by intrinsic processes which are not initiated by the ingrowth of nerve fibres from other regions. In subsequent stages these independent areas are interconnected by nerve fibres to form organized systems. Motor co-ordination centres may reach an advanced stage of differentiation before they have acquired any connexion with sensory tracts

---

[1] C. J. Herrick, 'Development of the brain in Amblystoma in early functional stages', *Journ. Comp. Neur.* **67**, 1937.

—in other words, they begin to mature in advance of their functional expression.

It seems probable that in the embryo two processes of differentiation occur, each predominating at a different phase of development. First of all, proliferating centres attract nerve fibres, and subsequently ingrowing nerve fibres induce or accelerate cell proliferation in the central nervous system. That the peripheral impulse is the more important as a primary factor causing the development of nerve centres is usually considered to be the explanation of evolutionary changes in the nervous system. For example, the atrophic condition of the visual centres of normally blind animals is commonly taken to be the secondary result of a degeneration of the peripheral mechanism of the eye.

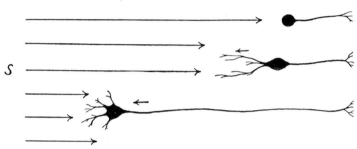

Fig. 125. Diagram illustrating the principle of neurobiotaxis. The source (S) of the stimuli influencing the neurons is represented by arrows. In response to these stimuli, dendrites grow out towards their source. This is followed by a migration of the cell body of the neuron in the same direction, which results in a corresponding lengthening of the axonal process.

One of the theories which have been elaborated to explain the underlying principles of neural differentiation was conceived by a Dutch anatomist, Ariëns Kappers, in 1907. He applied the term *neurobiotaxis* to a phenomenon which he believed to be of fundamental importance in determining the arrangement of nerve cells and their processes in the nervous system.

Stated briefly, the principle of neurobiotaxis expresses the tendency of nerve cells to move towards the source of the impulses by which they are predominantly affected. If, for example, we imagine a group of nerve cells *a* which are mainly activated by stimuli reaching them from a direction *b*, the cells will tend gradually to shift their position towards *b* (Fig. 125). This migratory tendency is manifested in embryological as well as in evolutionary development, and it may lead to a displacement of nerve cells some considerable distance away from the site of their primary differentiation.

Kappers first enunciated this principle on the basis of comparative anatomical observations. He noted that the nuclei of the cranial nerves in different vertebrates show marked variations in their relative position and extent in the brain stem. These variations, he believed, are the result of an actual displacement of the nerve cells in relation to the degree of development and varying position of fibre tracts with which they are functionally associated. Perhaps the most striking example of such a migration is the

nucleus of the seventh cranial (facial) nerve. In lower vertebrates this innervates branchial musculature and is therefore respiratory in function. It
lies in topographical relation to fibre paths which link up in one functional
system all the nuclei of the cranial nerves which have a similar distribution.
In terrestrial vertebrates, the functional significance of the seventh nerve
is completely changed in association with the fact that the musculature of
the second branchial arch becomes modified to form the muscles of the
face. The action of these muscles is now largely influenced by sensory
stimuli which arise from the region of the face, and which reach the brain
stem by the fifth cranial (trigeminal) nerve. Further, in mammals, the
activity of the seventh nerve nucleus also comes under cortical control by
way of descending motor paths (pyramidal tract) which have their origin in
the cerebral cortex. With this remarkable change-over of functional relations, the nucleus of the seventh nerve shifts from its original dorsal position in the brain stem to a ventral position where it comes to lie adjacent
to the nucleus of the fifth cranial nerve and the pyramidal tract. Even in
the human brain, the route which the seventh nerve nucleus has followed
in its evolutionary migration is still marked out by the tortuous path which
its fibres follow in order to reach the surface of the brain.

We have noted that, in its initial differentiation, the neural tube in the
embryo is organized on a fairly simple plan, with a dorsal lamina containing
the sensory neurons, and a ventral lamina containing the motor neurons,
and that, while this arrangement persists relatively unaltered in the spinal
cord, it becomes obscured in the brain stem. The disturbance of the primitive pattern in the brain has been explained as, in part, the result of the
neurobiotactic displacement of neurons from one lamina to the other.
Groups of nerve cells may also migrate over considerable distances in a
cranial or caudal direction under the same influence. An example of this is
the sensory nucleus of the fifth nerve, a portion of which extends right down
into the upper end of the spinal cord where it is brought into relationship
with the upper cervical nerves with which it is functionally associated.

Kappers attempted to explain his theory of neurobiotaxis on the basis
of changing electric potentials in the developing and functioning neurons,
and he argued that the morphology of the nerve cell and its processes, as
well as its final topographical position in the nervous system, is the product
of directional forces of a galvanotropic nature. However, although there is
some experimental evidence that growing nerve fibres may be influenced
by directional forces of an electrical nature (p. 372), the theory of neurobiotaxis, in spite of an extensive literature in which it has often been invoked
in explanation of nervous organization, remains a theory.[1]

It may be noted that the 'neurobiotactic' shifting of nerve cells is in many
cases partly due to a relative rather than a real displacement. As the result
of differential growth in the embryo, neurons tend to become greatly elongated in order to reach the tissues which they supply. This elongation mainly
affects the axonal process (which is specialized for conduction over long

[1] For general accounts of neurobiotaxis and its implications, see C. U. A. Kappers,
'On structural laws in the nervous system', *Brain*, **44**, 1921; 'Differences in the effect of
various impulses on the structure of the nervous system', *Irish Journ. Med. Sc.* Sept. 1934.

distances) and consequently the cell body, *relatively* to the termination of the axon, becomes apparently approximated more and more to the synaptic connexions of its dendrites. In other words, it *appears* to migrate bodily towards the source of its stimulation. But it is also the case that active migrations of developing nerve cells play a considerable part in the morphogenesis of the central nervous system. This phenomenon has been clearly demonstrated by placing experimental lesions in the developing brain of chick embryos with the intention of destroying groups of nerve cells before or during the process of their migration, or obstructing their migration route.[1]

7. PERIPHERAL NERVES

In man, thirty-one pairs of nerves take origin from the spinal cord in serial order, extending from the first cervical to the first coccygeal segments. Twelve pairs of cranial nerves also issue from the brain, but these are all specialized in some degree, and further reference to them may therefore be deferred for the moment. The spinal nerves are typically formed by the union of ventral and dorsal roots which are related respectively to the ventral and dorsal horns of grey matter in the cord.[2]

The ventral roots are composed of motor fibres predominantly concerned with the innervation of somatic musculature. In the thoracic and upper lumbar regions they also contain visceral efferent fibres which, by way of sympathetic ganglia, transmit motor impulses from the spinal cord to visceral or unstriped muscle and visceral efferent fibres of the parasympathetic system that emerge from certain of the cranial and sacral nerves (see p. 389). The dorsal roots consist of afferent or sensory fibres whose cell bodies are situated in the dorsal root ganglia. According to the nature of the stimuli which affect them, the end organs with which these sensory fibres are connected are divided into two main groups, *exteroceptors* which are affected by stimuli having their origin in the external environment, and *interoceptors* which are stimulated by changes in the body itself. Interoceptors, again, may be divided into *visceroceptors* which are related to visceral organs, and *proprioceptors* which are found in somatic structures such as muscles, tendons, and joints.

The arrangement whereby motor and sensory nerve fibres are segregated into separate roots is sometimes spoken of as the 'Bell-Magendie law'. From time to time doubt has been raised as to whether the 'law' is completely valid, for the suggestion has been put forward that certain types of efferent fibres may leave the spinal cord by way of the dorsal roots. On the physiological side, viscero-motor activities (such as vasodilatation) have been observed to follow stimulation of the peripheral cut end of dorsal roots, and numerous anatomists have sought for the morphological basis of these responses.[3]

[1] W. Harkmark, 'Cell migrations from the rhombic lip to the inferior olive, the nucleus raphae, and the pons. A morphological and experimental investigation on chick embryos,' *Journ. Comp. Neur.* **100**, 1954.

[2] The first cervical nerve quite commonly has no dorsal root—in other words, it may be entirely motor in function.

[3] See the discussion of this question by J. C. Hinsey, 'The functional components of the dorsal roots of spinal nerves', *Quart. Rev. Biology*, **8**, 1933.

There seems little doubt that in the lowest chordates (e.g. *Amphioxus* and *Petromyzon*) viscero-motor nerve fibres do leave the spinal cord by the dorsal roots, and there is some evidence, also, that they do so in amphibia and reptiles. In regard to mammals, however, the evidence is on the whole inconclusive. Most observers who have studied the problem have convinced themselves that, after section of a dorsal root between the ganglion and the spinal cord, all the nerve fibres in the proximal stump undergo degeneration, showing that they have their origin in the cells of the dorsal root ganglion. In direct contradiction to this conclusion was the statement of Kuré and his fellow workers, on the basis of similar experimental work, that as many as 40 per cent of the fibres in a dorsal root may be efferent fibres with an intraspinal origin. It has been pointed out, however, that in these experiments insufficient attention was paid to the possibility that regenerated or incompletely degenerated fibres may be mistaken for normal fibres. The same authors observed chromatolysis of cells in the dorsal horn of the spinal cord after section of the dorsal roots, and regarded this as further evidence of efferent fibres in the latter. Here again, the objection has been raised that this cell degeneration may have been the result of incidental interference with the blood supply of the cells by way of vessels which enter the spinal cord along the dorsal roots, or it may be an example of transneuronal degeneration (see p. 402).

In spite of the contradictory evidence, the *onus probandi* still remains with those who hold the view that a proportion of the dorsal root fibres in mammals are efferent in function. Numerical computations have shown that the number of nerve fibres in a thoracic dorsal root approximately equals the number of cells in the ganglion, which is what might be expected if all the fibres are sensory.[1] Indeed, the studies of Duncan and Keyser have led them to conclude that, in the cat, a 'one to one' ratio between the number of dorsal root ganglion cells and the number of dorsal root fibres is present in all the spinal nerves.

The union of a dorsal and ventral root forms a mixed nerve in which the motor and sensory fibres become mingled and rearranged so that they are carried towards their objective in single trunks. The peripheral course and the mode of branching of these trunks seem to be determined largely by convenience for their distribution. Like blood vessels and lymphatics, peripheral nerves tend to follow the most direct routes and paths of least resistance. The main trunks are therefore found along fascial planes in association with the main vessels. Nerves, however, are much more constant and stable than blood vessels, and the migration or displacement of any structure or organ during embryonic or evolutionary development usually leads to a corresponding displacement of the nerve which supplies it. So much is this the case, that the source of the nerve supply is regarded as one of the most reliable criteria for establishing the homologies of structures whose changed position and appearance in different animals may otherwise obscure their identity.[2]

---

[1] D. Duncan and L. L. Keyser, 'Further determinations of the numbers of fibres and cells in the dorsal roots and ganglia of the cat', *Journ. Comp. Neur.* **68**, 1938.
[2] See p. 151 for a reference to the constancy of nerve supply in muscles.

The structure of a peripheral nerve. A peripheral somatic nerve consists essentially of bundles of nerve fibres (mostly medullated) which are enclosed in a sheath of connective tissue, the *epineurium*. This is composed mainly of collagenous fibres and a small proportion of elastic fibres, and processes extend from it into the substance of the nerve, separating the fibres into secondary bundles. Lastly, the individual fibres are themselves embedded in delicate connective tissue which is termed the *endoneurium*.

The epineurium is most conspicuous in the larger nerve trunks, where it forms a relatively tough and thick fibrous sheath. It becomes more tenuous in the smaller trunks, and in the terminal branches it is absent altogether. Within the sheath, loose connective tissue forms perineurial spaces which are of considerable practical importance, since they provide channels through which infective processes can extend along the course of a nerve. They are particularly significant in the case of certain of the cranial nerves close to the point where they leave the skull.

The perineurial spaces of the olfactory nerves, for example, become continuous through the openings in the cribriform plate of the ethmoid with the subarachnoid space on the surface of the brain. Solutions of dyes or particulate suspensions introduced into the subarachnoid space inside the skull can track down along the nerves into the nasal cavities, while solutions dropped into the nasal cavities may also, on occasion, reach the brain along the same route. It is easy to understand, therefore, how meningitis may develop as a complication of nasal infections.

The vascular supply of peripheral nerves. All large and medium-sized peripheral nerves are supplied with blood vessels which are derived from vascular branches in the immediate neighbourhood. In each large nerve trunk in the human body there is embedded an axial artery which is reinforced at intervals of 3–5 cm. by a succession of arterioles; when the nerve divides the axial vessel divides with it. In the smaller branches, the axial vessel lies immediately adjacent to the nerve, bound up in the same connective tissue sheath. Experimental evidence has shown that functionally the blood supply to a nerve is of considerable importance. If the supply is interrupted the passage of nerve impulses is affected and ultimately a complete 'nerve block' may ensue. The large calibre fibres are first affected, and finally the fine fibres. In an experimental study, however, Adams has found that the interruption of a number of local nutrient arteries to a nerve may be effected without harmful results, since the longitudinal anastomosis within the nerve trunk is usually sufficient to ensure an adequate blood supply from more distant sources.[1] Lymphatic vessels are also present in the epineurium. It may be further noted that the epineurial sheath is itself supplied by fine nerve fibres (*nervi nervorum*) which are usually derived from branches of the main nerve trunk; these fibres are probably mainly vasomotor in function, innervating the blood vessels of the nerve trunk.

Nerve plexuses. It has been noted that the segmental arrangement of the spinal nerves is secondarily imposed upon the central nervous system by the myotomes or muscle segments in the embryo. Where the musculature

[1] W. E. Adams, 'The blood supply of nerves', *Journ. Anat.* **76**, 1942, and **77**, 1943.

retains its primitive segmental arrangement the corresponding spinal nerves in their peripheral course are distributed separately in an orderly segmental manner. This is the case in the thoracic region where the intercostal nerves each extend out in their own intercostal spaces to supply the serially arranged intercostal muscles and corresponding areas of skin. Such nerves may be termed uni-segmental.

In the case of the limbs, the primitive segmental arrangement of the musculature becomes greatly disturbed, and each muscle may derive its nerve supply from several segments of the spinal cord.[1] In order to adapt themselves to these more complicated patterns of musculature, the spinal nerves which supply the limbs undergo an extensive resorting of their constituent fibres as soon as they leave the spinal cord, so as to form mixed pluri-segmental nerves which may most conveniently carry in one bundle the fibres destined to innervate single muscles or groups of muscles. Motor fibres may also be commingled with sensory fibres which are required to supply an area of skin in a corresponding region of the limb. This is the basis for the formation of the limb plexuses, which are therefore essentially a provision for convenience of distribution.

Besides the large limb plexuses, which can be readily displayed by the ordinary process of dissection, it should be recognized that within single nerve trunks the individual fasciculi of nerve fibres commonly rearrange themselves in a plexiform manner as they extend towards their peripheral distribution.[2] While in the motor and sensory roots of the spinal nerves the component fasciculi run in parallel series and remain separate, in almost all peripheral mixed nerves (as can be shown in teased preparations) the fasciculi interconnect with each other and interchange fibres. Such intraneural plexuses occur even in uni-segmental nerves, and are probably largely concerned with the regrouping of motor and sensory fibres in accordance with their ultimate distribution.

The functional segregation of fibres in the course of a main nerve trunk has some practical significance. In the proximal part of its extent, fibres destined for a similar objective are still distributed diffusely through the nerve so that, for example, as much as one-third of its cross-sectional area can be divided without any marked sensory or motor loss in its territory of supply. On the other hand, individual muscles or groups of muscles may be selectively paralysed by partial section of a nerve trunk in the more distal part of its course.

The formation of plexuses in peripheral nerves is strictly comparable with the arrangement and grouping of fibres in the central nervous system to form circumscribed tracts. This process of *fasciculation*, as it is called, has been 'explained' on the basis of the principle of neurobiotaxis (p. 375), for it is clear that if nerve cells exposed to stimuli from similar sources tend to arrange themselves in common groups, their axonal processes will necessarily come into close topographical relation with each other. From

---

[1] As examples, it may be noted that, in the arm, the deltoid is related to the fifth and sixth cervical muscle segments and the flexor digitorum sublimis to the seventh and eighth cervical and first thoracic segments.

[2] J. E. A. O'Connell, 'The intraneural plexus and its significance', *Journ. Anat.* **70**, 1935–6.

the embryological point of view, also, it may be noted that nerve cells which undergo differentiation at the same time in any particular region will tend to be exposed to the same 'neurobiotactic' influence, and thus will send out their axonal processes in the same direction.

Nerve components. In a typical spinal nerve there are four functional systems of fibres which innervate the general tissues of the body. These systems can be listed as follows:

1. *General somatic afferent fibres,* concerned with general cutaneous and deep sensibility over the whole body.
2. *General visceral afferent fibres,* conveying impulses from mucous membranes, blood vessels, &c.
3. *General visceral efferent fibres,* distributed to unstriped musculature and glands.
4. *General somatic efferent fibres,* supplying general skeletal or striped musculature.

We have noted that the first and last of these categories are related respectively to the alar and basal laminae of the neural tube (which form the dorsal and ventral horns of grey matter in the spinal cord), while the visceral components are related to an intervening area of the neural tube alongside the sulcus limitans (which in the thoracic region of the cord forms a small lateral horn of grey matter). In the cranial nerves similar components of a general functional nature are present, but, in relation to the specialized structures and organs which are developed in the head region, other special categories are also to be recognized. These are as follows:

1. *Special somatic afferent fibres,* which innervate highly specialized sense organs, such as the retina,[1] and the cochlea and semicircular canals.
2. *Special visceral afferent fibres,* which supply specialized visceral sense organs, such as the taste buds in the tongue.
3. *Special visceral efferent fibres,* which supply the branchial musculature and its derivatives. Although it is of the striped variety and not distinguishable from skeletal muscle in its minute structure, this musculature has a different developmental origin. It is visceral in the sense that it is differentiated from the mesodermal wall of the pharynx and in the adult serves alimentary functions (at least in part).
4. *Special somatic efferent fibres,* innervating skeletal musculature which subserves specialized functions, e.g. the tongue muscles, and the extrinsic muscles of the eye.

The size of fibres in peripheral nerves. It will be noted by reference to Fig. 118A that the size of the fibres which make up a peripheral nerve shows a considerable variation. In a mixed nerve the diameter of myelinated fibres ranges from $1\,\mu$ to about $20\,\mu$, and there are in addition large numbers of unmyelinated fibres (that is to say, fibres in which the lipoid film coating the surface of the axon is not detectable with ordinary histological stains such as osmic acid). The finest unmyelinated fibres are beyond the limits of resolution with the light microscope and can only be defined by

---

[1] It should be noted that the 'optic nerve' is really a brain tract which has been drawn out from the general substance of the brain. On the other hand, the bipolar cells and their processes in the retina may be regarded as the equivalent of the posterior root fibres and ganglion cells of a spinal nerve.

electron microscopy; in the case of the olfactory nerves they may have a diameter of only 0.1μ, or even less. Indeed, it has now become clear that counts of nerve fibres in individual nerves, that have from time to time been published in the past, are not really valid, for they have not included those fibres which are beyond the limits of resolution with the light microscope. As already noted, the diameter of a nerve fibre and the thickness of its medullary sheath are related to the rate at which it conducts impulses. Thus it will be realized that in a mixed nerve impulses travel at different velocities, and it has been found in mammals that these range from over 100 metres per second for the largest fibres to 10 metres per second or less for the very fine myelinated fibres and the unmyelinated fibres. Since the normal functioning of the nervous system depends on a correct timing of the arrival of impulses at their destination, any alteration of the normal velocities of conduction is likely to interfere seriously with the efficiency of sensory and motor mechanisms. This may be a matter of some importance in the restoration of function after injuries to peripheral nerves. The regenerating nerve fibres in any case take some time to acquire their normal diameter and medullation (and therefore their normal velocities). If regeneration is long delayed, however, the endoneurial tubes in the peripheral stump of a cut nerve may shrink, and it has been suggested that this prevents the newly growing nerve fibres from expanding to their normal size. If this is so they will not recover their normal rates of conduction, and functional recovery will be to that extent impaired.

It is interesting to note that during the normal growth of the body the velocity of nervous impulses becomes progressively adjusted. Hursh found that in the growing kitten the axon diameter and conduction rate in the nerves of the hind-limb increase in direct proportion to the length of the limb. Thus the total conduction time remains constant as the animal grows.[1]

8. THE SPECIFICITY OF NERVE FIBRES

The afferent fibres in the posterior roots of the spinal nerves convey impulses to the spinal cord which are initiated by a variety of stimuli (e.g. impulses underlying sensations of touch, pressure, pain, and temperature, and proprioceptive and visceroceptive impulses). We have now to consider how far these fibres are *individually* concerned with specific types of sensation, that is to say, whether there are distinct categories of fibres subserving each sensation.

As previously noted (p. 323), there is a certain amount of experimental evidence for the view that, in addition to non-specific nerve endings in the skin and subcutaneous tissues, there are specific nerve endings which provide the receptors for stimuli of touch, pressure, temperature, or pain. It seems probable, therefore, that there are different sets of afferent fibres in peripheral nerves which are concerned with transmitting the impulses related to different sensory modalities. The frequency with which impulses

[1] J. B. Hursh, 'Conduction velocity and diameter of nerve fibres', *Amer. Journ. Phys.* **127**, 1939.

are transmitted along a nerve fibre depends on the strength of the stimulus, and it is possible on this basis to recognize variations in the intensity of a conscious sensation. In fibres of different calibre, also, the velocity with which the impulse travels varies. But numerous studies on the nature of a nervous impulse have failed to detect any *qualitative* difference in the impulse itself, by whatever kind of stimulus it is initiated. In other words, impulses conducted by any nerve fibres are always fundamentally of the same kind. It follows, therefore, that the *type* of sensation which is experienced depends on the nature of the receptor with which the nerve fibre is connected and on the connexions which it makes in the brain. In other words, some nerve fibres may be presumed to conduct impulses initiated by a specific type of stimulus, and the primary sorting-out of sensory stimuli must depend on a system of peripheral analysers. As it has already been pointed out, such analysers may be represented in certain areas of the skin by end organs of different morphological types, and there is considerable evidence to show that a similar system of analysers exists in the organs of special sense. For example, the sensory cells in different parts of the cochlea are stimulated by vibrations of different frequency, so that the recognition of variations in the pitch of a sound depends on a peripheral analysis in the internal ear. The rods and cones in the retina are contrasted in their reaction to light stimuli, and colour discrimination must ultimately depend on a differential sensitivity of the retinal receptors. Electrophysiological evidence strongly indicates that the olfactory epithelium contains sensory cells which are differentially sensitive to odorous substances, so that the quality of a smell is determined in the first instance by a system of peripheral analysers in the nose.

It has been stated in the previous section that peripheral nerves in general contain a spectrum of fibre sizes, and that the different sizes are associated with different conduction velocities. There is also evidence that they are in part associated with nervous impulses related to different types of sensation. Thus the large, rapidly conducting afferent nerve fibres transmit impulses from proprioceptors in muscles, and touch and pressure receptors, while the small, slowly conducting fibres are believed to be predominantly concerned with painful sensation. The evidence for this is derived from a variety of sources, including the study of the action potentials in sensory nerve fibres under various conditions and the anatomical study of the innervation of different types of sensory receptor in the skin and subcutaneous tissues.[1] That painful sensations are mediated in part by finely medullated and non-medullated fibres has also been claimed on the basis of experiments in which section of these fibres in the posterior roots of spinal nerves (without interfering with the other fibre components) was found to abolish responses to painful stimuli.[2]

In addition to the evidence cited above for a functional specificity of

[1] See E. D. Adrian, *The Basis of Sensation*, 1928; *The Mechanism of Nervous Action*, 1932; J. Erlanger and H. S. Gasser, *Electrical Signs of Nervous Activity*, 1937; H. H. Woollard, G. Weddell, and J. A. Harpman, 'Observations on the neurohistological basis of cutaneous pain', *Journ. Anat.* **74**, 1940.
[2] S. W. Ranson, 'Unmyelinated fibres as conductors of protopathic sensation', *Brain*, **38**, 1915.

peripheral nerve fibres, it is known that in the spinal cord afferent fibres become segregated (though not necessarily with mathematical precision) in accordance with the type of sensation which they mediate, and this may be taken to presuppose a corresponding functional differentiation of individual fibres in the peripheral nerves. However, even if this should prove to be the case in the normal individual, the specificity of nerve fibres may not be of such an order that it prevents them during the process of regeneration of a cut nerve from establishing connexions with types of end organs with which they were not previously related. In the regeneration of a sensory peripheral nerve which has been cut, it is theoretically possible that an individual fibre which previously mediated one type of sensation may become connected with a receptor which responds to a different type of sensory stimulus. This possibility may perhaps be of significance in regard to persistent disorders of sensation following peripheral nerve lesions.

In the case of motor nerve fibres there appears to be a general lack of specificity in regard to the muscles which they supply. It is a recognized surgical procedure to construct cross anastomoses of motor nerves in order to overcome muscle paralysis in cases where regeneration of the proper nerve has not occurred. Thus facial paralysis may be treated by anastomosing the central end of the hypoglossal or spinal accessory nerve (which is cut for the purpose) to the peripheral end of the degenerated facial nerve. When regeneration is complete, the patient can sometimes be educated to regain a certain degree of control over his facial musculature. A somatic motor nerve can also be anastomosed with a visceral motor nerve with the establishment of functional continuity, as in the classical experiments of Langley and Anderson[1] in which the central end of the severed phrenic nerve was united to the peripheral end of the cut cervical sympathetic trunk.

Certain of Detwiler's experiments on limb-bud transplantation in amphibian larvae have a bearing on the problem of the selectivity of growing nerve fibres, and therefore require mention here. He found that if the limb rudiment is shifted caudally for only a short distance it tends still to receive its nerve supply from its normal spinal segments—indeed, the brachial nerves may grow a considerable distance out of their usual course in order to reach the grafted appendage. However, there are definite limits to this possibility, for if the limb is transplanted to a distance of more than three segments behind its normal level, it now receives its nerve supply from new segments of the spinal cord which are not normally concerned with its innervation. In other words, it appears that growing nerve fibres in the embryo are probably capable of a limited power of selectivity in regard to the general region of their distribution.

A curious phenomenon has been noted in experimental transplantation of limbs in amphibian larvae which suggests some sort of functional specificity in motor nerve fibres in these forms. Weiss found that, even if a transplanted limb receives its motor innervation from quite an unusual source, its muscles still contract in sequence synchronously with the

[1] J. N. Langley and H. K. Anderson, 'The union of different kinds of nerve fibres', *Journ. Physiol.* **31**, 1904.

homologous muscles of a normal limb. The explanation of this phenomenon is quite obscure, though attempts have been made to explain it as a kind of 'resonance' effect. Except that the basis of this 'resonance' is known to reside in the nervous system, however, the mechanism which underlies it has not been analysed. Evidently it betokens some kind of functional reorganization in the spinal cord.

On the sensory side, experiments have demonstrated a remarkable example of the specific relationship between afferent fibres and their connexions in the central nervous system. Sperry[1] found that if the fibres of the optic tract are cut in the brain of a newt, they undergo regeneration and grow back into visual centres. In the scar formed at the site of the lesion, the regenerating fibres become disarranged and mixed up in a most complicated plexiform manner, but after traversing the scar they appear to sort themselves out again so that each group of fibres finally effects a functional connexion with its appropriate group of nerve cells. This is demonstrated by the fact that the animal regains its normal localizing optokinetic responses on stimulation of the retina. This 'selective patterning of synaptic connexions' can hardly be attributed to a mechanical guidance of the fibres, and it is suggested that it depends on differential biochemical affinities between the neuronal elements.

Sperry's observations provide an example of what has been termed the *sorting principle in sensory analysis*, that is to say, the sorting and resorting (or 'unshuffling') of afferent fibres in accordance with the category of impulse which they convey. Such a process has been shown to occur in the visual nervous system of the higher Primates.[2] It has also been strikingly demonstrated by studies of the visual pathways in the frog, for these showed that in the optic nerve 'axons follow sinuous paths constantly shifting their position . . . neighbouring axons in the nerve may come from as widely separated retinal areas as possible', but the fibres 'congregate again at their terminals, not only according to their points of origin but also according to their functions'.[3] A similar resorting of fibres occurs in cutaneous nerves; for example, experiments have demonstrated that sensory fibres derived from different circumscribed areas of the skin become mixed up in a random disarrangement in the main trunk of a sensory nerve, but on approaching the spinal cord they become 'unshuffled' to form a series of microbundles each of which again is made up of fibres related to a localized cutaneous area.[4] It is to be presumed that this process of resorting (the morphogenetic basis of which has yet to be elucidated) is of essential importance for effecting a high degree of sensory discrimination.

[1] R. W. Sperry, 'Orderly patterning of synaptic associations in regeneration of intracentral fibre tracts', *Anat. Rec.* **102**, 1948.

[2] W. E. Le Gros Clark, 'The sorting principle in sensory analysis as illustrated by the visual pathways', *Ann. Roy. Coll. Surg.* **30**, 1962.

[3] H. R. Maturana, J. Y. Lettvin, W. S. McCulloch, and W. H. Pitts, 'Anatomy and physiology of vision in the frog', *Journ. Gen. Physiol.* **43**, 1960.

[4] G. Weddell, D. A. Taylor, and C. M. Williams, 'The patterned arrangement of the spinal sensory nerves to the rabbit's ear', *Journ. Anat.* **89**, 1955.

9. THE CELLULAR ARCHITECTURE OF THE NERVOUS SYSTEM

In sections of the central nervous system which have been stained with basic dyes, such as methylene blue or neutral red, the cyto-architecture (that is to say, the disposition and arrangement of the nerve cells) is clearly displayed. While in such sections it can be seen that many nerve cells are scattered diffusely in a general groundwork of interlacing fibres, so that the extent of their distribution is difficult or impossible to define topographically, others which are concerned with common functions of a clearly defined nature are usually collected in fairly circumscribed and compact groups. The nuclei[1] of the individual cranial nerves, for example, form isolated groups which stand out conspicuously in microscopical preparations. In many cases, also, a nucleus is subdivided into secondary groupings as an anatomical reflection of an intrinsic functional differentiation. As instances of this, we may note that the nucleus of the seventh cranial (facial) nerve is composed of several clumps of cells, each concerned with the innervation of a different group of facial muscles, and that the motor neurons in the anterior horn of the spinal cord are similarly disposed in groups in accordance with the muscle groups which they supply.

Although the topographical separation of nerve cells into isolated formations is commonly related to functional differentiation, this is not always the case. Collections of cells otherwise homogeneous in their formation may be incidentally split up into component groups by fibre tracts which happen to pass among them, and in this case the topographical differentiation has apparently no functional significance.

In the higher functional levels of the brain, a common arrangement is found in which nerve cells are disposed in a series of one or more stratified layers. This lamination is best developed in the cortex of the cerebrum and cerebellum, and is also found in certain other masses of grey matter (such as the lateral geniculate body, the inferior olivary nucleus, and the dentate nucleus of the cerebellum).

The significance of lamination is probably related to the complexity of the fibre connexions by which the constituent cells of the laminae are linked up with other parts of the brain. Where these fibres are very numerous they would not find sufficient space to pass in order to reach the centrally placed cells of a single, compactly arranged, large group. Consequently, as a matter of practical convenience, the cells are spread out in a comparatively thin stratum so as to present a relatively large area to which fibres can have ready access from one or both sides. The more numerous and the more intricate the fibre connexions of the stratum, the thinner is the latter likely to be. It follows from this that the expansion of such a laminar formation of cells can hardly take place by an increase in its thickness; on the contrary, it can only expand by increasing its surface

---

[1] The use of the term 'nucleus' for a group of nerve cells which can be defined topographically has a certain inconvenience since 'nucleus' is essentially a cytological term. In spite of suggestions for an alternative nomenclature, however, it is hardly feasible to discard a term which has been so generally accepted for common use.

extent. This leads to the formation of convolutions, which are so characteristic of the cerebral and cerebellar cortex (and certain other structures) of more highly developed brains. That a laminated structure such as the cerebral cortex grows in quantity by surface extension in the form of convolutions, rather than by increasing its thickness, is perhaps partly to be explained also by convenience of blood supply. By remaining relatively thin, the whole width of the cortex maintains a position close to the blood vessels in the pia mater from which it derives its vascular supply.

It is interesting to note that, when cortical formations become folded in the course of their expansions, the pattern of folding is not entirely haphazard. Fissures, or sulci, tend to appear along the boundary lines separating cortical areas which differ in structure (and also in function), so that some of them come to form important topographical landmarks whereby the limits of these areas can be defined by surface inspection of the brain. For example, in the human brain a conspicuous fissure, the central sulcus, sharply defines the boundary between the motor and sensory areas of the cerebral cortex. In other words, the convolutional pattern of the brain has a direct relation to the distribution and extent of its component functional areas. The reason for this is to be found in the engineering principle that if a strut is exposed to longitudinal compression, the maximum stresses develop in the region where there is a change of thickness or mechanical strength, and it is here that the bending is most likely to occur. In the same way, it is to be expected that when, during its rapid growth, the cerebral cortex is exposed to lateral compression, it will tend to bend or fold where cortical areas of different width and texture adjoin each other.[1]

Besides the fact that the intrinsic structure of the central nervous system shows a topographical differentiation by the segregation of cells into compact groups and laminae, a further refinement of cyto-architectural organization is introduced by morphological differences in the cells themselves. Adjacent groups and laminae may be contrasted by the cytological characters of the cells which compose them. In the cerebral cortex, for example, there may be six or more separate layers each containing cells of a different type, varing in size, shape, dendritic patterns, and arrangement. Again, in the thalamus (a large sensory correlation centre in the fore-brain) a number of separate nuclei can be defined, not only by their spatial grouping, but also on the basis of their cell types. In many cases this cytological differentiation can be definitely correlated with functional differences, as in the cerebral cortex where one layer is mainly concerned with the reception of impulses from lower levels of the brain, another with the emission of efferent impulses, and yet another with associative functions.

It may be accepted as an axiom that differentiation in cyto-architecture is in general directly related to functional differentiation. The segregation of cells of a uniform type to form an isolated and compact group is indicative of the uniformity and specificity of their function, while a scattered arrangement of cells of diverse types is equally a reflection of the diffuseness

[1] W. E. Le Gros Clark, 'Deformation patterns in the cerebral cortex', *Essays on Growth and Form*, Clarendon Press, 1945.

and diversity of their function. In other words, localization of function is associated with the topographical localization of nerve cells. This principle has important reference to the study of sensory mechanisms of the brain, for it has become clear that sensory discrimination (that is to say, the ability to distinguish one quality of sensation from another in each sensory modality) is ultimately made possible by different spatial patterns of excitation in the cerebral cortex. Nerve fibres from sense organs, as we have just noted, become segregated on their arrival in the central nervous system in accordance with the source of the impulses which they carry, and each group terminates in localized (and in some cases very sharply defined) groups of nerve cells before they are relayed to the cerebral cortex. It thus appears that the so-called 'lower sensory centres' are more than just 'relay stations'; they are 'sorting stations' where impulses are sorted out in such a way that they may then be projected on to the cerebral cortex to form the various patterns of excitation on which conscious discrimination depends. It is probable, also, that sensory discrimination depends in part on temporal patterns of stimulation, as the result of differential rates of conduction in nerve fibres transmitting impulses from the different sensory receptors.

In many parts of the central nervous system the descriptive anatomy of cyto-architectural features has outstripped physiological studies of their functional significance, and, indeed, some neurologists have doubted the utility of subdividing the cellular areas of the brain into so many separate entities as anatomists in the past have done. With the progress of physio-logical research, however, it is being demonstrated more and more that these entities are often related to different functional systems, i.e. that they are not just meaningless units of interest only to the descriptive morpho-logist. This is the case, for instance, with the cerebral cortex where a number of different areas, which have been defined anatomically on the basis of the contrast in their cyto-architecture and their different fibre connexions, are now known to serve correspondingly different functions, and in the thala-mus and hypothalamus where well-defined groups of cells have been shown to have equally well-defined functional implications.

10. THE FIBRE ARCHITECTURE OF THE NERVOUS SYSTEM

The arrangement of myelinated fibres in the brain and spinal cord may be studied in sections stained by the Weigert-Pal technique (p. 400), while the pattern made by unmyelinated fibres can be demonstrated by methods of silver impregnation. Grey matter is everywhere permeated by neuropil, which is an exceedingly intricate network of interlacing fibres, and its den-sity varies in different situations (Fig. 126). Such a diffuse arrangement of fibres is a fundamental characteristic of correlation and co-ordination centres of the nervous system, and provides a basis for the rapid dispersion of impulses in all directions over a wide area.

Fibres which are concerned with reactions of a more localized and precise nature tend to collect themselves into circumscribed bundles or *tracts*, and, in general, the more specific the function of a group of fibres, the more sharply defined and compact is the tract which they form. The tendency of fibres conveying similar impulses to be segregated in anatomically well-defined tracts is termed *fasciculation*, and we have seen that this process may be explicable on developmental grounds (p. 380).

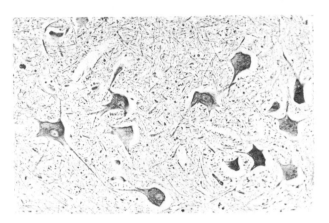

Fig. 126. Microphotograph showing nerve cells embedded in a mass of neuropil (stained with protargol). × 120.

The fibre architecture of grey matter in the central nervous system usually reflects—as a negative image—the cyto-architecture, for cell groups are commonly outlined by 'capsules' of fibres which skirt their margins as they pass by. It may happen, however, that in their course fibre tracts penetrate cell masses with which they have no connexion. Such 'fibres of passage' break up the outline of cell groups and, in sections of normal material, they may easily be mistaken for terminal fibres. In the cerebral cortex, the differentiation of cortical areas based on cyto-architecture is reproduced also in the fibre architecture, so that it is possible to delineate several of these areas in sections which have been stained by the Weigert-Pal method (which stains the myelinated fibres).

11. THE AUTONOMIC NERVOUS SYSTEM

In rather marked contrast to the centralized organization of the brain and spinal cord, concerned so obviously with the somatic reactions of an animal to changes in its external environment, are the peripheral ganglia and plexuses through which visceral reactions are controlled. As these peripheral nervous structures provide in some cases a regulatory mechanism for local activities which can occur independently of the control of the central nervous system, they are commonly grouped under the term autonomic nervous system'. Since this term was first employed, however, it has been realized that, with few exceptions, the control of visceral activities usually involves the central nervous system as well as peripheral ganglia—in other words, it is not really possible sharply to distinguish a visceral nervous system from a somatic nervous system on anatomical or physiological grounds. Thus the definition of the 'autonomic nervous system' has been extended by some writers to include not only the peripheral nervous basis of visceral functions, but also the central nervous mechanisms which are concerned in these reactions.

As already indicated, the visceral and somatic activities of the body are very closely interlocked, both being represented at all functional and anatomical levels of the nervous system, but it may be accepted that, broadly speaking, somatic nervous mechanisms are concerned with the regulation of an animal's overt behaviour in relation to the external environment, while the autonomic nervous system is concerned with regulating the internal environment, including nutritional processes, temperature, vaso-motor, viscero-motor, and secretory and excretory activities.

On anatomical and physiological grounds, the peripheral autonomic system is divided into two categories—*sympathetic* and *parasympathetic*. They control activities which are often mutually antagonistic—hence most visceral structures are innervated from both sources.

In general, it may be said that sympathetic reactions tend to be 'mass reactions', widely diffused in their effect, and that they are directed towards the mobilization of the resources of the body for the expenditure of energy in dealing with emergencies or emotional crises. For example, the activity of the sympathetic system produces a vasoconstriction of cutaneous vessels resulting in an increased blood supply to the muscles, the heart, and the brain, which may be called upon to accomplish additional work. Further, the heart is accelerated and the general blood pressure raised, while other visceral activities of less immediate importance are slowed by inhibition of peristalsis and the closure of sphincters.

On the other hand, the effects of parasympathetic activity are usually discrete and isolated, and are directed rather towards the conservation and restoration of resources of energy in the body. The heart is slowed, the pupils are constricted, and processes of digestion and assimilation are promoted by increased glandular secretion and enhanced viscero-motor activities.

In brief, the sympathetic and parasympathetic systems represent respectively the katabolic and anabolic aspects of autonomic activity. Such a statement, however, is to be regarded merely as a convenient generalization, for, in fact, there are many visceral activities in which the sympathetic and parasympathetic systems co-operate to the same end. We may now briefly review the general organization of the sympathetic and parasympathetic systems so far as their peripheral components are concerned.[1]

The sympathetic system. The peripheral ganglia of the sympathetic system are represented by the sympathetic chain which extends alongside the vertebral column in continuity from the base of the skull to the lower end of the sacrum, and by certain collateral or paravertebral ganglia situated in close relation to the aorta in the abdominal cavity (e.g. the coeliac or

[1] For a detailed general account of the anatomy of the autonomic system, see G. A. G. Mitchell, *Anatomy of the Autonomic Nervous System*, E. and S. Livingstone, 1953.

solar plexus). These ganglia contain nerve cells (predominantly multipolar in type) enclosed in nucleated capsules, all of which are efferent in function. The ganglia of the sympathetic chain are arranged segmentally, though the segmental arrangement is to some extent obscured in the cervical region by the fusion during embryological development of adjacent ganglia. They are linked together by a longitudinal strand of nerve fibres (Fig. 127).

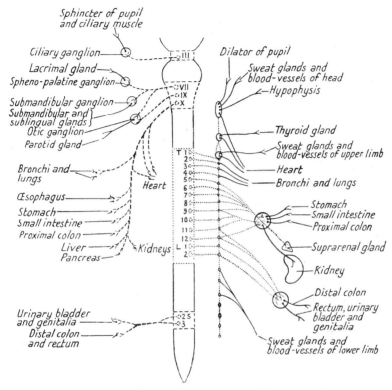

FIG. 127. Diagram illustrating the general arrangement of the autonomic nervous system. Preganglionic fibres of the parasympathetic system are shown in broken lines, and those of the sympathetic system in dotted lines. Postganglionic fibres of both systems are shown in continuous lines. (After Cunningham.)

The part which sympathetic ganglia play in the organization of the autonomic nervous system may be indicated by reference to a diagram representing the main connexions of a ganglion of the sympathetic chain in the thoracic region (Fig. 128). Arising from small cells in the lateral horn of grey matter in the spinal cord are efferent sympathetic fibres which, after emerging with the ventral root of the spinal nerve, leave the latter to join the ganglion. These fibres are finely medullated and are termed *preganglionic fibres*. Together they form a white strand which can be discerned macroscopically and which is called the *white ramus communicans* of the ganglion. Many of the preganglionic fibres terminate by making synaptic connexions with the cells in the ganglion, and the axons of these cells, which are non-medullated, are called *postganglionic fibres*. Some post-ganglionic fibres may run more or less directly to the tissues which they are destined to innervate (e.g. unstriated muscle in the walls of viscera or blood vessels, or gland cells), while others form a fasciculus termed the

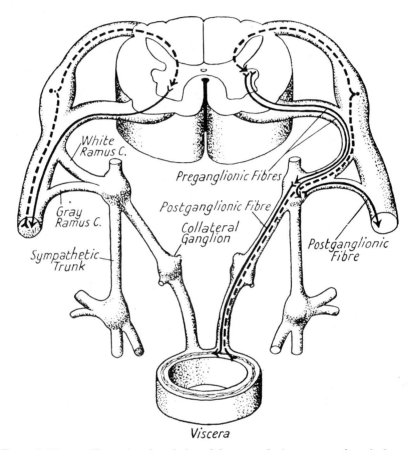

FIG. 128. Diagram illustrating the relation of the sympathetic system to the spinal nerves and spinal cord. The main visceral fibres are shown on the right side and, for contrast, the somatic fibres on the left. Afferent fibres are indicated by interrupted lines and efferent fibres by continuous lines. (After Cunningham.)

grey ramus communicans which rejoins the corresponding spinal nerve, and are then carried to their destination along its peripheral branches.

Preganglionic fibres emerging in a ventral root do not necessarily end in the corresponding segmental ganglion of the sympathetic chain. They may pass without interruption *through* the ganglion and ascend or descend to terminate in ganglia at different segmental levels, or they may continue more peripherally to end in a synaptic junction in one of the paravertebral ganglia. The position of the synapse in the peripheral pathway of a sympathetic impulse can be ascertained in many instances by applying Langley's nicotine technique. This depends upon the fact that the application of a weak solution of nicotine blocks the passage of an impulse at a synaptic junction, while having no effect on an uninterrupted nerve fibre.

The sympathetic outflow from the spinal cord in man is limited to segments extending from the first thoracic to the second or third lumbar.[1] It follows from this that white rami communicantes are only found in connexion with the sympathetic ganglia of these segments, while grey rami

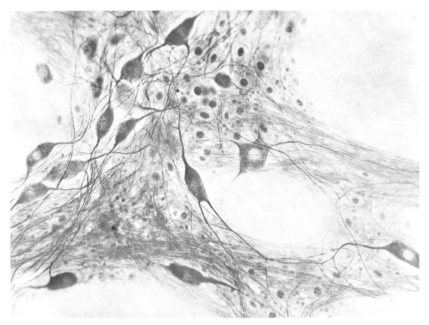

FIG. 129. Microphotograph of Auerbach's (myenteric) plexus from the small intestine of a cat, stained by Bielschowsky's method. A part of the sympathetic chain had been extirpated 21 days previously. A considerable proportion of the sympathetic postganglionic fibres have therefore undergone degeneration, exposing to better advantage the ganglion and their processes. × 350 approx. (From a preparation by C. J. Hill.)

communicantes are found issuing from all the ganglia. It should be noted, also, that a single preganglionic fibre may effect a synaptic connexion with several cells in a ganglion, thereby allowing for a diffusion of the impulses which leave the spinal cord.

It will be observed that the peripheral course of a sympathetic impulse involves two neurons—a preganglionic fibre and a postganglionic fibre. There is apparently only one exception to this arrangement, and that occurs in the nerves which supply the specialized glandular tissue which secretes adrenalin. This tissue (which is commonly called *chromaffin tissue* because of the affinity which its cells show for chromic salts) is mainly aggregated

[1] According to O. Gagel, the *cells of origin* of the preganglionic fibres of the sympathetic system in man extend in the lateral horn as far up as the seventh cervical segment ('Zur Histologie und Topographie der vegetativen Zentren im Rückenmarke', *Zeitschr. f. d. ges. Anat.* **85**, 1928).

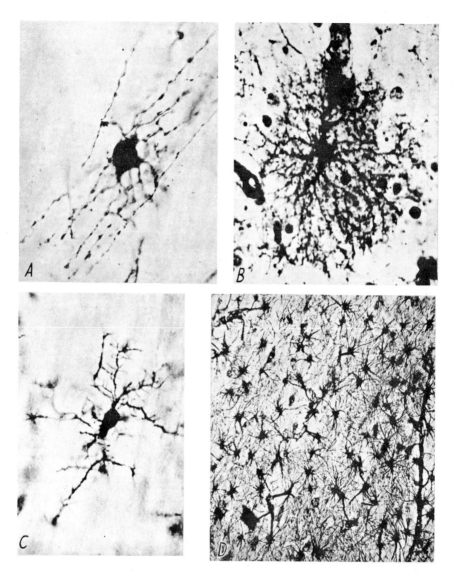

Fig. 130. Microphotograph showing different types of neuroglia. *A.* Oligodendroglia. Note the disposition of the processes of this cell, which are arranged along adjacent nerve fibres. *B.* Protoplasmic astrocyte. *C.* Microglia. *D.* Fibrous astrocytes. (Photographs lent by the late Professor del Rio Hortega.)

in the medulla of the adrenal glands. Preganglionic fibres end directly in relation to the cells of chromaffin tissue, and it is evident, therefore, that these cells are the morphological equivalents of postganglionic fibres. The significance of this has become clear since, with the development of the humoral theory of nervous transmission, it has been shown that on stimulation of sympathetic fibres adrenalin (together with noradrenalin) is actually produced at the terminals of the postganglionic neurons, and provides the medium through which the ultimate effects of sympathetic stimulation generally are brought about.

Efferent sympathetic impulses are initiated by the activities of lateral horn cells in the spinal cord, either as the result of reflex stimulation, or of stimulation from descending paths in the central nervous system which may have their origin at different functional levels, including the cerebral cortex. It is thus apparent that the whole sympathetic system is ultimately under the dominating control of the central nervous system. Nevertheless, it is possible for certain visceral activities to be mediated entirely by peripheral mechanisms. Such local mechanisms, however, are very limited, and are probably confined to the enteric ganglionated plexuses and to axon reflexes.

The peripheral afferent paths of the visceral nerves are similar to those of the somatic nervous system. That is to say, they are represented by afferent fibres which take origin as sensory nerve endings (which in this case are situated in visceral structures, blood vessels, &c.), have their cell body in a posterior root ganglion, and enter the spinal cord by the posterior root.

The parasympathetic system. In its general organization the peripheral part of the parasympathetic resembles the sympathetic system in that it is made up of pre- and postganglionic fibres which are connected by synapses in ganglia. It differs in the fact that it is limited to certain cranial nerves and sacral segments, and that the ganglia are more peripherally situated, usually being found actually in the viscera which they innervate in the form of rather diffuse ganglionated plexuses. Such are the enteric plexuses (of Auerbach and Meissner) in the wall of the alimentary canal, and the cardiac plexuses in the heart wall. The preganglionic neurons thus have usually a relatively long course, and this, together with the fact that their impulses are not diffused over a large number of postganglionic neurons at synaptic junctions, provides the anatomical basis for more discrete and isolated reactions, in contrast to the mass reactions characteristic of the sympathetic system.

The cranial part of the parasympathetic system (or the cranial autonomic system) is found in relation to the third cranial nerve, and the seventh, ninth, and tenth nerves. The parasympathetic components of these nerves innervate visceral structures ranging from the sphincter pupillae and ciliary muscle of the eye and the salivary glands, to the heart, bronchi, and the alimentary tract and its derivatives as far distally as the middle portion of the transverse colon. The sacral contribution (or the pelvic autonomic system) is composed of preganglionic fibres which arise in the second, third, and fourth sacral segments of the spinal cord and are distributed to peripheral plexuses in the lower end of the alimentary tract and the pelvic viscera.

It will be observed that, so far as the outflow of fibres from the cerebro-spinal axis is concerned, the whole autonomic nervous system is confined to the brain stem, the thoracic and upper lumbar segments of the spinal cord, and certain sacral segments. In other words, it shows obvious gaps in the segments from which the large limb plexuses take their origin. That this relation is significant is shown by the fact that variations in the segmental origin of the limb plexuses are always associated with corresponding variations in the segmental extent of the preganglionic outflow of the autonomic system. It may be inferred, therefore, that this outflow is inhibited by the local augmentation of somatic neurons related to the development of the limbs.

12. THE 'CONNECTIVE TISSUES' OF THE NERVOUS SYSTEM

We have noted that in a peripheral nerve the individual axons are enveloped by the cytoplasm of Schwann cells, and the whole nerve is invested by a sheath of vascular connective tissue from which processes and septa extend in among component fasciculi. Moreover, while the vascular connective tissue is of mesodermal origin, the Schwann tissue is ectodermal, being derived during embryonic development from the cells of the neural crest.

In the central nervous system equivalent tissues are found. In the first place, the whole brain and spinal cord are enclosed within a layer of tough fibrous tissue, the dura mater, which contains blood vessels and is comparable to the epineurium of peripheral nerves.

Under the dura mater are finer membranes (*leptomeninges*) comprising the arachnoid and pia mater. These two membranes are composed of essentially the same tissue—flattened mesothelial cells supported in a matrix of interlacing fibrils—and are commonly referred to as the *pia-arachnoid*. There is some evidence to support the conception that this tissue is derived from the neural crest and that it may be regarded as the morphological equivalent of the Schwann cells of peripheral nerve fibres. Harvey and Burr[1] found that if in the larval amphibian, *Amblystoma*, a portion of the cerebral hemisphere without any associated neural crest cells is transplanted to another part of the body, differentiation proceeds without the formation of leptomeninges. If, on the other hand, a portion of the mid-brain together with neural crest cells is similarly transplanted, leptomeninges are subsequently developed in relation to the transplant. These embryological experiments, ingenious as they are, have not carried conviction in the minds of some anatomists, so that the conclusions drawn from them must be regarded as still doubtful. It is a matter of extreme difficulty to investigate by ordinary embryological methods a problem such

---

[1] S. C. Harvey and H. S. Burr, 'The development of the meninges', *Arch. of Neurol. and Psych.* **15**, 1926.

as this, for at an early stage neural crest cells become so intimately mingled with mesenchymal cells that they cannot be readily distinguished.

It has already been noted (p. 57) that under conditions of irritation the cells of the pia-arachnoid are capable of being set free and becoming phago-cytic; in other words, they can become converted into macrophages like certain mesodermal elements of connective tissue elsewhere. This be-haviour certainly suggests a mesodermal origin.

The arachnoid itself forms a continuous sheet investing the whole of the central nervous system (resembling in this way a serous membrane), and beneath it are spaces containing loose trabecular tissue. In the meshes of this tissue circulates the cerebrospinal fluid. The latter is formed mainly within the ventricles of the brain (from the choroid plexuses) and, after being distributed over the surface of the brain and spinal cord, is filtered back into the venous system.

When the nervous system first begins to be developed in the embryo, it is composed entirely of ectodermal tissue. Very soon, however, it is invaded by capillary blood vessels which carry in with them tissues of mesodermal origin. From the latter are differentiated the muscular and fibrous coats of the vessels themselves, and also cells of a special type which become scattered throughout the nervous tissue. These are the microglial cells which will be discussed below.

Neuroglia. In microscopical sections of the central nervous system stained by ordinary methods, there are always to be seen among the nerve cells and their processes numerous small, deeply stained nuclei (Fig. 112). These are the nuclei of specialized 'connective-tissue' elements termed collectively *neuroglia*. The cytoplasm of neuroglial cells is extremely difficult to stain, and can usually only be demonstrated by particular histological methods such as silver impregnation.

Neuroglia is composed of several types of cell. Their relation to each other is not always clear, and their functional significance is still less certain. Nevertheless, the recognition of these different types is of considerable practical importance, since they are responsible for forming certain kinds of tumour which are met with clinically, and the nature of a tumour and the prognosis in regard to its growth or operative removal can often be determined by reference to its cellular basis. Except for microglia, neuro-glial tissue is ectodermal in origin, being derived from the same embryonic cells which also give rise to neural elements.

Microglia (Fig. 130 c). This consists of small cells with oval nuclei and numerous branching cytoplasmic processes, found predominantly in the grey matter of the central nervous system. They have been studied very intensively by Hortega, who has adduced strong reasons for believing that, unlike other neuroglial elements, they have a mesodermal origin.[1] In embryonic development they appear later than other types of neuroglia, at a time when the meningeal and vascular organization of the brain is well established. Moreover, they appear first at the surface of the brain, imme-diately below the pia, and only later do they penetrate more deeply to reach

[1] P. del Rio Hortega, 'Microglia', *Cytology and Cellular Pathology of the Nervous System*, ed. by W. Penfield, 1932.

the walls of the ventricular cavities. Lastly, they are amoeboid and phagocytic. Their mobility has been actually observed in tissue cultures of living
cells, and Hortega has described in detail how they migrate actively towards the site of a lesion in the brain, and take part in reparative processes
by removing the débris of diseased or injured cells.

Wells and Carmichael[1] have shown that microglial cells bear a very close
resemblance to the wandering phagocytic cells which are present in cultures
of ordinary connective tissue, and that they do not appear in cultures of
embryonic nervous tissue in which no mesodermal elements are present.
They found, also, that microglial cells react to vital dyes (e.g. trypan blue)
in the same way as cells of the macrophage or reticulo-endothelial system
(see p. 55). There seems little doubt, therefore, that microglia is to be
regarded as a derivative of the macrophage system. Its mesodermal origin is
indicated by some authorities by using the alternative term *mesoglia*.

Astrocytes. Scattered throughout the grey and white matter of the
central nervous system are stellate cells with a small cell body and elaborate
branching processes. They are divided into two categories, *protoplasmic
astrocytes*, whose cytoplasm is rather granular in appearance and has complicated irregular processes (Fig. 130 B), and *fibrous astrocytes*, which have
fewer, longer and straighter processes, and contain in their cytoplasm fine
fibrils (Fig. 130 D). Protoplasmic astrocytes predominate in grey matter,
while fibrous astrocytes are found mainly in white matter.

A characteristic of astrocytes is their intimate relation to capillary blood
vessels. Some of their processes end in flattened expansions closely applied
to the capillary endothelium, and are often termed 'sucker-processes'.
This suggests an important functional relationship, and it is in fact supposed by some histologists that astrocytes play a nutritional role as inter-
mediaries between the vascular system and nervous elements. In general,
however, astrocytes may be regarded as supporting structures, equivalent
in this respect to connective tissue in other parts of the body.

Following lesions or degenerative processes in the central nervous system, astrocytes undergo marked proliferation, leading to a hyperplasia
which is termed *gliosis*. Astrocytes thus play an important part in the process of repair, providing for the formation of a special type of cicatricial
tissue. It has been suggested that the granular inclusions of protoplasmic
astrocytes betoken a secretory process, but the evidence for this has as yet
no certain foundation.

Oligodendroglia (Fig. 130 A). The cells of this type of neuroglia are
much smaller than astrocytes, and their processes are less elaborate and
less numerous. Transitional forms, however, are not uncommon. Oligo-
dendroglial cells are found in grey and white matter, and they commonly
have a very intimate relation to nerve-cell bodies and nerve fibres. On these
grounds it is suggested that they are functionally equivalent to the Schwann
cells of peripheral nerve fibres and to the satellite cells found in relation to
nerve cells in peripheral ganglia. It seems probable, therefore, that oligodendroglia subserves a nutritive and protective function, and it is possibly

[1] A. Q. Wells and E. A. Carmichael, 'Microglia: an experimental study by means of
tissue culture and vital staining', *Brain*, **53**, 1930.

also concerned with the deposition and maintenance of the myelin sheaths in white matter. In regard to the latter suggestion, it is stated by some histologists that the differentiation of oligodendroglia cells in the course of development coincides in time with the onset of myelinization.

Insulating functions have been ascribed to neuroglial tissue generally. In this connexion it has been noted that it is relatively more abundant in areas of dense neuropil, where fibres carrying different types of impulse are mingled together in a close network, whereas it is less profuse in well-defined tracts whose fibres have a common functional significance.

Ependyma. Lining the ventricles of the brain and the central canal of the spinal cord is a columnar ciliated epithelium termed *ependyma*. Although this does not really come into the category of 'connective tissue', its consideration here is convenient because of its close morphological relation to neuroglia.

From the bases of the ependymal cells fine processes extend peripherally. In the developing neural tube these reach to the surface, where they come into relation with the pial tissue. This connexion usually becomes broken as the wall of the neural tube increases in thickness, but it persists in certain regions (e.g. in the roof of the mid-brain, and in the spinal cord where the central canal is in close proximity to the antero-median fissure). In the ventricles of the brain, the ependyma covering the choroid plexuses (invaginated tufts of vascular pia mater) becomes modified to form a true secretory epithelium which is concerned in the production of cerebrospinal fluid.

The histogenesis of ectodermal neuroglia. As already described (p. 358), when the neural tube of the embryo is first formed its wall consists of a few layers of cells which by rapid proliferation increase to many layers. The cells which line the lumen of the tube comprise the germinal layer. From this layer, cells migrate peripherally and differentiate into two categories, neuroblasts or embryonic nerve cells and spongioblasts or embryonic neuroglial cells. The latter are distinguished by their small cell bodies, and by their branching processes which at first appear to be interconnected to form a true syncytium. They provide a matrix or scaffolding to support the growing nerve fibres and nerve cells. Finally, the cells of the germinal layer which remain *in situ* ultimately form the ependymal epithelium. The differentiation of spongioblasts into the various types of ectodermal neuroglia has been followed in detail by histological studies. Intermediate forms in the development of astrocytes and oligodendroglia have been recognized, and described as *astroblasts* and *oligodendroblasts*.

13. ANATOMICAL METHODS EMPLOYED IN THE STUDY OF THE NERVOUS SYSTEM

The functional significance of any group of nerve fibres in the brain and spinal cord is directly related to the connexions which they establish with sensory or motor end organs, or with other parts of the central nervous system. One of the primary aims of the neurological anatomist, therefore, is to trace the origin, course, and terminal distribution of these conducting elements. With a detailed knowledge of their interconnexions, it would be theoretically possible to infer the functional implications of each part of the nervous system and to obtain a much more complete picture of the physiology of the brain as a whole than is at present available. The anatomist, it may be noted, is concerned with more than the elucidation of the purely morphological structure of the nervous system; it also falls within his province to determine how far it is possible to relate specific functions to specific structural elements. Functional localization in the nervous system is the common ground of the anatomist and physiologist in neurological studies.

Methods employed in the anatomical investigation of the nervous system can be divided into two main categories, those involving the direct study of normal material and those involving the study of experimental and clinical material. These methods may now be briefly discussed.

The study of normal material. (*a*) *Gross Dissection.* It is possible by ordinary dissection of the brain (particularly with the use of blunt dissecting instruments) to trace several of the better defined fibre tracts throughout the whole or part of their course. In this way, for example, the optic tract can be followed to its termination in the lateral geniculate body and superior colliculus, the optic radiations can be traced to the visual cortex, and certain motor and sensory tracts in the brain stem, such as the pyramidal tract and medial fillet, can be displayed. The method, however, has very considerable limitations. At the best it can only give a general indication of the origin and destination of the more compact tracts of fibres, and even then their precise connexions must be verified by experimental studies.

(*b*) *Histological methods.* The disposition of nerve cells in the central nervous system can be studied in microscopic sections stained with basic dyes such as methylene blue, toluidin blue, thionin, neutral red, cresyl violet, &c. With proper treatment these dyes ordinarily leave the fibres unstained since they select out the Nissl granules, the chromatin of the nucleus, and the nucleolus.

Myelinated fibres can be selectively stained by the Weigert-Pal technique, the principle of which depends on the fact that, after treatment with a suitable oxidizing agent (e.g. potassium bichromate), the myelin acts as a mordant towards haematoxylin. The sections are first over-stained with the haematoxylin, and subsequently differentiated. The stain is thereby removed from all tissues except the myelinated fibres, which hold it very tenaciously.

By the careful study of Weigert-Pal sections it is possible to observe in some detail the relation of nerve fibres to areas of grey matter in the central nervous system. Particularly is this the case if the material is counterstained with a suitable dye so that the individual nerve cells and the myelinated nerve fibres are shown up in the same sections. The Weigert-Pal method is often used as a routine procedure in the microscopical examination of the nervous system, but it has certain disadvantages when applied for the purpose of elucidating the details of fibre connexions. It demonstrates only those fibres which are provided with myelin sheaths, and, since myelinated fibres may lose their sheaths some distance before they actually reach their destination, their precise termination cannot always be demonstrated with certainty. Again, in the more diffuse fibre-tract systems, bundles of fibres may interweave in a complicated fashion with fibres of adjacent systems so that in actual practice it is extremely hazardous to try to establish their continuity from their origin to their termination. Lastly, fibres of passage (which may be passing *through* a collection of nerve cells to reach some other objective) are often distinguished only with great difficulty, if at all, from terminal fibres.

Non-myelinated fibres can be demonstrated in suitably prepared sections of the nervous system by treatment with silver salts. In this method the axons are displayed not as the result of being stained in the ordinary sense

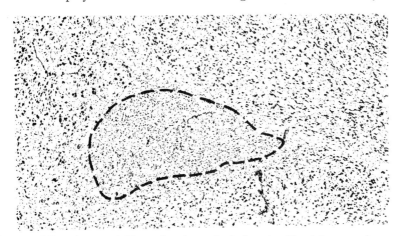

FIG. 131. Microphotograph illustrating the method of tracing fibre interconnexions in the central nervous system by retrograde cell degeneration. The section is taken from the ventral nucleus of the thalamus of a monkey two months after a local ablation of part of the 'face area' of the sensory cortex. The operation has led to a sharply circumscribed patch of atrophy of the nerve cells whose axons terminate in the cortical area. × 25.

of the term, but because they become impregnated by an extremely fine deposit of reduced silver oxide. In successful preparations the fine terminal arborizations of individual nerve fibres can be demonstrated with great precision, so that their relations to nerve cells with which they come into synaptic contact are accurately determined.

One of the difficulties of most silver methods is that they tend to display all the details of the fine neuropil of the nervous system, and it is usually difficult to trace the course of individual fibres in a matrix of other intertwining fibres. For this reason the Golgi technique and its modifications are of special value, for by this method individual and isolated cells together with their processes are picked out here and there. This capricious reaction makes it much easier to study in detail the connexions of representative neurons in any collection of nerve cells and fibres. In principle it involves treatment with silver nitrate or certain other metallic salts after preliminary hardening in potassium bichromate with or without osmic acid.

The methods of investigating the intrinsic structure of the nervous system to which brief reference has been made above rarely allow of any final conclusions regarding the origin and termination of fibre paths. Rather, they provide the means for a preliminary survey of the field of inquiry as a basis for subsequent experimental work.

The study of experimental and clinical material. (*a*) *Chromatolysis and cell atrophy.* It has been noted (see p. 336) that, if an axon is cut, its cell of origin undergoes characteristic changes involving the breaking up and disappearance of its Nissl granules, a swelling of the cell body, and a displacement of its nucleus to an excentric position. If regeneration does not occur (as in the case of lesions inside the central nervous system), the cell body and its processes may eventually undergo complete atrophy.

This process of retrograde cell degeneration provides a most valuable technique for studying the origin of groups of nerve fibres. If, for example, it is required to determine from which particular group of cells (e.g. in the thalamus) fibres to a specific area of the cerebral cortex are derived, this may be done by destroying the cortical area in an experimental animal, allowing the latter to survive for a few weeks, and then studying serial sections of the thalamus which have been stained with methylene blue. The site of origin of the fibres which have been interrupted by the cortical lesion is marked out by a zone of cell atrophy (Fig. 131). The cell origin of peripheral motor nerves may similarly be studied by cutting the latter, and searching for evidence of chromatolysis in the spinal cord or brain stem a week or two after the operation.

In some cases it happens that, following the section of a nerve fibre, the cell with which it is in synaptic relation undergoes atrophy. This phenomenon of *transneuronal degeneration* is uncommon, but where it does occur it provides a method by which the termination of nerve fibres can be defined with the greatest precision. If, for example, the optic nerve is cut, within a few weeks the cells of one of the main lower visual centres (the lateral geniculate body), in relation to which the optic fibres end, undergo profound atrophy. It has been found, moreover, that a small retinal lesion is followed by the appearance of a localized patch of cell atrophy in the lateral geniculate body, and in this way it has been possible to show that each portion of the retina is projected on to a separate portion of the geniculate body. A more rapid and severe type of transneuronal degeneration has been described in certain cell groups in the hind-brain, following lesions of the cerebral cortex. In this case, the affected cells are reported actually to undergo necrosis and fragmentation within a few days.[1] Retrograde transneuronal degeneration has also been observed in experimental material, its degree of severity being a function of the age of the animal at the time of operation and of the post-operative survival period. This distant retrograde reaction of nerve cells has been used to good effect in tracing the more remote afferent connexions of certain areas of the cerebral cortex.[2] It is not clear why transneuronal degeneration occurs in some neural paths and not in others. Possibly it is an indication of an extreme specificity of function.

[1] A. Torvik, 'Transneural changes in the inferior oilve and pontine nuclei in kittens', *Journ. Neuropath. Exp. Neurol.* **15**, 1956.
[2] W. M. Cowan and T. P. S. Powell, 'An experimental study of the relation between the medial mamillary nucleus and the cingulate cortex', *Proc. Roy. Soc.* B, **143**, 1954.

(b) *Wallerian degeneration.* When myelinated nerve fibres within the central nervous system are interrupted by experimental section or by disease, the myelin sheaths distal to the lesion undergo complete degeneration and ultimately disappear. This process is termed Wallerian degeneration. If a part of the brain or spinal cord containing completely degenerated fibres is sectioned and stained by the Weigert-Pal technique, the affected tracts are displayed because, in the absence of myelin, they do not take up the stain. Such a negative picture allows well-defined and compact groups of fibres to be followed throughout a great part of their course, but it does not permit the tracing of individual fibres, or diffusely arranged fibre tracts.

(c) *Marchi degeneration.* Of all experimental anatomical methods, the Marchi technique has been most widely used in the past for studying myelinated fibre paths in the central nervous system. It has the great advantage over the Wallerian method that it provides a positive instead of a negative picture in microscopical preparations, for it enables fibres undergoing degeneration to be stained while normal fibres are left unstained.

The Marchi technique depends on the fact that, during the course of their degeneration, myelin sheaths become broken up into granules of unsaturated fatty acids. If these are now treated with a weak solution of osmic acid, the latter is reduced to black osmic oxide, so that the degenerating fibres are stained a dense black. Preliminary treatment with potassium bichromate is sufficient to oxidize the small proportion of unsaturated fatty acids in the myelin sheaths of normal fibres, and the latter do not therefore reduce the osmic acid in the same way. In principle, therefore, the course and termination of any particular fibre tract can be determined by interrupting it with an experimental lesion, allowing the animal to survive for a fortnight or three weeks, preparing serial sections through the appropriate region of the central nervous system, staining with osmic acid, and following the degenerating fibres through the serial sections where they are marked out by a train of black granules.

Valuable as it is, the use of the Marchi method requires extreme care in order to obtain satisfactory results. For several reasons, also, the interpretation of Marchi material is not easy. In the first place, certain parts of the white matter in the brain have a tendency to produce a diffuse deposit of black osmic granules even when normal; this is termed a *pseudo-Marchi reaction*. Granules of degenerating myelin may also be carried along the perivascular spaces of blood vessels and simulate the course of degenerating fibres. Again, it may not always be easy to distinguish between degenerating terminal fibres and degenerating fibres of passage, though in the latter case the Marchi granules are confined to definite fasciculi of fibres rather than being scattered diffusely in the ground substance among nerve cells.

In order to trace Marchi degeneration with certainty, it is always necessary to have serial sections through which the osmic staining in any particular system of fibres can be traced *in continuity* to their ultimate destination. In any case, of course, the Marchi staining only displays the course of a degenerating fibre so far as the latter is invested by a myelin sheath. If a fibre loses its sheath some distance before it ends, the actual site of its termination cannot be determined by this method. Lastly, the Marchi technique is not likely to lead to accurate results in the case of extensive lesions, for the latter may cause a general trauma of the brain or spinal cord which leads to the formation of an osmic deposit in fibre tracts other than those which have been directly injured.

(*d*) *Degeneration of axon terminals.* It has been mentioned (p. 368) that, at a synaptic junction, the terminal arborizations of an axon may end in knob-like thickenings closely applied to the body of a nerve cell or its dendrites. These terminal boutons become thickened and swollen a few days after the axon is sectioned, and stain deeply with silver nitrate. Finally they break up into granular masses which may appear very conspicuous on histological examination (Fig. 132). The technique used for demonstrating this terminal degeneration (Glees' method) is much superior to the experimental methods which have hitherto been mainly employed for the study of fibre connexions in the central nervous system, for it makes possible the definition of the actual synaptic configuration of fibre terminals. The method has been widely used in recent years, for example, in the study of the terminal connexions of fibres of the optic tract and the olfactory tract, and also for establishing the interconnexions of some of the intrinsic fibre systems of the brain. Another technique for demonstrating axonal degeneration by silver impregnation has been devised by Nauta and his colleagues. This method has the great advantage that it stains the degenerating fibres deeply while leaving normal fibres unstained. On the other hand, it is less effective than Glees' method for demonstrating the finest terminal ramifications of the degenerating fibres.[1] While terminal degeneration can be satisfactorily established by Glees' method in the case of the main terminal connexions of a fibre tract, it is not so reliable as an index for locating synaptic connexions in the case of individual isolated fibres. We still await a technique by which degenerating terminals can be differentially stained so decisively that a *single* fibre can be traced to its ultimate destination without any possibility of doubt.[2]

(*e*) *Degeneration of peripheral nerve endings.* Motor and sensory nerve endings undergo degeneration after section of peripheral nerves. Histological studies of the changes which the endings show provide an obvious method for tracing the precise terminations of peripheral fibres, or for ascertaining the root value of specific motor and sensory nerves, particularly when the course of the latter may be obscured from direct anatomical investigation by a plexiform arrangement.

(*f*) *Functional changes following neural lesions and experimental stimulation.* The anatomical course and relations of motor and sensory tracts can often be studied indirectly by reference to the effects of impulses which they

[1] For a general account of the silver techniques of Glees and Nauta, see P. Glees and W. J. H. Nauta, 'A critical review of studies on axonal and terminal degeneration', *Monats. f. Psych. v. Neurol.* **129**, 1955; D. H. L. Evans and L. H. Hamlyn, 'A study of silver degeneration methods in the central nervous system', *Journ. Anat.* **90**, 1956.

[2] To illustrate the care which is needed in the interpretation of the degeneration of isolated nerve cells and fibres in experimental material, it may be noted that merely the operative exposure of the brain or spinal cord, even without effecting any direct lesions in the nervous tissue, can lead to chromatolytic changes in occasional cells or degenerative changes in occasional fibres.

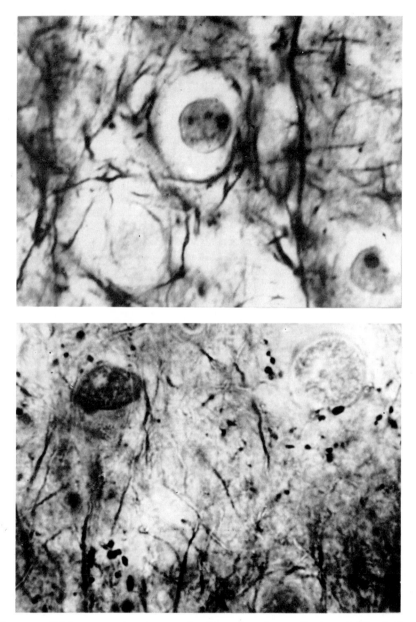

Fig. 132. High-power microphotographs illustrating the application of Glees' silver technique to the study of fibre connexions in the central nervous system. Above are shown cells and pericellular plexuses of fibres in a normal nucleus of the hypothalamus (at the base of the brain). Below is a microphotograph of the corresponding region a few days after effecting a small lesion of the cerebral cortex. The fibres which have their origin in the cortex have become broken up into conspicuous beaded fragments, and others have disappeared altogether. × 1500.

carry, in so far as interruption of the tracts abolishes these effects while stimulation may accentuate them. These methods are particularly valuable when dealing with neural paths which are diffuse, or which involve a series of neuronal relays, and thus are not readily demonstrable by more direct anatomical procedures. In order to illustrate such functional methods, we may cite a number of examples.

It is known that a large motor tract (the pyramidal tract) takes origin from an area of the cerebral cortex and passes down into the spinal cord where its fibres terminate at different segmental levels. Electrical stimulation of different points in the cortical area gives rise to movements involving different groups of muscles, and it is possible by this means to determine the point of origin of the component parts of the pyramidal tract system. Similar information can be provided by observing which groups of muscles are paralysed as the result of localized lesions in the cerebral cortex (either produced experimentally or observed clinically). Further, the course followed by the fibres conveying these motor impulses can be inferred to a considerable extent by making lesions in different parts of the brain stem and spinal cord and observing which particular lesions prevent the effects of cortical stimulation. Similar methods have been employed from time to time in order to trace out the anatomical course of neural paths carrying visceromotor, vasomotor, pilomotor, sudomotor impulses, &c., by stimulating functional 'centres', central or peripheral paths (or by interrupting the latter), and observing the corresponding functional changes and their distribution.

On the afferent side, observations on sensory changes associated with lesions of the central nervous system have helped to determine the course and termination of sensory pathways. For example, lesions in the visual mechanism of the brain have provided anatomical information regarding the precise path followed by retinal impulses, and the site of their termination in the cerebral cortex. The site of termination in the cerebral cortex of sensory tracts conveying impulses from different regions of the body has also been studied experimentally by applying to localized spots of the cortex a weak solution of strychnine hydrochloride. This leads to a sensitization which manifests itself in a hyperaesthesia of certain areas of the skin.[1]

An ingenious method was evolved by McSwiney and his collaborators for studying the anatomy of visceral afferent pathways by using pupillary dilatation as an index of visceral afferent nerve activity.[2] Stimulation of afferent visceral fibres leads to immediate dilatation of the pupil as the result of inhibition of the nucleus of the third cranial nerve in the midbrain. By making lesions in different dorsal spinal roots and in different parts of the spinal cord, and observing which particular lesions abolish the pupillary reaction, the route followed by the afferent fibres under investigation can be mapped out with considerable precision.

[1] J. G. Dusser de Barenne, 'Experimental researches on sensory localization in the cerebral cortex of the monkey', *Proc. Roy. Soc.* B, **96**, 1924.
[2] See, for example, W. A. Bain, J. T. Irving, and B. A. McSwiney, 'The afferent fibres from the abdomen in the splanchnic nerves', *Journ. Physiol.* **84**, 1935.

The course taken by afferent impulses has also been studied by recording the action currents of nerve fibres (after amplification with a cathode-ray oscillograph) following peripheral stimulation. This technique has been facilitated by the use of barbiturates for inducing anaesthesia in experimental animals, for these drugs damp down the 'spontaneous' electrical activity of the brain sufficiently to permit the detection of the small electrical changes which occur when a peripheral sense organ is stimulated. For example, if in an anaesthetized animal one of the tactile vibrissae of the snout is moved and at the same time the cerebral cortex is explored with electrodes leading off to an amplifier and oscillograph, the precise point in the cortex where the impulse arrives can be accurately defined. In this way the representation of different parts of the body in the cerebral and cerebellar cortex, and in certain deeper structures of the brain, has been worked out in some detail.[1]

An indirect anatomical method of a similar kind has been devised by Dusser de Barenne and his collaborators. It may be termed the method of strychninization, for it involves the local application of a weak solution of strychnine hydrochloride to different parts of the grey matter of the brain, and the recording of the electrical changes induced by this chemical stimulation. Experimental evidence shows that the strychnine acts on the cell bodies of the neurons, causing a discharge of impulses, and that the electrical changes are produced at the axon terminals.[2] By this method, therefore, it is possible to determine the origin and termination of many intricate fibre connexions of the brain whose course can hardly be followed by direct anatomical study. The method of strychninization has been successfully employed in the delineation of the complicated association systems which link up the different areas of the cerebral cortex, and the diffuse pathways by which the cortex is connected to deep-lying structures in the brain such as the basal ganglia. Since it has recently been demonstrated that some groups of nerve cells in the brain are resistant to the action of strychnine, negative results obtained by this method are evidently inconclusive.

(g) *Comparative anatomy.* Information regarding the anatomical significance of different elements in the brain can sometimes be derived from comparative studies. For example, the atrophy or disappearance of certain nervous structures in animals which are normally blind suggests that they are related to the visual system. The enlargement of certain cell groups in an animal which has a well-developed sense of smell suggests that they may be concerned with the reception of olfactory impulses. Such observations usually provide only a general indication in regard to functional localization, but they can sometimes supply negative evidence of considerable value.

[1] See E. D. Adrian, 'Localization in the cerebrum and cerebellum', *Brit. Med. Journ.*, 29 July 1944.
[2] J. G. Dusser de Barenne and W. S. McCulloch, 'Physiological delineation of neurons in the central nervous system', *Amer. Journ. Physiol.* **127**, 1939.

14. THE VASCULAR SUPPLY OF NERVOUS TISSUE

During embryonic development the tissues of the central nervous system become vascularized at an extremely early stage. In the chick, for example, Williams found capillary vessels invading the neural tube at the second day of incubation, while at twelve days arterioles and venules begin to appear.[1] When the vessels first enter the tube, they are apparently attracted towards the germinal layer where the cells are undergoing rapid division, and which is therefore presumably a region of high metabolic activity. Subsequently, as the proliferative activity of the germinal layer spreads throughout the developing brain following on cell migration, the vessels send out sprouts to these new areas.

In the adult the relative vascularity of different parts of the nervous system bears a broad relation to their metabolic level. Grey matter is richly vascular while white matter is poorly supplied with capillaries. This corresponds with the fact that, in general, grey matter when active consumes more oxygen than white matter. It has been suggested that the oxidation associated with nervous activity occurs in dendrites and synaptic structures rather than in the nerve-cell bodies themselves,[2] and in relation to this it has been observed that the degree of vascularity tends to vary rather with the number of synaptic connexions than with differences in the number and density of nerve-cell bodies.[3] For example, the vascularity of the cerebral cortex, which contains numerous synapses, is out of all proportion greater than that of the sensory ganglion of the trigeminal nerve in which synapses do not occur in any significant number. Neuropil, also, is richly supplied with capillary vessels. But this relationship is not consistent; thus, the glomerular formations in the human olfactory bulb are apparently devoid of capillary vessels, and yet they are the site of exceedingly intricate synaptic interconnexions.

Craigie's work has shown clearly that relative vascularity depends on neural activity as well as on the density of nerve cells.[4] All parts of white matter are not equally vascular, and in grey matter motor nuclei are in general less rich in capillaries than sensory and correlation centres. Moreover, the vascularity of grey matter increases rapidly with the onset of functional activity during development.

It will be appreciated from what has been said that the vascular architecture of the brain reproduces very closely the cellular architecture. In sections of a well-injected brain, therefore, cell collections and cell laminae are outlined by the disposition of capillary plexuses. In the most vascular regions the capillary network may be so fine that individual cells are encircled by each mesh. Indeed, endocellular capillaries are sometimes found

[1] R. G. Williams, 'The development of vascularity in the hind-brain of the chick', *Journ. Comp. Neur.* **66**, 1937.

[2] E. G. Holmes, 'Oxidations in central and peripheral nervous tissue', *Biochem. Journ.* **24**, 1930.

[3] H. S. Dunning and H. G. Wolff, 'The relative vascularity of various parts of the central and peripheral nervous system of the cat and its relation to function', *Journ. Comp. Neur.* **67**, 1937.

[4] E. H. Craigie, 'On the relative vascularity of various parts of the central nervous system of the albino rat', ibid. **31**, 1920.

actually piercing the cytoplasm of a nerve-cell body. This is particularly the case with certain cell collections in the hypothalamus (in the neighbourhood of the pituitary gland at the base of the brain), which are believed to possess secretory functions.[1]

The human brain is well supplied with blood vessels. Two pairs of relatively large arteries (the internal carotid and vertebral arteries) carry blood to it, and it is drained by veins which ultimately open into large venous sinuses in the dura mater. These anatomical facts are correlated with the observation that the blood flow through the brain in a given time is considerably higher than through most other parts of the body, including skeletal muscle.

On the surface of the brain and spinal cord the arteries anastomose very freely, allowing an even distribution of blood to all parts and also an equalization of pressure in all the vascular channels. On the other hand, within the substance of the central nervous system anastomoses are much less free. Indeed, from the point where a branch comes off a main artery in order to enter nervous tissue it functions as an end artery (see p. 195). A study of the vascular architecture of the brain shows, however, that there are no end arteries in the strict anatomical sense, since adjacent vessels do anastomose through a capillary bed and also to some extent by direct arteriolar connexions.[2] It has been shown experimentally, however, that these anastomoses are not adequate for the formation of a collateral circulation, and if one arterial channel is blocked, the territory of nervous tissue which it vascularizes undergoes rapid necrosis.[3] It follows from this that interruption to the vascular supply by lesions such as thrombosis or embolus must always lead to a complete destruction of the corresponding part of the brain. In any case, nerve cells are particularly susceptible to the effects of anaemia. Gomez and Pike[4] found that, if the brain is temporarily deprived of blood (at normal body temperature) by clamping the main vessels, the small pyramidal cells of the cerebral cortex are fatally injured in eight minutes. Other cells are susceptible to lesser degrees, and sympathetic ganglion cells are the most resistant. On the other hand, neuroglia cells seem not to be affected at all by a temporary anaemia of corresponding duration.

The intracranial circulation offers a series of interesting problems, for, as Munro originally pointed out and Kellie later on confirmed, the brain is enclosed within a rigid container and, since the substance of the brain is practically incompressible, the quantity of the blood circulating within the cranium would therefore appear at all times to be almost constant. This thesis is known as the *Munro-Kellie doctrine*.

[1] E. Scharrer and R. Gaupp, 'Neuere Befunde am Nucleus supraopticus und Nucleus paraventricularis des Menschen', *Zeitschr. f. g. Neur. Psych.* **148**, 1933.

[2] It is interesting to note that in marsupials and certain amphibia true end arteries do exist. In these animals, each small artery entering the cerebral cortex makes a loop with a corresponding vein, and affects no anastomosis at all with neighbouring vessels. G. B. Wislocki and A. C. B. Campbell, 'The unusual manner of vascularization of the brain of the opossum', *Anat. Rec.* **67**, 1937.

[3] S. Sunderland, 'The production of cortical lesions by devascularization', *Journ. Anat.* **73**, 1938.

[4] L. Gomez and F. H. Pike, 'The histological changes in nerve cells due to total temporary anaemia of the central nervous system', *Journ. Exp. Med.* **2**, 1909.

It has been shown, however, that there are reciprocal volume changes between the blood and the cerebrospinal fluid, varying in magnitude with the position of the body. Weed[1] has suggested, indeed, that the chief function of the cerebrospinal fluid may be to allow a prompt reciprocal adjustment in response to changes which occur in the amount of blood in the arteries and veins within the skull. Certainly any rise in arterial or cerebrospinal fluid pressure will tend to compress the thin-walled cerebral veins and to obstruct the venous return from the brain. In order to obviate this tendency, the cerebral veins effect unusually strong anastomoses with each other. In some cases, also, their mode of termination by opening into the dural venous sinuses in a direction opposite to the blood flow in the latter suggests a mechanism for maintaining the pressure in them and thus preventing their easy collapse.

The large terminal veins inside the skull are represented by intradural venous sinuses. These are contained between layers formed by rigidly held processes of dura mater, and are thus protected from the effects of any sudden rise in intracranial pressure.

The nervous control of the circulation in the central nervous system has been discussed in a previous chapter (see p. 223).

15. THE EVOLUTION OF NERVOUS ORGANIZATION

The history of the evolution of the nervous system is the history of progressive centralization. In lowly invertebrates of a radially symmetrical type (i.e. coelenterates), the whole nervous system is composed of a diffuse network of nerve cells and their processes, forming a ganglionated plexus which is dispersed throughout the body of the organism. This regulatory mechanism mediates reactions of a diffuse type, but it is hardly capable of initiating and synthesizing rapid adjustments involving widely different types of effector, or of evoking individuated responses of an isolated and discrete nature.

When, during the course of organic evolution, axially symmetrical animals came into being, the head end became equipped with specialized receptor organs to enable them to respond rapidly to environmental changes which they encountered in their movements in search of food, &c. It seems that these receptors, in turn, provided a stimulus for the elaboration of nervous ganglia concentrated in the head region. Thus a primitive brain came into existence. The primary function of the brain was evidently to receive impulses from specialized sense organs and, on the basis of information derived from these, to dominate and control the activity of the rest of the nervous system. To facilitate this, a centralization of nervous tissue along the median axis of the animal occurred, forming a nerve cord which served to link up segmental ganglia with each other, and the whole ganglionated chain with the brain. In this manner a spinal cord was developed. In brief, as indicated above, the central nervous system provides a centralized mechanism which allows the interaction of diverse stimuli

[1] L. H. Weed, 'Meninges and cerebrospinal fluid', *Journ. Anat.* **72**, 1938.

arriving from receptors of all kinds, as well as the simultaneous co-ordination of widely separate effectors.

In all vertebrates the nervous activities of the body are mediated by a central nervous system consisting of a brain and spinal cord. In addition, there are peripheral ganglionated plexuses dispersed in the walls of visceral organs, which are to a limited extent capable of local regulating functions. These plexuses may be regarded as persistent representatives of the diffuse network which comprises the whole of the nervous system in lowly invertebrates.

In the central nervous system there can be recognized three main anatomical and functional levels. First there is the *segmental* organization, comprising a single segment of the spinal cord with its corresponding pair of spinal nerves and sympathetic ganglia. This provides the basis of segmental reflex activities, and may therefore be regarded as an elementary nervous mechanism for controlling the reactions of an individual segment of the body. As already noted, however, the segmental arrangement of the central nervous system in vertebrates is not a primary feature—it is secondarily imposed from a peripheral direction by the segmentally disposed muscle plates or myotomes (see p. 373).

Secondly, there is an *intersegmental* organization whereby the segmentally arranged neural mechanisms are linked together. This is represented anatomically by intersegmental association tracts of fibres, which course up and down in the spinal cord and brain stem connecting segmental mechanisms at different levels, and also by the longitudinal fibres of the sympathetic chain. It provides the basis for intersegmental reflex activities.

Lastly, superimposed on these lower functional levels is a *supra-segmental* organization which dominates the whole nervous system. This is represented by elaborate neural mechanisms such as the cerebellum in the hind-brain, the red nucleus and tectum in the mid-brain, and the thalamus, corpus striatum, and cerebral cortex in the fore-brain. These supra-segmental mechanisms are built up on the foundation of correlation centres which were first developed as terminal stations for specialized cranial nerves. Thus, the cerebellum was primarily a correlation centre through which impulses from the semicircular canals and lateral line system can exert their influence over lower neural levels, the tectum of the mid-brain similarly provides visual and auditory centres, while the fore-brain was at first predominantly an olfactory mechanism developed in relation to the termination of olfactory tracts.

During the course of vertebrate evolution a process occurs which involves a gradual shifting forwards of controlling centres from the lower levels of the spinal cord and brain stem to the fore-brain or prosencephalon. This process is termed *prosencephalization*, and it is particularly well illustrated by the progressive development of the cerebral cortex in the vertebrate series.

In lower vertebrates the cerebral cortex is almost entirely olfactory in its afferent connexions, and becomes highly organized in accordance with the fact that in these forms behavioural reactions are dominated by olfactory stimuli. In the early stages of mammalian evolution this elaborate organization was made use of to allow the co-operation at the same functional level of other sensory stimuli in determining adaptational reactions. The activities of correlation centres for visual, auditory, and general sensory stimuli were projected by fibre tracts on to the cerebral cortex so that in this highly intricate mechanism senses other than that of smell came to acquire a representation. Indeed, in the further evolutionary development of higher mammals (particularly in Primates) this process has continued to the point where the olfactory areas of the cortex have become completely overshadowed by non-olfactory areas. The whole cerebral cortex finally comes to form an integrating mechanism whose increasing complexity is manifested on the functional side in a greater ability to apprehend the nature of incoming stimuli and in a capacity for a wider range of adjustments to any environmental change.

It is a matter of considerable importance from the point of view of experimental neurology that the process of prosencephalization should be recognized in all its implications. Functional centres of control, which in lower mammals are situated in the mid-brain or hind-brain, become progressively shifted to the higher levels of the fore-brain in higher mammals. It is apparent, therefore, that the analysis of functional processes by experimental study in lower mammals cannot be directly transferred to the interpretation of the functions of corresponding parts of the brain in higher forms. Another consequence of prosencephalization depends upon the fact that, when a lower controlling centre is superseded by the development of a centre at a higher level, the former does not simply atrophy and disappear —it persists relatively unaltered anatomically, but its activities are subordinated to the controlling and inhibitory influences of the higher centres. Many of the results of experimental or pathological destruction of higher centres are therefore 'release' phenomena, due to the resultant uncontrolled activity of the corresponding lower centres.

The process of prosencephalization in mammals leads to a progressive increase in the volume of the cerebral hemispheres—the direct result of the increasing complexity of integrating mechanisms in the cortex and related structures. Apart from changes associated with an increase in the size of the body as a whole, nervous tissue is unique in that its efficiency as a system is directly related to its bulk as well as to its structural complexity. The advantage in the struggle for existence which accrues from a large brain is well shown in the later palaeontological history of mammals. In early Eocene times (about seventy million years ago) mammals were already well differentiated into divergent groups comparable with those which exist today. They were remarkable, however, for the astonishingly small size of their brains (as inferred from the study of endocranial casts). In their subsequent evolution all these groups showed a progressive and parallel enlargement of their cerebral hemispheres. This evolutionary growth of the brain was not confined to Primates, although it was most obtrusive in this Order, but was also manifested to a considerable degree in other groups such as carnivores, ungulates, &c. Even after the Primates themselves had become differentiated in the course of evolution into separate and divergent groups, each group continued independently to elaborate its cerebral equipment, so that large and well-convoluted brains of a similar organization were evolved to varying degrees in a parallel manner.

Man is contrasted with all other mammals in the relative size of his brain. In him prosencephalization has proceeded further, so that more bodily functions have come to be represented in the most highly organized levels of the cerebral cortex. Moreover, since the cerebral cortex is the anatomical substratum of processes involving voluntary control, its progressive elaboration in man is simply an expression of his supreme position in the animal world as arbiter of his own destiny. The range of his behavioural reactions, which is directly related to the complexity of the highest functional levels of the nervous system, has become extended to an almost unlimited degree. It may be said, indeed, that the cerebral cortex of the human brain is the organ of liberty of action and its correlate, liberty of thought. Herein many will see an inspiration for the present and a hope for the future.

INDEX OF NAMES

INDEX OF SUBJECTS

PRINTED IN GREAT BRITAIN
AT THE UNIVERSITY PRESS, OXFORD
BY VIVIAN RIDLER
PRINTER TO THE UNIVERSITY